Impacts of Globalization and Innovation in Mathematics Education

Evelyn Njurai
Kisii University, Kenya

Alphonse Uworwabayeho
University of Rwanda, Rwanda

A volume in the Advances in Educational Technologies and Instructional Design (AETID) Book Series

Published in the United States of America by
IGI Global
Information Science Reference (an imprint of IGI Global)
701 E. Chocolate Avenue
Hershey PA, USA 17033
Tel: 717-533-8845
Fax: 717-533-8661
E-mail: cust@igi-global.com
Web site: http://www.igi-global.com

Library of Congress Cataloging-in-Publication Data

CIP DATA PROCESSING

2024 Information Science Reference
ISBN(hc) 9798369328736 | ISBN(sc) 9798369346716 | eISBN 9798369328743

British Cataloguing in Publication Data
A Cataloguing in Publication record for this book is available from the British Library.

Advances in Educational Technologies and Instructional Design (AETID) Book Series

Lawrence A. Tomei
Robert Morris University, USA

ISSN:2326-8905
EISSN:2326-8913

MISSION

Education has undergone, and continues to undergo, immense changes in the way it is enacted and distributed to both child and adult learners. In modern education, the traditional classroom learning experience has evolved to include technological resources and to provide online classroom opportunities to students of all ages regardless of their geographical locations. From distance education, Massive-Open-Online-Courses (MOOCs), and electronic tablets in the classroom, technology is now an integral part of learning and is also affecting the way educators communicate information to students.

The **Advances in Educational Technologies & Instructional Design (AETID) Book Series** explores new research and theories for facilitating learning and improving educational performance utilizing technological processes and resources. The series examines technologies that can be integrated into K-12 classrooms to improve skills and learning abilities in all subjects including STEM education and language learning. Additionally, it studies the emergence of fully online classrooms for young and adult learners alike, and the communication and accountability challenges that can arise. Trending topics that are covered include adaptive learning, game-based learning, virtual school environments, and social media effects. School administrators, educators, academicians, researchers, and students will find this series to be an excellent resource for the effective design and implementation of learning technologies in their classes.

Coverage

- Adaptive Learning
- Collaboration Tools
- Digital Divide in Education
- Educational Telecommunications
- Higher Education Technologies
- Hybrid Learning
- Instructional Design
- Web 2.0 and Education

IGI Global is currently accepting manuscripts for publication within this series. To submit a proposal for a volume in this series, please contact our Acquisition Editors at Acquisitions@igi-global.com or visit: http://www.igi-global.com/publish/.

The (ISSN) is published by IGI Global, 701 E. Chocolate Avenue, Hershey, PA 17033-1240, USA, www.igi-global.com. This series is composed of titles available for purchase individually; each title is edited to be contextually exclusive from any other title within the series. For pricing and ordering information please visit http://www.igi-global.com/book-series/advances-educational-technologies-instructional-design/73678. Postmaster: Send all address changes to above address.

Titles in this Series

For a list of additional titles in this series, please visit:
http://www.igi-global.com/book-series/advances-educational-technologies-instructional-design/73678

Integrating Generative AI in Education to Achieve Sustainable Development Goals
Ruchi Doshi (Universidad Azteca, Mexico) Manish Dadhich (Sir Padampat Singhania University, India) Sandeep Poddar (Lincoln University College, Malaysia) and Kamal Kant Hiran (Sir Padampat Singhania University, India & Lincoln University College, Malaysia)
Information Science Reference • copyright 2024 • 545pp • H/C (ISBN: 9798369324400) • US $255.00 (our price)

Considerations and Techniques for Applied Linguistics and Language Education Research
Hung Phu Bui (University of Economics, Ho Chi Minh City, Vietnam)
Information Science Reference • copyright 2024 • 241pp • H/C (ISBN: 9798369364826) • US $245.00 (our price)

Comprehensive Sexuality Education for Gender-Based Violence Prevention
Mariana Buenestado-Fernández (University of Cantabria, Spain) Azahara Jiménez-Millán (University of Córdoba, Spain) and Francisco Javier Palacios-Hidalgo (University of Córdoba, Spain)
Information Science Reference • copyright 2024 • 342pp • H/C (ISBN: 9798369320532) • US $245.00 (our price)

Enhancing Curricula with Service Learning Models
Sharon Valarmathi (Christ University, India) Jacqueline Kareem (Christ University, India) Veerta Tantia (Christ University, India) Kishore Selva Babu (Christ University, India) and Patrick Jude Lucas (Christ University, India)
Information Science Reference • copyright 2024 • 328pp • H/C (ISBN: 9798369359334) • US $265.00 (our price)

IGI Global
PUBLISHER of TIMELY KNOWLEDGE

701 East Chocolate Avenue, Hershey, PA 17033, USA
Tel: 717-533-8845 x100 • Fax: 717-533-8661
E-Mail: cust@igi-global.com • www.igi-global.com

Table of Contents

Detailed Table of Contents

Today, two vectors shape the world-globalization and technology; the former control the economies, politics, and cultures, and the latter dictate how fast they happen or spread or shared within the global space. The two vectors have had a great impact in the field of mathematics education especially the mathematical objects of content and process. The content strand entails what is taught whereas the process strands entail how the content is taught. This chapter distils the effects of the two vectors on the process strand. The process strand includes problem solving, reasoning and proof, communication, connections, representations and use of mathematical tools. Globalization and technology have influenced the teaching of mathematics in a number of facets that include use of 21st Century teaching methodologies, influence of Artificial intelligence and mathematics education, ICT integration in the teaching of mathematics, communities of practice and assessment.

This chapter explores the impact of artificial intelligence (AI) on mathematics education, including teaching, research, and administration. As AI reshapes life, it also transforms how mathematics is taught and learned. The chapter reviews how AI enhances instructional methods, offering personalized learning and interactive ways to understand complex mathematical concepts. It also discusses AI's role in mathematics research, particularly in solving complex problems, and its influence on the research field. Additionally, the chapter covers AI in administrative tasks, like curriculum design and policy-making, using data analytics. It addresses challenges, including ethical issues and the need for educators to adapt. The chapter concludes with future prospects for AI in mathematics education, advocating for a balanced approach that leverages AI's potential while upholding the core principles of mathematical learning and inquiry.

Globalization has a significant bearing on mathematics teacher education and there is need for a comprehensive restructuring to meet the demands of the interconnected world. This study advocates for a paradigm shift that encompasses cross-cultural competencies, linguistic diversity, and a nuanced understanding of global educational trends. It further highlights the significant influence of virtual learning environments (VLEs) in responding to the challenges posed by globalization and emphasizes the synergy between VLEs and heutagogical practices. Drawing from the cited case studies and other existing literature, this chapter underscores the departure from conventional pedagogical methods towards a learner-cantered, self-determined approach. The chapter concludes by discussing the far-reaching implications for policy, practice, and theory, urging stakeholders to recognize the benefits of heutagogy in preparing mathematics teachers for the complexities of the twenty-first century teaching and learning.

Mathematics is a very important human venture that finds its applications in all spheres of life. Despite this importance attached to it, the interest and performance in it is dismal. This is as a result of how it is taught. There is every need for teaching it better, thus the need for innovative and powerful approaches. This is this concept is "used to identify the knowledge and capabilities required to accommodate the mathematical demands of private and public life, and to participate in society as informed, reflective, and contributing citizens." Therefore, letting the students make sense of the mathematical content they learn can have a significant positive impact on multiple dimensions of their lives. Math enhances our capacity to deal with these kinds of day-to-day challenges. There is a need to develop innovative learning environments that benefit all learners. The second theme is centered on how both tasks and tools can be used to support powerful pedagogical practices. Finally, the third theme reviews the challenges of developing innovative mathematical learning environments.

This chapter will give a thorough overview of the current state of mathematics education in the public universities in Burundi. The chapter examines how instructors in Burundi's universities are adjusting to the changing nature of education, with a particular focus on creative teaching approaches and practices. Understanding, identifying, and addressing the unique challenges and opportunities in the context of mathematics teaching in Burundi's higher education institutions are the main goals and concerns of this chapter.

This chapter is based on the systematic review that investigates assessment methods in mathematics education in the global era where pertinent skills to address specific problems are at the centre of assessment purposes. Of 442 records obtained from Google Scholar, 50 studies published between 2013 and 2023 were reviewed. These studies, sourced from journals, conference proceedings, and book chapters indexed by Scopus, were analysed using document and thematic analysis. The review aimed to identify the assessment methods applied, the specific skills measured, and how these methods are implemented. Findings revealed that formative assessment, including performance-based, peer, oral, observation, written tests, portfolios, self, and open-ended methods, is central to classroom practices. These methods assessed mathematical and generic skills such as problem-solving, reasoning, thinking, representations, conceptual understanding, and feedback generation. The review provides insights into these methods and their implications for mathematics education in the globalized era.

The teaching of mathematics globally at university level, has created an environment that favours the performance of men over women. Men tend to perform better than women at university level. The gender gap has been explained by using biological and social cultural theories. It is also noted that at doctoral level the number of women who earn degrees is very minimal and the women graduates tend to be viewed as less competent when compared with men. The study employs poststructural feminism, socialcultural, and situated learning theories to explore the gender gap in mathematics. Gender stereotypes seem to have a major impact on performance of women in mathematics.

From 2012 to 2018, pupils taking the Kenya Certificate of Secondary Education (KCSE) routinely performed poorly in Mathematics on a national level in Kenya. In KCSE Test data from 2012 to 2018, it was noted that probability was one of the mathematical concepts that the majority of pupils found challenging. A novel approach that advocates for the incorporation of computer assisted learning methods like simulation has been put out to address the issue. In this study, learner achievement in probability in Mathematics will be compared to computer-based simulation (CBS) effect in public secondary schools in Kisii County. The objective of the study will be to determine the effect of computer simulation (innovation) on students' achievement in probability by sex (mathematics education).

Editorial Advisory Board

Preface

In an era of ever-emerging issues, globalization and innovation have been instrumental in educational progression. Globalization is a dynamic and multifaceted construct, and thus education, particularly mathematics education, cannot be assumed to remain unchanged during these emergent times. Globalization explores how culture, language, global migration, equity, equality, access, inclusion, ethical issues, technological advancements, climate change education, gender issues, technology, collaborations and exchange programs, pedagogical policies and practices, theories and the process of theorizing, models, and teacher training, among others, influences and impacts on the world and its people. Influence and impact in mathematics education include curriculum development, implementation, assessment, proficiency, administration, and research. With globalization, classrooms have become more diverse: different cultures bring varied perspectives on and about mathematics, which can influence teaching strategies and curriculum development. Educators now strive to incorporate culturally relevant examples and teaching methods into mathematics education to cater for diverse student populations. Globalization also affords space for comparative studies of mathematics education across the globe, thereby benchmarking for best practices and learning from challenges. Reporting of issues in mathematics education through local and international media influences the perceptions and understandings of the report consumers on the subject.

In the recent past, there have been increased innovations in mathematics education pertaining to instruction, assessment, and resources necessary in classrooms. While the innovations may help to improve students' classroom participation and performance in mathematics, there are issues of incorporating these innovations into policies, curricula, and actual practice. The innovations may also conflict with some cultural identities.

The publication "Impacts of Globalization and Innovation in Mathematics Education" delves into the multifaceted impacts of globalization and innovation in the field of mathematics education. The objectives of the book are to:

- Inform on global trends: With the cross-cutting research, the book informs on global, regional, and local trends in mathematics education that have been influenced by globalization and innovation.
- Provide insights: The book provides knowledge and insights into how globalization and innovation have impacted mathematics education.
- Provide analysis: An in-depth analysis of local, regional, and global practices in and out of the mathematics classrooms are provided, hence identifying best practices, challenges, and opportunities.
- Inform future research: The cross-cutting findings from the chapters shall help identify research gaps, contestations, and even agreements on the impacts of globalization and innovation in mathematics education, thus prompting further research.

The book will contribute to knowledge development in mathematics education in the light of globalization and innovation. It is intended for mathematics and science researchers, educators, postgraduate students, and teachers. Curriculum developers, investors in mathematics projects, innovators, and industry professionals will also find the book valuable.

Organization of the Book

The book is organized in **Eleven** chapters and further into **Four** themes. The themes are **Pedagogy** (Chapters 1-5), **Assessment** (Chapter 6), **Gender** (Chapter 7-8), and **Equity** (Chapter 9-11). The content in each chapter builds on one or more theme(s) of concern from another chapter. The book directly addresses SDG4 and indirectly SDG1 and SDG5 hence addressing global concerns.

The first Chapter sets the pace of the book by discussing the impact of globalization and innovation on pedagogies in mathematics classrooms. With mathematics education being a discipline which borrows from a variety of other disciplines, the author notes that the impact of globalization and innovation is evident in mathematics classrooms. The chapter details the impact on methodologies of teaching mathematics with the emergence of new pedagogies and assessment methods and techniques. Further, the chapter unpacks theories and models informing use of technology in mathematics learning. The chapter concludes by noting the impact of globalization and innovations in the mathematics classroom to include the changed role of a teacher to being a facilitator, methods of instruction being more learner centered, use of ICT in promoting conceptual understanding and expanded means of sharing of knowledge from local to global contexts and vice versa.

Building on AI, Chapter two presents a discussion on redefining mathematics education in the age of Artificial Intelligence with an eye on challenges and opportunities afforded by AI. The chapter delves into the transformative impact of AI technological innovations on mathematics education, encompassing instruction, research, and administration. It examines the evolving role of AI in enhancing instructional methodologies, exploring the intersection of AI and mathematics research and a focus on how AI-driven data analytics are employed to refine curriculum design, assessment strategies, and educational policy-making in mathematics education. The author critically analyses the challenges and proposes future directions for integrating AI into mathematics education, emphasizing the need for a balanced approach that harnesses the potential of AI while maintaining the fundamental values of mathematical learning and inquiry.

Chapter three presents a discussion on mathematics teacher education in a globalised age with a focus on heutagogy and Virtual Learning Environments (VLEs) in context. The author advocates for a paradigm shift that encompasses cross-cultural competencies, linguistic diversity, and a nuanced understanding of global educational trends. The chapter further highlights the significant influence of VLEs in responding to the challenges posed by globalization and emphasizes the synergy between VLEs and heutagogical practices. The chapter underscores the departure from conventional pedagogical methods towards a learner-centered and self-determined approach. The chapter concludes by discussing the far-reaching implications for policy, practice, and theory, urging stakeholders to recognize the benefits of heutagogy in preparing mathematics teachers for the complexities of the twenty-first century teaching and learning.

In chapter four, the author argues for innovative pedagogical practices in mathematics education. The author notes that innovation and technology have the potential to revolutionize the way mathematics is taught and learned. By giving some ways in which this can happen, the author explores interactive learning practices, problem based learning and AI among others.

Chapter five titled Innovative Pedagogical Approaches and Practices for Mathematics Education in Public Universities of Burundi provides overview of the current state of mathematics education in public universities in Burundi. The chapter examines how instructors in Burundi's universities are adjusting to the changing nature of education, with a particular focus on creative teaching approaches and practices. Insights into how cutting-edge instructional strategies are being applied in mathematics classrooms and integration of technology and digital tools into mathematical instruction is examined. The chapter also analysis difficulties that arise when putting innovative pedagogies into practice and possible solutions are discussed. The findings of the study are intended to inform educators, scholars, and policymakers who aim to improve the quality and relevance of mathematics education in Burundi. By examining both successful practices and challenges, the chapter contributes to the current conversation on educational innovation and development within the unique context of Burundi universities.

Chapter six provides insights on assessment in mathematics education by focusing on methods and skills assessed in the era of globalization. This chapter investigates assessment methods in mathematics education in the global era where pertinent skills to address specific problems are at the centre of assessment purposes. From the empirical study, findings revealed that formative assessment, including performance-based, peer, oral, observation, written tests, portfolios, self, and open-ended methods, are central to classroom practices. These methods assessed mathematical and generic skills such as problem-solving, reasoning, thinking, representations, conceptual understanding, and feedback generation. The review provides insights into these methods and their implications for mathematics education in the globalized era.

Chapter seven introduces matters of gender in the book by exploring globalisation gender sensitivity in the teaching of mathematics in higher education. Research shows that male students tend to surpass female students in performance in mathematics at university level. The number of females who engage in mathematics in higher education also declines as they move up the academic spectrum. The chapter, therefore, discusses matters of gender sensitivity and identity and further the theories that underpin gender sensitivity. The chapter further presents the global trends on gender equality and use of technology in mathematics education.

Chapter eight addresses matters of gender and innovations in mathematics education. In teaching and assessment of mathematics, probability is one of the mathematical concepts that the majority of students find challenging. The chapter sought to address this concern by determining the effect of computer simulation on students' achievement in the concept of probability by gender. In this study, learner achievement in probability in Mathematics was compared to Computer-Based Simulation (CBS) in public secondary schools in Kenya. The findings showed that the use of CBS being an effective teaching strategy for both male and female learners.

Chapter nine addresses matters of equity and inclusion in mathematics education in with particular focus on strategies for supporting diverse learners in the context of India. The chapter examines the disparities in mathematics performance based on gender and social categories, utilising data from the National Achievement Survey of India. It explores the significance of equity and inclusion in mathematics education and pedagogical approaches aimed at fostering equity and inclusivity. The findings of the study show that equity and inclusion in mathematics education necessitate a comprehensive strategy that encompasses policy reforms, allocation of resources and infrastructure, teacher professional development, cultural sensitivity, and the establishment of inclusive learning environments. The findings of the study offer strategies to support learners from diverse backgrounds and help educators, policymakers, and educational administrators to establish an equitable and inclusive mathematics education.

Chapter ten is titled Math Achievement: How Does Income Influence Children's Third Grade Math Achievement? Drawing from family investment models, the chapter examined parental monetary investment in essential mathematics materials and mathematics tutoring as mediators of the association between family income and mathematics achievement during third grade. In this empirical study, data were collected from third graders and their mothers in Bangladesh. Results showed that there were significantly positive and direct relations between low-income, fewer monetary investments, and poor mathematics outcomes during third grade, after accounting for child and maternal characteristics. When mediators were included in the model, the estimated significant pathways from low-income to poor mathematics outcomes were decreased, but the associations were still significant. The results suggested that parental monetary investment in mathematics tutoring and number word was more important than other materials in children's early primary mathematics achievement. Future directions for research and policy implications are discussed.

Chapter eleven addresses language use in mathematics education. Titled Foundational-Level Mathematics Teaching Using Learners' Mother Tongue, the chapter provides a summary and interpretation of study findings on the use of mother-tongue (MT) in teaching of mathematics in South Africa. Past findings that show that the use of MT as a language of instruction in mathematics leads to higher-quality instruction and improved learner performance and that it is easier for teachers to teach mathematics in their MT. However, most educators struggle with the transition of content and concepts from English to MT. For a long time, children in township schools in South Africa went to class learning a language they were unable to comprehend. As a result, this article focuses on how to use MT in a math classroom. The chapter address the concern whether mathematics educators had a sufficient grasp of mathematical language to adequately instruct their learners. The chapter also discusses the challenges that educators experience while using MT to teach mathematics concepts. The findings of the study inform policy development of Language of Learning and Teaching (LoLT) not just for South Africa but also other context.

In Conclusion

In conclusion, the profound impacts of globalization and innovation in mathematics education cannot be overstated. This book, "Impacts of Globalization and Innovation in Mathematics Education," offers a comprehensive exploration of these influences and impacts, reflecting on the dynamic and evolving landscape of mathematics education worldwide.

Our aim has been to provide readers with understanding of how globalization has transformed educational practices, fostering diversity, inclusivity, and intercultural exchange in mathematics classrooms. Similarly, we have delved into the myriad ways innovations are reshaping teaching methodologies, assessment strategies, and resource utilization, while also acknowledging the challenges and cultural tensions that may arise.

Through the diverse perspectives and in-depth analyses presented in this volume, we hope to inform and inspire educators, educational administrators, innovators, curriculum developers and implementers, researchers, and policymakers. By highlighting global trends, sharing insights, and providing rigorous analysis, this book aspires to serve as a valuable resource for those dedicated to advancing mathematics education in our interconnected world.

We anticipate that the findings and discussions contained herein will spark further research and dialogue, helping to identify new pathways for improvement and collaboration. As the field of mathematics education continues to evolve, it is our sincere hope that this book will contribute to a deeper understanding and appreciation of the complex interplay between globalization, innovation, and educational practice.

We extend our heartfelt gratitude to the contributors, whose scholarship and dedication have made this publication possible. It is our collective effort that brings this book to fruition, and we are confident that it will serve as a significant contribution to the ongoing discourse in mathematics education.

Evelyn Njurai
Kisii University, Kenya

Alphonse Uworwabayeho
University of Rwanda, Rwanda

Section 1

Pedagogy

Chapter 1
Impact of Globalization and Innovation on Pedagogies in Mathematics Classrooms

Benson Njoroge Wanjiru
https://orcid.org/0009-0003-4607-2317
Mount Kenya University, Kenya

ABSTRACT

Today, two vectors shape the world-globalization and technology; the former control the economies, politics, and cultures, and the latter dictate how fast they happen or spread or shared within the global space. The two vectors have had a great impact in the field of mathematics education especially the mathematical objects of content and process. The content strand entails what is taught whereas the process strands entail how the content is taught. This chapter distils the effects of the two vectors on the process strand. The process strand includes problem solving, reasoning and proof, communication, connections, representations and use of mathematical tools. Globalization and technology have influenced the teaching of mathematics in a number of facets that include use of 21st Century teaching methodologies, influence of Artificial intelligence and mathematics education, ICT integration in the teaching of mathematics, communities of practice and assessment.

DOI: 10.4018/979-8-3693-2873-6.ch001

BACKGROUND

Globalization may be termed as the coming together of world economies, politics and cultures (Volle, 2023). Theodore Levitt, a German- born American Economist is credited with the term globalization. In 1873, he wrote an article titled, "Globalization of the Markets" (Levitt, 1983). The concept was widespread in 19th Century with the advent of Industrial revolution. This has been escalated with innovation of modern technologies in the communication and transportation as well as adoption of liberal trade policies by counties in the world. The concept may be viewed as "removal of territories-geographical boundaries-the world being seen as a global village. The social activities which could be political, economic or cultural takes place in disregard of geographical location; meaning that people worldwide can connect and share experiences and information without being physically present in the same location or time. Modern technologies have enabled people to connect worldwide enhancing interconnections and interdependence of countries in the world.

Today, two vectors shape the world-globalization and technology (Levit,2008). The former control the economies, politics and cultures and the latter dictate how fast they happen or spread or shared within the global space. These two vectors have had impact on all sectors of the economy; education being one. Mathematics education being a discipline which borrows from a variety of disciplines such as psychology, sociology, philosophy, history, semiotics, cognitive science, post-modernism has not been spared. Technology is one of the drivers of modern innovations. Technologically driven innovations have changed the nature of teaching and learning in Mathematics classrooms (BECTA, 2023). New pedagogies have emerged; new mathematical concepts have been developed; new ways of solving global problems using mathematical models have emerged; efficient ways of sharing mathematical knowledge in communities of practice have been created; assessment techniques have been revolutionised. Innovations in mathematics educations as result of globalization and technology are endless.

A fundamental question then is, "What are the pre-existing problems in mathematics education that have called for the mathematics educators to embrace technology as a driver of improving and globalizing mathematics education"?. The importance of mathematics cannot be overemphasized. Mathematics is applied in everyday life activities, in the workplace, in the industry, as tool for other subjects, for further education and training, and in research. However, students' achievement in mathematics has been low (Kihwele & Mkomwa, 2023). Mathematics is an abstract subject and most teachers approach the teaching using traditional methods (Sandhu & Grover, 2017). This has resulted in students losing interest and as a result low performance. In addition, tradition methods of 'talk and Chalk' and 'drill and skills' does not afford the learners opportunities for acquisition of higher order skills like critical and analytical thinking, problem-solving, logical reasoning and decision making. The resultant is learners with instrumental rather than relational understanding. In response to this, mathematics educators have come up with innovations mostly driven by technology to bridge the gaps. In addition, in this era of information sharing and collaborations, mathematics educators share their innovations. The twin vectors of globalization and technologically-driven innovations have had an impact on the pedagogies in mathematics education. This chapter dissects the impact of innovations and globalization on pedagogies in Mathematics classrooms.

Impact of Globalization and Technology on teaching methodologies in Mathematics

Innovations in Technology has had great impact in the mathematics classrooms. They revolutionized the teaching methodologies of the teachers of mathematics. Before, we delve into revolution of the teaching methodologies, it is important to understand the characteristics of the 21st century mathematics classroom. The 21st Century classroom is a collaborative environment in which learners can acquire competencies that are required in daily life, the workplace, industry and in research and teachers are facilitators of their learning. Globalization and technology have rendered the traditional classroom ineffective and mathematics teachers have adopted and have to adopt to new teaching strategies. The 21st century classroom is more learner-centric and the teachers take the role of facilitators and guides instead of being reservoirs of knowledge. This entails using a variety of instructional methods and following pedagogical approaches aided by ICT. Saxena (2013) expounds on the characteristics of the 21st century classroom. These include student-centric, contains computing devices, involves active learning, adaptive learning, invitational environment, performance-based assessment and collaborative learning among others. This implies that the methodologies adapted for the mathematics classroom must facilitate learning and not mere exposition. Beswick *et al* (2023) point out three major international trends in instructional discourses within the mathematics classrooms; greater advocacy in student-centric pedagogies, promotion of reasoning and problem-solving and increased interest in regular assessment of learners' progress rather computational and procedural skills.

Theories Supporting Use of Technology in the Learning of Mathematics

This book chapter shall limit itself to the Cognitive Theory of Multimedia Learning (CTML) by Richard Mayer and TPACK model

Cognitive Theory of Multimedia Learning (CTML) by Richard Mayer

Mayer and Moreno's (1999) Cognitive Theory of Multimedia Learning (CTML) posits that for meaningful learning to take place information should be presented in both textual and graphical formats rather than textual format only (MC Graw Hill, 2029). Mayer states that the brain takes in and processes information using two channel based on how the information is presented. Hence, the Dual Coding Theory (DCT) assumption. The first channel is for visually represented material and the second is for auditory represented material. According to this theory, knowledge is represented and manipulated through two cognitive channels: visual-pictorial and auditory-verbal. CTML asserts that the words and visuals that we choose to use during the instructional process are important and impactful. These two channels are used to process information in the working memory. A leaner must coordinate and monitor five cognitive processes in the working memory in order to learn in a multimedia environment. These includes selecting relevant words from a multimedia message; relevant visuals from a multimedia message; building connections among certain words to create a coherent verbal model and pictorial models and integrate these models and with prior message (Malamed, 2023). This is shown in Figure 1.

Figure 1. Cognitive theory of multimedia learning (CTML)

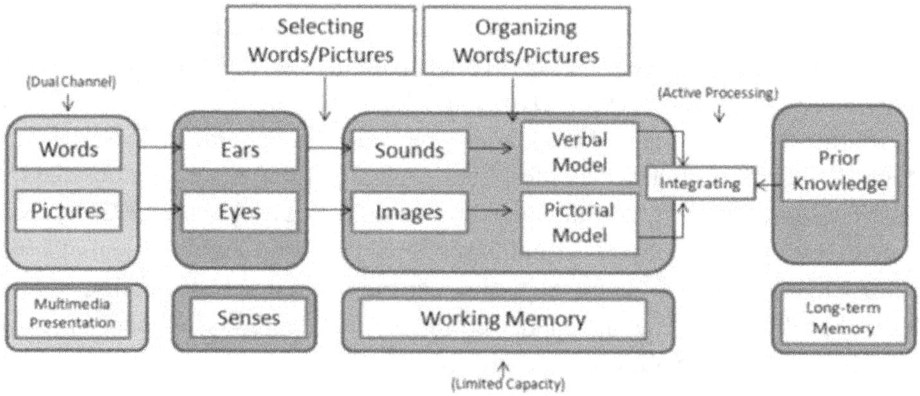

(Malamed, 2023)

The implications of the Model for the Mathematics classroom are that a teacher should:

(i) present content and instruction in both textual and pictorial formats.

(ii) These texts and graphics should be aligned.
(iii) Use narration whenever possible but should not use narration and on screen text at the same time.
(iv) Avoid using images, audios, transitions and other elements that are not related to the content as they act as distractors.
(v) Use conversation style during instruction
(vi) break down mathematics lessons into small sessions of 10-15 minutes.

TPACK Model

Mishra and Kohler (2006) developed the TPACK framework. The model postulates that in order to effectively integrate knowledge into instruction, teachers need knowledge in three domains or areas: Technological Knowledge (TK), Pedagogical Knowledge and Content Knowledge (CK). Sinha (2023) gives a teacher's guide to the TPACK technology Integration model. TPACK (Technological, Pedagogical Content Knowledge) framework proposed by Mishra & Koehler (2006) builds upon the foundation of Pedagogical Content Knowledge formulated by Shulman. It offers a comprehensive approach to teaching and learning in the digital era as involves the integration of Technology, Pedagogy and Content. This model acknowledges the dynamic relationship between Technological Knowledge (TK), Pedagogical Knowledge and Content Knowledge (CK). Technological Knowledge (TK) involves the educator's understanding of the tools, resources and applications available for use. This includes knowledge on hardware, software, multimedia tools, educational platforms, and digital resources. Pedagogical Knowledge (PK) is the art and science of teaching. This involves the tutors' understanding on psychology of teaching and learning, effective instructional strategies, classroom management techniques, assessment techniques and ability to create a conductive learning environment. Content Knowledge (CK) refers to the teacher's deep understanding of the subject matter being taught and ability to present concepts in a manner that is accessible and meaningful to the learners. TPACK framework emphasizes the intersection and integration of the 3 components as in Figure 2.

Figure 2. TPACK framework

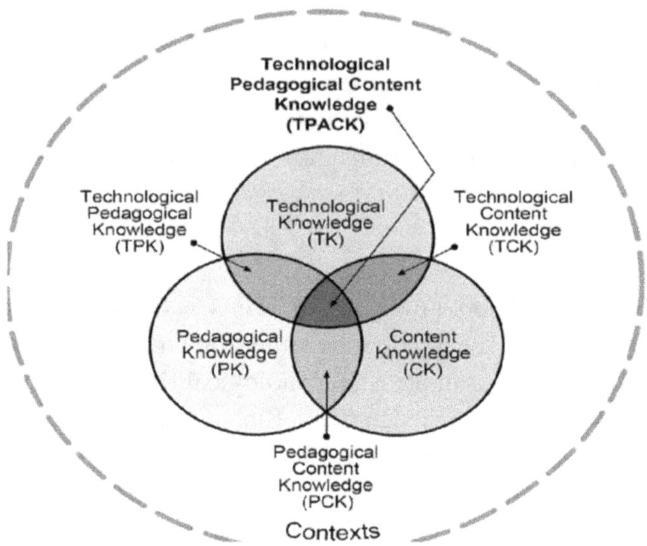

(tpack.org)

The interaction between the three primary knowledge domains give rise to three secondary domains. They entail Pedagogical Content Knowledge (PCK), Technological Content knowledge (TCK) and Technological Pedagogical Knowledge (TCK). According to Mishra & Koehler, PCK is the "knowledge of pedagogy that is applicable to specific content" (P.1027). On the other hand, TCK is the "Knowledge about the manner in which technology and content are reciprocally related" (P.1028). Finally, there is the TPK, that is, "knowledge of existence, components, and capabilities of various technologies as they are used in teaching and learning settings, and conversely how teaching might change as the result of using particular technologies" (p.1028). The interaction of these domains birth TPACK- technological pedagogical content knowledge. This is the ideal knowledge needed by a teacher to effectively integrate ICT in teaching.

The knowledge needed to use technology in the mathematics for conceptual understanding is Technology Pedagogical Content Knowledge, or TPACK. (Rakes, 2022). The TPACK model guides teachers on how to use digital technology to deepen learners conceptual understanding in Mathematics. The CK is the 'what—subject matter; the teacher must be well versed with mathematical knowledge acquired during pre-service or in-service or by engaging with colleagues in communities of practice. This involves the conceptual understanding of the content strand: Numbers

& Operations, Algebra, Geometry, measurement and probability and data analysis. The PK is 'the how'- Methodology or pedagogy. A teacher must be able to use a repertoire of strategies to facilitate content-CK. This entails the process category of the mathematics objects. The TK is the knowledge of the technology in education, that is, the knowledge of ICT tools and aids are need to facilitate learning.

The TPACK model encourages teachers to explore how technology can be leveraged to transform instruction and engage learners in deep learning experiences.

An Example of Application of TPACK Framework in Integrated Lesson Plan in Mathematics

County: Kenya
Curriculum: CBC Curriculum
Strand: Geometry
Sub-strand: Common Solids

Specific Learning Outcome: By the end of the sub-strand, the learner should be able to use IT devices or other materials to draw models and nets of solids in different situations

Technological Knowledge (TK)

- The learner will use IT devices to draw models and nets of solids
- The teacher to use authentic assessment tools to monitor learners progress and provide feedback. These include tools like Mentimeter, Quizizz, Polls, google forms, Socrative, Kahoot, e-portfolios and Warm-ups and Exit tickets among others.

Pedagogical Knowledge (PK)

- the learner will engage in hands-on-activities collaboratively in small group, using digital devices to solve problems related to common solids.
- The teacher to use authentic assessment tools to monitor learners progress and provide feedback.
- The learners/Groups to share their experiences with the entire class.

Content Knowledge (CK)

- Learners collaboratively Sketch nets of cubes, cuboids, cylinders, pyramids and cones in different situations
- will apply mathematical reasoning to solve geometry problems using visual representations.

ICT Integration in Mathematics Education

The ability of the teacher to explain concepts is facilitated by technology as the he or she can model abstract ideas and concepts. Unlike the traditional setting where the teacher relied on drawings and hand-made models, modern technology offers a wide range of instructional tools. This enables the teacher to model realistic problems in the class setting. Mathematics uses problem solving to create contexts that simulate the real-life situation. This is the interface that is bridged by technology. ICT tools and software are being used to enhance student's visualization of concepts in topics like geometry and trigonometry and modelling power to solve problems. Research has shown that technology enhances deeper conceptual and relational understanding of mathematical concepts and algorithms. Students using ICT tools such as calculators and computers are able to operate at a higher of abstractions.

There are several functions that ICT play in learning mathematics. Key among these is in Interaction and communication. This may involve the use of *flipped learning technologies* which enable learners to watch videos in any topic well in advance before the teacher goes to the Class. This enables them to prepare themselves to take part in discussion in the main classroom. This will enhance understanding of mathematical concepts. *Video conferencing technologies* are other interaction and communication tools. These tools allow both learners and the facilitators to active participants by engaging in discussion while presenting a topic. The class discourses mirror the brick-and-mortar classes safe this is occurring virtually. One more function in this category is the use of *social media*. These are websites and software applications that enables both teachers and learners to communicate, share information, ideas, personal message on internet using computer and mobile devices.

The Learning management system (LMS) software used as platforms that helps to organise, manage, deliver and track courses and programmes. LMS is the digital version of face-to-face classroom. They can be used to deliver mathematics content to learners that separated physically from the facilitators or places there are inadequate teachers. These platforms have been used extensively by institutions of higher learning to facilitate online classes for learners in Virtual and blended modes of study. The pedagogies adopted here include synchronous method of instructional delivery. Here, the teaching and learning takes place in an environment where the teacher of mathematics and learner are available remotely at the same time in the same platform. This happens in real-time with a set class schedule. This takes the form of video conferencing, teleconferencing, live chatting, and live-streamed lessons, scheduled radio and television lessons. The second instructional strategy is asynchronous method. The learners interact with the tutor in the same platform but at different times and or locations. This strategy does not require real-time interaction.

ICT can be used in *simulation*, where learners model or role-paly situations for the purpose of learning concept or evaluation their level of understanding of concepts. Simulations can be range from simple scenarios to complex, highly written or even interactive games. They have the potential to enhance comprehension, increase initial learning and retention as well as transfer of skills and knowledge effectively.

Mobile apps can also be used in ICT integration in the teaching of mathematics. Mobile apps are education software designed for specific use in the teaching-learning process. In the market currently, we have mobile apps like Kahoot, Edmodo, Moodle, among others. Teachers of mathematics with ICT skills of coding can also develop *mobileapp* for collaboration, communication, assessment and for uploading additional resources that the learners can make references to.

Teachers and learners can also use *audio conferencing*. This is a two way-speech channel of communication over the telephone or a live radio broadcast where both the leaners and tutor are separated by distance. On the hand, there teacher and the learner can engage in a visual communication via video conference-synchronous learning. This can be enabled using software like Microsoft Teams, Zoom, Google Meet among others that facilitate video conferencing. The users connect in real time through audio or video communication over the web. The teachers can explain concepts and the learners can collaborate with peer. This mirrors a classroom only that there is no brick and mortar.

Example of ICT Integration in the Curriculum

Level: GRADE (Kenya)
Strand: Numbers
Sub-strand: Fractions

Specific Learning Outcome: By the end of the sub-strand, the learner should be able to play games of operations on fractions using IT devices

Link from the internet: https://www.mathmammoth.com/practice/order-operations #num=2&operations=add,sub,mult,div,exponents,parens

Figure 3. Hexingo – order of operations game

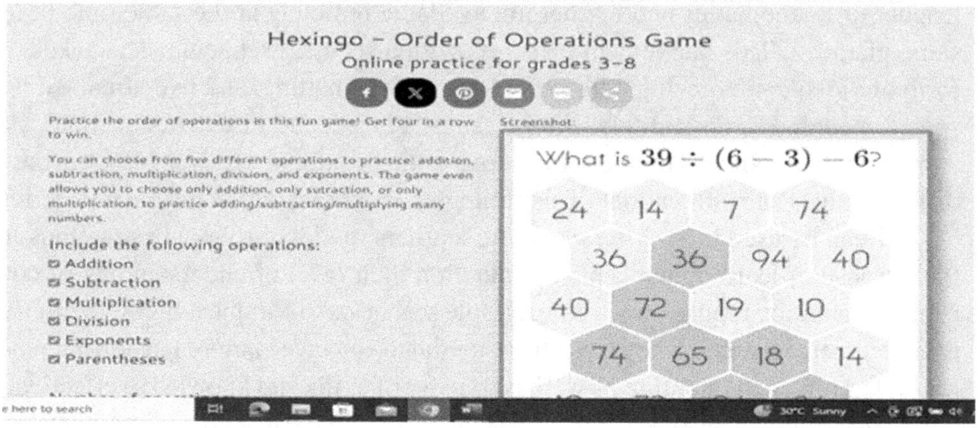

(https://www.mathmammoth.com/practice/order-operations)

In the above game, the learner selects the number of operations, and a problem is generated. He/she is able to then work out the solution and key in the answer. The system is able to provide immediate feedback.

Learners can use ICT tools to develop content in a learning area. The teacher can involve learners in using ICT tools to take pictures of observations, take videos of demonstrations in project work in mathematics or create 2D and 3D art forms using graphic design tools such as procreate. They can also create and send SMS or social media messages to their facilitator or their peers. They can also develop Power Point presentations on different topics in the curriculum designs or syllabus/course.

There are many Open Education Resources (OER) over the internet for use in teaching and learning mathematics be accessed using ICT devices. These are teaching, learning and research materials in digital or print media that can be accessed without limited restrictions at no cost for adaptation and redistribution to others (UNESCO, 2019).

Some of the OER that can be used include:

a) General Collections
   i) Discovery Education (2020) (see https://www.discoveryeducation.com/teachers/) has content in Mathematics, and other subjects, interactive games, puzzles, programs and contest.
   ii) **KICD** Open Educational resource www.kec.ac.ke
   iii) **Khan Academy**https://www.khanacademy.org/
   iv) Mason OER Metafinder

Searches across notable OER collections including MERLOT, OASIS, and OER Commons

v) MERLOT: Mathematics

Searches 90,000+ records with the option to search other OER collections and the web

vi) OASIS: Math

Searches 380,000+ records from 90+ reputable OER sites

vii) OER Commons: Mathematics

Searches 50,000+ records of curated, high-quality OERs

viii) National Science Digital Library: Mathematics

Educational resources for teaching and learning in the STEM field

b) Textbooks
   i) American Institute of Mathematics
      - A collection of free open textbooks covering various mathematic topics
   ii) LibreTexts: Mathematics
      - Next-generation textbook for postsecondary education
   iii) Openstax: Mathematics
      - Textbooks and additional resources on introductory courses
   iv) Open Textbook Library: Mathematics
      - Textbooks used at higher education institutions
   v) Open Textbook Store: Mathematics
      - Open Textbook Store offers a selection of free, online mathematics textbooks
   vi) Teaching Commons: Physical Sciences & Mathematics
      - Textbooks, course sites, syllabi, course materials, lectures and more

Books and Journals

i) arXiv.org
   - E-prints in physics, mathematics, computer science, and more

ii) E-Books Directory: Mathematics

- Books, documents, and lecture notes

iii) InTechOpen: Mathematics

- Books in the science, technology, and medicine fields

iv) Impact of Globalization and Technology on Assessment in Mathematics

Courses and Related Materials

i) Khan Academy: Math

- Practice exercises, instructional videos, and reference materials

ii) MIT Open CourseWare: Mathematics

- Course content from MIT

iii) Saylor Academy: Mathematics

- Courses at college and professional levels

ICT Tools in Mathematics Education

ICT tools available for Mathematics classrooms may include but not limited to:

i) Devices. They include computers, Televisions, Radios, Digital Cameras, LCD Projectors Interactive boards, Smartphones, Data storage devices (CDs, DVDs, Memory sticks among others
ii) Productivity Tools. These include tools like Word, Excel, PowerPoint, Microsoft 365 Office,
iii) Searching Tools that include tools like e-library, Google Scholar etc.
iv) Collaborative Tools among these are WhatsApp, Goggle Doc, etc.
v) Web Based Tools –Kahoot and Mentimeter for gamification, assessment and evaluation
vi) Learning Management Systems. These include google classroom, Moodle, Edmodo, Sakai, Canvas, KICD Cloud LMS.
vii) Conferencing/Interactive Tools. This category include tools like Zoom, Google Meet, Webex, Microsoft Teams Google Hangouts, Skype, Meta platform(Facebook Live), YouTube live among others.
viii) Meeting Tools (i.e., Attendify, Calendar, etc.)
ix) Assessment Tools. Some of the tools include Google Forms, Exam.net,
x) Creating Resources (i.e. Recording, Editing, Wikis, Podcasting etc.
xi) Cloud storage-google drive, OneDrive

Artificial Intelligence and Mathematics Education

Artificial Intelligence (AI) has the potential to revolutionize the learning of Mathematics. Sinha (2023) posits that AI has the potential to change the pedagogies in Mathematics Education by providing personalized learning experiences, improving student engagement and enhancing problem-solving skills. Mathematics teachers can create adaptive and dynamic learning environments that cater to the individuals needs and learning styles of students. Figure 3 demonstrate the many parts of the AI system.

Figure 4. Artificial intelligence

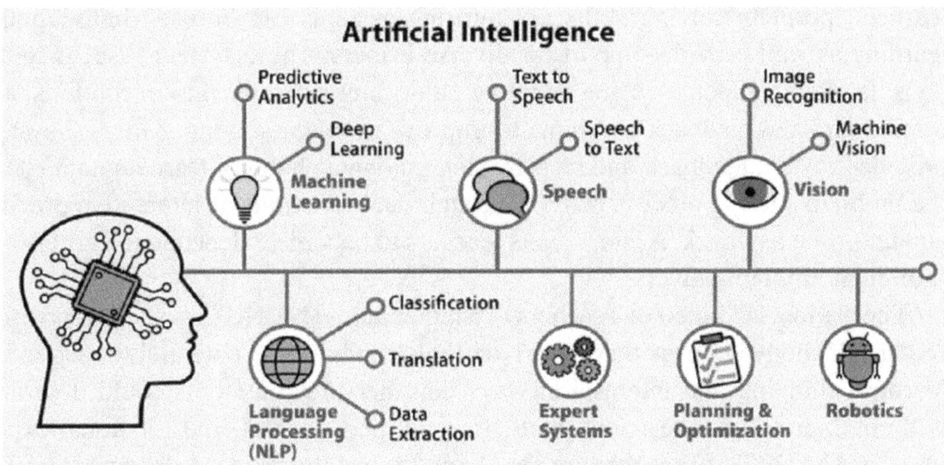

(https://www.engineering-admission.in/direct-admission-in-bangalore-for-ai-and-ml/)

In a traditional class, the teacher progresses at a predetermined pace and these often leaves some students behind. AI- powered platforms like *Khan Academy* and *Smarttick* can bridge such a gap by utilizing machine learning algorithms. AI can be utilised in offering help to a learner, for instance who is experiencing difficulties in Geometry can be provided with more practice exercises with specific explanations on the areas of difficulty. The gifted learner can be provided with more advanced areas to explore, advanced exercises testing higher thinking skills or project work activities. AI technologies used utilised to enhance learners' participation in the learning process. For instance, Virtual Reality (VR) and augmented reality (AR) applications can be used to create visual images of abstract mathematical concepts in Mathematics. Learners can to explore 3-deimensional geometric shapes, manipulate objects to understand spatial relationship, and visualize complex functions in a dynamic virtual space. These hand-on learning activities make mathematics more concrete removing the abstraction's currently being taught in classrooms enhancing conceptual understanding of mathematical ideas.

Teachers of Mathematics can use the AI tools to model real-life situations exemplifying concepts students are learning. Dykema (2023) cautions on the on the implications that AI has on what we teach. Questions requiring computational skills are area solved with AI tools like *ChatGPT* or *photoMath*. The implication is that teachers of mathematics must reconsider the types of questions they asking in mathematics. implies that we must reconsider the types of questions we are asking in mathematics (Dykema, 2023). The primary goal of teaching and learning mathematics is problem-solving. Learners should be able to solve complex unfamiliar problems in a variety of settings. AI technologies can be used to effectively develop learner's problem-solving skills. Ai tutoring systems can be used individualize learning as well provide support for diverse leaners. These systems use AI to analyze learners' responses using adaptive algorithms, identify misconceptions, and offered targeted feedback and hints to improve problem-solving skills. AI system provided instant feedback and step-by-step guidance, helping learners understand the problem solving process better. Through dual functions of interactive practice and real-time feedback, AI empowers learners to become independent learners and proficient problem-solvers.

The National Council of Teachers of Mathematics (NCTM) has spelt its position on AI. The body believes that AI driven tools can have the potential to respond to learner's thinking and interests in ways no other previous tools could. Learners will continue to need teachers' mathematical, pedagogical, and relational expertise (NCTM, 2023). Teachers are also likely to benefit from AI-driven tools. The teachers' role will be creating links between prior knowledge, new knowledge and shared knowledge. teachers must keep abreast with developments in AI so as be ready to assist the learners. NCTM takes the position that AI tools do not replace the need to teach Mathematics or problem solving. Further, NCTM contends that AI tools encourage teachers to reimagine teaching and assessment. AI tools excel at recognizing and solving computational problems. Finally, AI tools have the potential to personalize learning. They can be used to provide personalized learning experiences as well as assessment. AI can also be used to globalize the teaching, learning and research in mathematics education. AI can be used in collaboration of learners across the globe. Teachers of mathematics have the opportunity to use AI of mathematics to form communities of practice and engage remotely with peers. AI can be use in global research in mathematics concepts, applied mathematics in field like medicine, engineering, marine, aviation among others.

Gamification in Mathematics Education

Best (2019) defines gamification as "overlaying a set of rules and rewards on educational activities: The value of games and gamification in mathematics can be overemphasized. Mathematics can be made more engaging and fun by the use of games. There is software such as *Mathletic*s which transforms multiplication into a competitive game. Learners can engage with gaming activities. The activities are designed in a manner that engage students with mathematics learning through such features such as competition among peers, point earning system, a personal avatar that they can update with in-game points and animated progress bars (Best, 2019).

Gamification is a suitable teaching strategy when the teacher wants to rekindle his/her learner's enthusiasm for mathematics, to develop procedural practice and fluency and also make real-worlds connections. Gamification has benefits of encouraging creative conjecture making, fostering and integrating learners into collaborative environments of teamwork and promoting learning on own initiative. Obviously, there certain drawbacks of gamification as a learning tool. Top on the list is addiction as games are addictive in nature. Learners are likely to focus on obtaining the individual achievements rather than understanding the mathematical concepts. Yet still, learners may lack the applicability of the concepts learner as they lack of transferability of the concept to real life situations.

Globalization, Technology, and Assessment in Mathematics

Assessment is another area in mathematics education that has been revolutionized by globalization and innovations. Computer based assessment may take the form of Assessment ***As***, assessment **For** and Assessment **Of** Learning. The tools used in Assessment For Learning include Google Forms, Menti meter, Vocaroo, Kahooit!, Aswer Garden, Quizlet, Flipgrid, Flex Cli, Slido, Ted-Ed among others(Rao, 2020).

Google Docs have also been used to serve as electronic learners books. The learners' wok out exercises and the teachers are able to grade them allowing powerful *assessment FOR learning*. The teachers can be to give feedback; address conceptual errors and share good examples instantaneously to the class or individually. This saves a lot of time in the feedback process. The learners also work on several books. The learners have a forum to ask interactive questions and prompt through additional resources or scaffolds. This affords the opportunity not to overload the learner in an online environment. There are softwares that use AI to proctor assessment in mathematics. The teacher can use AI to carry out any form of assessment in mathematics with the help of software. This can help save trees and indeed help reduce the effects of climate challenges. This can also save money by reducing or elimination the need for physical testing centres.

Challenges of Using ICT in Mathematics Education

While AI had the potential of revolutionizing the teaching and learning of mathematics, there are certain challenges and ethical issues that need to be addressed. There are also challenges that come with AI technologies. Teachers need to be ready to validate and verify the correctness of the AI generated exercises and solutions to prevent misconceptions and misleading learning outcomes. The AI systems must also be designed in a way they safeguard learners' privacy and innovations. Additionally, the issue of data privacy and security arises when implementing AI in education. It is essential to safeguard students' personal information and ensure that AI systems adhere to strict privacy regulations.

CONCLUSION

Globalization and Technology have changed the role of the teacher from Expert and knowledge transmitter to that of a facilitator, coach and contractor. Innovations that are technologically have revolutionised the pedagogies in the mathematics classrooms. The methods are more learner centred, friendly and self-paced. Integration of ICT in the classroom has promoted conceptual understanding of mathematics. AI can be used to proctor examinations. Globalization have also influenced the sharing of knowledge among the teacher educators. Teaching of mathematics can be done synchronously and asynchronously. The twin vectors of globalization and innovations that are technologically motivated have led to the evolution of new pedagogies for use in mathematics classrooms, new knowledge in mathematics, discoveries in mathematics and applied field and in collaborative research. As the technologies continue to advance and more innovations done, we expected more sophisticated learning platforms. Technology and globalization can facilitate collaborative learning by connecting peers globally enabling them exchange ideas, solve problems together and develop a global perspective to issues leading to global peace. The pedagogies in mathematics education have been globalised and revolutionized by globalization and technologically driven innovations.

REFERENCES

BECTA. (2003). *ICT Research. What the Research Says about Using ICT in Maths*. British Educational Communications and Technology Agency.

Beswick, K., Rawlings-Sanaei, F., & Tuohilampi, L. (2023). Interactive Mathematics Teacher Activities. In Manizade, A., Buchholtz, N., & Beswick, K. (Eds.), *The Evolution of Research on Teaching Mathematics. Mathematics Education in the Digital Era* (Vol. 22). Springer. 10.1007/978-3-031-31193-2_5

Levitt, T. (1983). The globalization of markets. *Harvard Business Review*, 93–94.

Malamed, C. (2023). *Applying Richard Mayer's Cognitive Theory of Multimedia Learning*. The e-learning coach. https://theelearningcoach.com/learning/learning-theory-and-multimedia/

Mayer, R. E. (2014). Incorporating motivation into multimedia learning. *Learning and Instruction*, 29, 171–173. 10.1016/j.learninstruc.2013.04.003

Mishra, P., & Koehler, M. (2006). Technological pedagogical content knowledge: A framework for teacher knowledge. *Teachers College Record*, 108(6), 1017–1054. 10.1111/j.1467-9620.2006.00684.x

Rakes, C. R., Stites, M. L., Ronau, R. N., Bush, S. B., Fisher, M. H., Safi, F., Desai, S., Schmidt, A., Andreasen, J. B., Saderholm, J., Amick, L., Mohr-Schroeder, M. J., & Viera, J. (2022). Teaching Mathematics with Technology: TPACK and Effective Teaching Practices. *Education Sciences*, 12(2), 133. 10.3390/educsci12020133

Rao, K.U, Prema, V and Subramanya, K. N. (2020). Plickers: An ICT Tool for Formative Assessment and Feedback. *Journal of Engineering Education Transformations, 33.*

Saxena, S. (2013). *Top 10 Characteristics of a 21st Century Classroom.* EdTech Review. https://www.edtechreview.in/author/saomya-saxena/

Sinha, S. (2023). *Artificial Intelligence and Mathematics Education: Revolutionizing Learning for the Digital Era.* LinkedIn. https://in.linkedin.com/in/sanjeev-sinha-3474533a?trk=article-ssr-frontend-pulse_publisher-author-card

UNESCO. (2019). *Open educational resources*. UNESCO. https://www.unesco.org/en/open-educational-resources

Yeh, C., Cheng, H., Chen, Z., Liao, C., & Chan, T. (2019). Enhancing achievement and interest in mathematics learning through Math-Island. *Research and Practice in Technology Enhanced Learning*, 14(5), 1–19. 10.1186/s41039-019-0100-9

KEY TERMS AND DEFINITIONS

Globalization: This is the coming together of world economies, politics and cultures, such that the world is viewed as global village.

Innovation: This is the process of developing new pedagogies in mathematics education which are driven by technology and globalization.

Chapter 2
Redefining Mathematics Education in the Age of Artificial Intelligence:
Challenges and Opportunities

Mohamad Zreik
https://orcid.org/0000-0002-6812-6529
Sun Yat-sen University, China

ABSTRACT

This chapter explores the impact of artificial intelligence (AI) on mathematics education, including teaching, research, and administration. As AI reshapes life, it also transforms how mathematics is taught and learned. The chapter reviews how AI enhances instructional methods, offering personalized learning and interactive ways to understand complex mathematical concepts. It also discusses AI's role in mathematics research, particularly in solving complex problems, and its influence on the research field. Additionally, the chapter covers AI in administrative tasks, like curriculum design and policy-making, using data analytics. It addresses challenges, including ethical issues and the need for educators to adapt. The chapter concludes with future prospects for AI in mathematics education, advocating for a balanced approach that leverages AI's potential while upholding the core principles of mathematical learning and inquiry.

DOI: 10.4018/979-8-3693-2873-6.ch002

INTRODUCTION

Artificial intelligence (AI) has profoundly impacted various aspects of daily life, revolutionizing how we interact with technology and each other (Elliott, 2019). The influence of AI is evident across multiple domains, ranging from everyday conveniences to complex professional environments. In the realm of consumer technology, AI has redefined user experience. Smartphones, home assistants, and personalized recommendation systems on streaming platforms exemplify how AI tailors interactions based on individual preferences and behaviours. These advancements not only enhance convenience but also create more intuitive user experiences.

In healthcare, AI's impact is transformative. Advanced algorithms assist in diagnosing diseases, predicting patient outcomes, and personalizing treatment plans. AI-driven tools analyse vast datasets, identifying patterns that might elude human observation (Zreik, 2024a). This capability is pivotal in developing new treatments and understanding complex medical conditions.

The business sector has also been reshaped by AI. From automating routine tasks to providing insights through big data analysis, AI enhances efficiency and decision-making. In finance, AI algorithms assist in risk assessment, fraud detection, and algorithmic trading, profoundly affecting how financial institutions operate (Cheng et al., 2021).

Transportation and logistics have witnessed significant changes due to AI. Autonomous vehicles, ranging from cars to drones, are becoming increasingly prevalent, promising to revolutionize how goods and people are transported. AI optimizes routes and logistics operations, significantly reducing costs and improving service quality (Abduljabbar et al., 2019).

In education, AI tools offer personalized learning experiences, adapting content and pace to individual learner needs (Delgado et al., 2020). This personalization has the potential to bridge educational gaps and provide equitable learning opportunities.

AI's impact extends to the realm of entertainment and media, where it is used to create more engaging content and interactive experiences. In video games, AI enhances realism and responsiveness, creating more immersive experiences (Ravyse et al., 2017).

AI plays a crucial role in environmental sustainability. It aids in monitoring climate change, managing natural resources, and optimizing energy consumption (Kaack et al., 2022). AI-driven analysis contributes to more informed decisions regarding environmental policies and conservation efforts (Nova, 2023).

The advent of AI has initiated a significant shift in the traditional paradigms of teaching and learning mathematics. This shift is characterized by a move from conventional, one-size-fits-all teaching methods to more personalized, adaptive, and engaging approaches. One of the most notable changes is the personalization of learning. AI-driven educational platforms and tools can tailor learning experiences to individual students' needs and learning styles (Srinivasa, Kurni, & Saritha, 2022). These systems can adjust the difficulty level of problems, provide targeted feedback, and suggest resources based on a student's performance and progress. This approach contrasts with traditional methods where the pace and content are uniform, often not catering to the diverse needs of students in a classroom.

Interactive learning is another area where AI has made significant inroads. Traditional mathematics education often relies heavily on textbooks and teacher-led instruction. In contrast, AI-powered educational tools offer interactive simulations and problem-solving environments (Alam, 2023). These tools can make abstract mathematical concepts more tangible and understandable by allowing students to visualize and manipulate mathematical entities, fostering a deeper understanding and engagement.

Moreover, AI has changed the role of the teacher in mathematics education. In traditional settings, teachers often spend considerable time on administrative tasks like grading or preparing standard lesson plans. AI tools can automate these tasks, freeing teachers to focus more on addressing individual student needs, mentoring, and facilitating in-depth discussions (Alam & Mohanty, 2022). This shift enhances the teacher's role as a guide and mentor rather than just an information provider.

AI also encourages a more data-driven approach to teaching mathematics. Traditional methods may not effectively track individual student progress or identify specific areas of struggle. AI systems, on the other hand, can continuously collect data on students' performance, providing educators with insights into learning patterns, common misconceptions, and effective instructional strategies (Touretzky, Gardner-McCune, & Seehorn, 2023). This data-driven approach allows for more informed and responsive teaching methods. Furthermore, AI has expanded access to quality mathematics education. Traditional education systems can be limited by geographical, economic, and resource constraints. AI-driven online platforms and resources provide high-quality, accessible mathematical education to a broader audience, including those in remote or underserved areas (Miao et al., 2021).

AI IN MATHEMATICS INSTRUCTION

Role of AI in Enhancing Instructional Methodologies

The integration of AI in mathematics instruction has significantly enhanced instructional methodologies, leading to innovative teaching and learning experiences. AI's role in this transformation is multifaceted and impactful, as it addresses various aspects of the educational process. AI has enabled the development of adaptive learning systems. These systems leverage AI algorithms to analyse students' learning patterns, strengths, and weaknesses (Feldman, Monteserin, & Amandi, 2015). Based on this analysis, they adapt the instructional content to suit individual learners' needs. This means that students can learn at their own pace and in a way that is tailored to their specific learning style, which is a considerable shift from the one-size-fits-all approach of traditional teaching methods.

Another significant contribution of AI in mathematics instruction is the provision of instant feedback and support. In traditional classroom settings, students may need to wait for teacher feedback on assignments or tests. AI-powered tools, however, can provide immediate feedback on students' work, allowing them to understand and correct their mistakes in real-time (Pokrivcakova, 2019). This immediate feedback loop is crucial for learning complex mathematical concepts and problem-solving skills. AI also enriches the instructional content through the use of interactive tools and simulations. Complex mathematical theories and problems can be hard to visualize and understand through traditional teaching methods. AI-driven interactive models and simulations can make these abstract concepts more tangible and easier to comprehend (Sun, Guo, & Hu, 2023). These tools allow students to experiment with variables and see the real-time impact of changes, deepening their understanding of mathematical principles.

Moreover, AI facilitates personalized learning paths in mathematics education. By analysing a vast amount of data on individual student performance, AI systems can identify the most effective learning path for each student (Grivokostopoulou, Perikos, & Hatzilygeroudis, 2017). This can include recommending specific areas of focus, suggesting additional resources, or modifying the difficulty level of problems. This personalized approach ensures that each student can advance through the curriculum in a way that is most effective for their learning process. AI aids in the creation of a collaborative learning environment. AI-driven platforms can connect students with peers working on similar problems or concepts, fostering a collaborative and supportive learning community. These platforms can also connect students with tutors or mentors who can provide additional support, further enhancing the learning experience (Ouyang, Xu, & Cukurova, 2023).

Personalized Learning Experiences Through AI Tools

The application of AI in education, particularly in providing personalized learning experiences, marks a significant evolution in instructional methodologies. AI tools are uniquely capable of tailoring educational content to meet the individual needs, abilities, and learning styles of each student, a feat traditional educational approach often struggle to achieve. Personalized learning through AI tools begins with the system's ability to analyse a student's performance in real-time. AI algorithms can assess a student's responses to various tasks and questions, identifying patterns in their understanding and application of mathematical concepts (Grivokostopoulou, Perikos, & Hatzilygeroudis, 2017). This deep analysis allows the AI to pinpoint areas where a student excels and, more importantly, areas where they require additional support or instruction (Srinivasa, Kurni, & Saritha, 2022).

Based on this continuous assessment, AI tools can then adapt the curriculum to suit the student's specific learning needs. For instance, if a student struggles with a particular mathematical concept, the AI can provide additional resources, such as explanatory videos, interactive simulations, or practice problems tailored to that concept. This adaptability ensures that students are not left behind in areas where they may find difficulty. Moreover, AI tools can customize the pace of learning. In a traditional classroom, the pace of instruction is often dictated by a set curriculum or the average progress of the class. However, with AI, each student can progress through the material at their own pace (Zreik, 2024b). Advanced learners can move ahead to more challenging topics, while those who need more time can take it without feeling rushed or pressured.

Interactive elements provided by AI tools also contribute to personalized learning. Through gamification and interactive challenges, AI can make learning more engaging and enjoyable (Bennani, Maalel, & Ben Ghezala, 2022). This approach not only motivates students but also aids in the retention of mathematical concepts by applying them in varied and practical contexts. AI tools also facilitate personalized feedback, a crucial component in the learning process. Unlike generalized feedback in traditional settings, AI can provide detailed, specific feedback on each student's work (Bernius, Krusche, & Bruegge, 2022). This feedback can include explanations for why an answer is incorrect and guidance on how to approach similar problems in the future. Furthermore, AI-driven personalized learning systems can cater to diverse learning styles. Visual learners, for example, can be provided with more graphical representations of mathematical problems, while auditory learners might benefit from explanatory lectures or discussions (Yang et al., 2013).

Interactive Methods for Exploring Complex Mathematical Concepts

The integration of AI in mathematics education has introduced a range of interactive methods for exploring complex mathematical concepts, significantly enhancing the learning experience. These interactive methods, powered by AI, offer dynamic and engaging ways for students to understand and engage with intricate mathematical ideas, often difficult to grasp through traditional teaching methods. One prominent interactive method involves the use of AI-powered simulations and visualizations. Complex mathematical concepts, such as calculus, algebraic structures, or geometric theorems, can be abstract and challenging to conceptualize (Ritter & Urcid, 2021). AI-driven visual tools bring these concepts to life, allowing students to see and interact with mathematical models in real-time. For example, a student learning about parabolic curves can use these tools to adjust coefficients and parameters, visually observing how these changes affect the curve's shape. This visual and interactive exploration aids in deeper understanding and retention of mathematical principles.

Another interactive approach facilitated by AI is the use of gamified learning experiences. Gamification applies game-design elements in educational contexts, making learning more engaging and motivational (Bennani, Maalel, & Ben Ghezala, 2022). AI can customize these educational games to adapt to the learner's skill level, providing challenges that are neither too easy nor too hard. This adaptive difficulty ensures that students remain engaged and challenged throughout their learning journey. Gamified learning experiences in mathematics might include solving puzzles, embarking on quests requiring mathematical problem-solving, or competitive math-based challenges. AI also enables the creation of virtual laboratories or math workshops where students can experiment with mathematical concepts and theories. In these virtual environments, students can manipulate variables, test hypotheses, and see the outcomes of their experiments in a controlled, risk-free setting (Makransky, Borre-Gude, & Mayer, 2019). This hands-on approach to learning complex mathematical concepts encourages exploration, curiosity, and a deeper understanding of the subject matter.

Moreover, interactive AI tools often incorporate natural language processing (NLP) capabilities, allowing students to ask questions and receive explanations in a conversational manner (Slack et al., 2023). This interaction mimics a one-on-one tutoring session, where a student can receive immediate, personalized responses to their queries. This feature is particularly useful for complex mathematical concepts that require detailed explanation and discussion. Collaborative learning is another area where AI-driven interactive methods shine. These tools can connect students from different locations, enabling them to work together on mathematical problems and projects (Ashri, 2019). Such collaboration not only aids in learning from peers but also fosters skills like teamwork and communication, essential in the modern, interconnected world.

AI IN MATHEMATICS RESEARCH

Contribution of AI Algorithms to Solving Complex Mathematical Problems

The realm of mathematics research has been profoundly influenced by the advent of AI, particularly through the contribution of AI algorithms in solving complex mathematical problems. This influence marks a significant shift from traditional problem-solving methods, opening new avenues and perspectives in mathematical research. One of the primary ways AI contributes to mathematics research is through its exceptional data processing capabilities. AI algorithms, especially those based in machine learning and deep learning, can analyse vast datasets much more efficiently than traditional methods (Janiesch, Zschech, & Heinrich, 2021). This capability is particularly beneficial in areas of mathematics that deal with large-scale data sets or require the processing of significant amounts of information, such as statistical analysis, pattern recognition, and computational mathematics.

AI algorithms also excel in identifying patterns and correlations within complex datasets, a task that is often challenging and time-consuming for human mathematicians. This ability has led to breakthroughs in various branches of mathematics, including number theory, algebra, and geometry. By identifying previously unnoticed patterns or simplifying complex relationships, AI algorithms have provided new insights and solutions to longstanding mathematical problems (Deng, 2018). Moreover, AI has contributed to the field of mathematical modelling. By leveraging AI algorithms, researchers can develop more accurate and sophisticated models for a range of applications, from predicting financial market trends to modelling climate change (Nishant, Kennedy, & Corbett, 2020). These models can handle a level of complexity and nuance that was previously unattainable, providing more precise and reliable predictions and analyses.

In addition to problem-solving, AI is instrumental in hypothesis generation within mathematics research. AI systems can suggest new avenues of inquiry by analysing existing mathematical knowledge and identifying potential areas of interest (Cockburn, Henderson, & Stern, 2018). This approach can lead to the exploration of entirely new mathematical theories and concepts, pushing the boundaries of the field. AI also plays a crucial role in optimizing algorithms used in mathematical research. Through techniques like reinforcement learning and neural network training, AI can refine and improve the efficiency of existing mathematical algorithms, making them faster, more accurate, and capable of handling more complex tasks (Mahmud et al., 2018).

Furthermore, AI's contribution extends to the democratization of mathematics research. AI-powered tools and platforms make advanced mathematical research more accessible to a broader range of researchers, including those who may not have extensive resources or computational power at their disposal. This democratization leads to a more inclusive and diverse research community, fostering innovation and collaboration.

The Revolutionary Impact of AI on Mathematics Research

The impact of AI on mathematics research is nothing short of revolutionary, fundamentally transforming how research is conducted in this field. AI's influence extends across various facets of mathematical investigation, from enhancing computational methods to enabling new discoveries and innovations. One of the most significant impacts of AI on mathematics research is the acceleration of computational mathematics. AI algorithms, particularly those involving machine learning and deep learning, are capable of processing and analysing data at unprecedented speeds and scales (Zhou et al., 2017). This capability is crucial in fields like numerical analysis, algorithmic algebra, and statistical modelling, where the ability to quickly process large volumes of data can lead to faster and more efficient problem-solving. AI's computational power is not just about speed; it also allows researchers to tackle problems that were previously too complex or resource-intensive to solve (Chang et al., 2021).

AI is also playing a transformative role in the realm of predictive modelling within mathematics. By leveraging AI's predictive analytics, mathematicians can develop more accurate models for a wide range of applications (Agbehadji et al., 2020). These models have applications in various sectors, including finance, environmental science, and engineering, where they contribute to better decision-making and innovation. The accuracy and efficiency of AI-driven models represent a significant leap over traditional mathematical modelling techniques. Furthermore, AI has enabled new methods of pattern recognition and data analysis in mathematics research. Complex and intricate patterns within large datasets, which might be imperceptible to human analysis, can be identified and interpreted by AI algorithms (Jin et al., 2020). This capability has led to new insights and discoveries in areas such as graph theory, number theory, and geometric analysis, expanding our understanding of these fields (Wigderson, 2019).

AI also contributes to the development of new mathematical theories and concepts. By analysing vast amounts of mathematical data and existing literature, AI can generate hypotheses and suggest new areas of research (Cockburn, Henderson, & Stern, 2018). This aspect of AI as a tool for discovery is particularly exciting, as it opens up possibilities for exploring uncharted territories in mathematics, potentially leading to ground-breaking new theories. Collaboration between mathematicians and AI researchers has resulted in the refinement and development of more advanced mathematical tools and algorithms. AI's ability to learn and adapt has led to the optimization of existing mathematical algorithms, making them more efficient and powerful (Tang, Liu, & Pan, 2021). This collaboration is not just enhancing the tools available to researchers but also fostering a cross-disciplinary approach that enriches both fields.

The impact of AI on mathematics research extends to democratizing access to advanced research tools. AI-powered platforms and software make sophisticated mathematical tools available to a broader range of researchers, including those in academia, industry, and independent scholars. This democratization is breaking down barriers to entry in the field, leading to a more diverse and innovative research community.

AI IN ADMINISTRATIVE ASPECTS OF MATHEMATICS EDUCATION

Use of AI-Driven Data Analytics in Curriculum Design

The incorporation of AI into the administrative aspects of mathematics education, particularly in curriculum design, represents a significant advancement in educational planning and implementation. AI-driven data analytics is playing a crucial role in this area, offering a more nuanced and effective approach to curriculum development. One of the key contributions of AI in curriculum design is the ability to analyse extensive educational data. AI systems can process and interpret data from a wide range of sources, including student performance records, engagement metrics, and feedback from educators and learners (Zawacki-Richter et al., 2019). This comprehensive analysis enables a deeper understanding of how students interact with and respond to different mathematical concepts and teaching methodologies. Such insights are invaluable for designing curricula that are more aligned with students' learning needs and preferences.

AI-driven data analytics also facilitates the creation of dynamic and adaptable curricula. Traditional curriculum design often follows a static, one-size-fits-all approach, which may not effectively address the diverse needs of all students. With AI, however, curricula can be made flexible and responsive. Based on real-time data, the curriculum can be adjusted to incorporate new teaching methods, topics, or resources that better suit the evolving needs of students (Kinshuk et al., 2016). This adaptability ensures that the curriculum remains relevant and effective in fostering students' mathematical understanding. Another important aspect is the personalization of learning pathways within the curriculum. AI algorithms can identify individual student's strengths, weaknesses, and learning styles, and suggest customized learning paths within the broader curriculum framework (Tapalova & Zhiyenbayeva, 2022). This personalized approach ensures that each student is engaged with material that is challenging yet attainable, maximizing their learning potential.

AI-driven data analytics also aids in predictive analysis in curriculum design. By analysing trends and patterns in student performance and engagement, AI can predict which areas of the curriculum might need reinforcement or re-evaluation in the future (Evans, Kandiko Howson, & Forsythe, 2018). This predictive capability enables proactive adjustments to the curriculum, ensuring that it continuously meets the highest educational standards and effectively prepares students for both academic and real-world challenges. Furthermore, AI contributes to the alignment of the curriculum with educational standards and goals. By analysing the curriculum against established educational standards and benchmarks, AI ensures that the content, teaching methods, and assessment strategies are not only effective but also compliant with regulatory and accreditation requirements (Williams, 2023).

Improvement of Assessment Strategies Through AI

The integration of AI into the assessment strategies in mathematics education marks a significant improvement in evaluating student learning and performance. AI technologies offer sophisticated, efficient, and nuanced methods of assessment that surpass many traditional testing and evaluation practices. AI-driven assessment tools are capable of providing more personalized and adaptive testing experiences (Tan, 2023). Unlike standard tests that present the same questions to all students, AI-based assessments can adapt in real-time to a student's performance (Zreik, 2024c). For example, if a student answers a series of questions correctly, the AI system can present more challenging problems, ensuring that the assessment accurately gauges the student's understanding at different competency levels. Conversely, if a student struggles, the AI can adjust to offer less complex questions, helping to accurately pinpoint the student's current level of understanding without causing undue frustration.

Another significant improvement brought about by AI in assessment strategies is the provision of instant and detailed feedback. Traditional assessment methods often involve a delay between test taking and results, and the feedback may be limited to scores or grades without detailed explanations. AI-based assessments, however, can provide immediate feedback on each question or problem, explaining why an answer is correct or incorrect and offering hints or tips for improvement (Coviliac, 2022). This instant feedback is invaluable for student learning, allowing them to quickly understand and correct their mistakes. AI also enhances the ability to analyse assessment data comprehensively. AI systems can process large volumes of assessment data, identifying trends and patterns in student performance across various topics (Chen et al., 2022). This analysis can reveal insights into common areas of difficulty, effectiveness of teaching methods, and overall curriculum alignment. Such insights are crucial for educators and administrators to make informed decisions about curriculum adjustments and teaching strategies.

Furthermore, AI in assessment can reduce the bias and subjectivity often associated with traditional grading methods. AI algorithms, when well-designed, can provide consistent and objective evaluations of student responses, particularly in areas like mathematics where answers are often more clear-cut (Rothman, 2020). This objectivity ensures a fairer assessment process, where students are evaluated based on their understanding and skills rather than subjective criteria. AI-driven assessments also offer scalability and efficiency, especially beneficial in large educational settings or remote learning environments. AI systems can administer and grade assessments to a large number of students simultaneously, saving significant time and resources while maintaining consistency and accuracy in grading (Alam, 2022).

AI's Role in Educational Policy-Making

AI is increasingly playing a pivotal role in educational policy-making, providing data-driven insights and tools that are reshaping the landscape of educational strategies and decisions. The influence of AI in this domain is multifaceted, impacting various aspects of policy formulation and implementation. One key area where AI contributes to educational policy-making is through the analysis of large-scale educational data. AI algorithms are adept at processing vast amounts of data from various sources, including student performance metrics, teacher evaluations, and resource allocation (Rastrollo-Guerrero, Gómez-Pulido, & Durán-Domínguez, 2020). By analysing these datasets, AI can uncover trends and patterns that might not be visible through traditional analysis methods. These insights are invaluable for policy-makers, as they provide a comprehensive understanding of the educational landscape, including areas of success and those needing improvement.

AI also aids in the customization and optimization of educational policies. By leveraging predictive analytics, AI can forecast future trends in education, such as changes in student enrolment, the emergence of new learning needs, or the impact of technological advancements on education (George & Wooden, 2023). This predictive capability allows policy-makers to develop more proactive and targeted policies that are better equipped to address future challenges and opportunities. Moreover, AI can play a role in evaluating the effectiveness of existing educational policies. By monitoring and analysing the outcomes of implemented policies, AI systems can assess their impact on student learning, teacher performance, and resource utilization (Chen, Chen, & Lin, 2020). This evaluation is crucial for determining whether policies are achieving their intended goals and where adjustments may be necessary.

AI's role extends to enhancing equity and inclusivity in education. AI tools can identify disparities in educational access and outcomes across different demographics, regions, or school systems (Baidoo-Anu & Ansah, 2023). This analysis enables policy-makers to develop more equitable policies that address these disparities, ensuring that all students have access to quality education. Furthermore, AI can contribute to the personalization of education at a policy level. Policies informed by AI can promote individualized learning approaches, recognizing that students have diverse learning styles, abilities, and needs. By advocating for educational models that cater to these individual differences, policy-makers can foster more inclusive and effective educational environments.

AI in educational policy-making facilitates more efficient resource allocation. AI can optimize budget spending, staffing, and resource distribution by analysing where resources are most needed and where they can have the greatest impact (Valle-Cruz, Fernandez-Cortez, & Gil-Garcia, 2022). This optimization ensures that educational resources are used effectively and efficiently, maximizing the benefits for students and educators.

CHALLENGES AND ETHICAL CONSIDERATIONS

Ethical Concerns Related to AI in Education

The integration of AI into education, while offering numerous benefits, also brings forth various ethical concerns and challenges. These issues need careful consideration to ensure that the use of AI in educational settings is responsible, fair, and beneficial to all stakeholders. One of the primary ethical concerns is data privacy and security. AI systems in education often require collecting, processing, and storing vast amounts of personal data from students, including academic performance, learning habits, and sometimes even biometric data. Ensuring the security of this data and protecting it from unauthorized access or breaches is paramount. Additionally, there's the question of consent – how and to what extent students and parents are informed about the data collection and its usage.

Another significant concern is the risk of bias and discrimination in AI algorithms. AI systems are only as unbiased as the data they are trained on and the algorithms that drive them (Liang et al., 2022). There's a risk that these systems may perpetuate existing biases or create new ones, leading to unfair or discriminatory treatment of certain student groups. For instance, if an AI system is trained on data that does not adequately represent a diverse student population, it might be less effective or fair for underrepresented groups. The issue of transparency and explainability in AI-driven educational tools is also a major ethical consideration (Klimova, Pikhart, & Kacetl, 2023). It's essential for educators and students to understand how AI tools make decisions and provide recommendations. A lack of transparency can lead to mistrust and can make it difficult for educators to intervene or make informed decisions based on AI insights.

There's also the concern of over-reliance on AI in educational settings. While AI can be a powerful tool for enhancing learning and teaching, there's a risk that it might be seen as a replacement for human teachers or that it could diminish the human elements of teaching, such as empathy, understanding, and personal connection. Ensuring that AI is used as a complement to, rather than a replacement for, human educators is crucial. Moreover, AI's impact on student autonomy and agency is a concern. There's a risk that highly personalized and predictive AI tools might over-prescribe learning paths and choices, leaving little room for students to explore, make mistakes, and learn independently. Balancing personalization with the need to foster independent learning and critical thinking skills is vital.

Accessibility and equity issues arise with the implementation of AI in education. There's a risk that AI-driven educational tools could widen the digital divide, with students in under-resourced schools or regions having less access to these advanced technologies (Brokensha, Kotzé, & Senekal, 2023). Ensuring that AI tools are accessible and beneficial to all students, regardless of their socio-economic background, is a critical ethical consideration.

Necessity for Educators to Adapt to AI Advancements

The necessity for educators to adapt to advancements in AI is becoming increasingly important in the modern educational landscape. As AI continues to revolutionize various aspects of teaching and learning, educators must evolve their skills and understanding to effectively integrate this technology into their teaching practices. Educators need to develop a foundational understanding of AI and its applications in education. This involves not only grasping the basic concepts of AI but also understanding how AI tools and platforms can be utilized to enhance teaching and learning. With a solid understanding of AI, educators can make informed decisions about which tools to use and how to integrate them into their lesson plans and teaching strategies.

Adapting to AI also means educators need to embrace new roles in the classroom. AI can take over certain tasks such as grading and personalized content delivery, allowing teachers to focus more on higher-level teaching activities like facilitating discussions, mentoring students, and providing individualized support (Brokensha, Kotzé, & Senekal, 2023). Educators must be prepared to shift from traditional teaching methods to more of a facilitator or guide role, where they support and augment the AI-driven learning process. Another important aspect of adapting to AI advancements is the continuous professional development in this area. The field of AI is rapidly evolving, with new tools and applications emerging regularly. Educators must stay current with these advancements to effectively incorporate them into their teaching. This ongoing learning can be achieved through professional development courses, workshops, and self-directed learning.

Critical to adapting to AI in education is the development of digital literacy skills among educators. This encompasses not only the ability to use AI tools but also an understanding of digital ethics, data privacy, and cybersecurity. As AI systems handle a large amount of sensitive student data, educators must be knowledgeable about protecting this data and using AI tools responsibly. Furthermore, educators need to cultivate a mindset of experimentation and adaptability. Integrating AI into education often requires trial and error, and what works well in one educational setting may not work in another. Educators should be prepared to experiment with different AI tools and approaches, assess their effectiveness, and adapt their strategies accordingly.

Educators must advocate for equitable access to AI tools and resources. As AI becomes more integral to education, it's essential to ensure that all students, regardless of their background, have access to these technologies. Educators should work towards eliminating the digital divide and ensuring that AI is used in a way that is inclusive and equitable.

CONCLUSION AND FUTURE DIRECTIONS

The transformative impact of AI on mathematics education is undeniable. AI has ushered in a new era of teaching and learning, research, administrative efficiency, and policy-making. It has empowered educators with tools to create personalized and engaging learning experiences, aided researchers in solving complex mathematical problems, optimized administrative aspects of education, and informed educational policies with data-driven insights. However, this transformative journey is not without its challenges and ethical considerations, including data privacy, bias, transparency, and issues of accessibility and equity.

As we look to the future, there are several promising directions for the integration of AI in mathematics education. Firstly, continued research and development in AI-driven educational tools and platforms will lead to even more sophisticated and effective learning experiences. These tools will become increasingly adaptive, responsive, and capable of catering to diverse learning needs. Moreover, AI's role in educational research will expand, leading to more breakthroughs in mathematical theory and practical applications. AI algorithms will continue to contribute to mathematical modelling, problem-solving, and data analysis in ways that were previously inconceivable.

In the administrative aspects of education, AI-driven data analytics will become more integral in curriculum design, assessment strategies, and policy-making. The optimization of resource allocation and the personalization of educational pathways will further enhance the efficiency and effectiveness of educational systems. Addressing the ethical concerns related to AI in education will also be a paramount focus. There will be efforts to ensure data privacy and security, mitigate bias and discrimination, enhance transparency and explainability, and promote equitable access to AI technologies. Furthermore, educators will play a crucial role in shaping the future of AI in mathematics education. Continuous professional development, digital literacy, and adaptability will be essential for educators to harness the potential of AI effectively.

REFERENCES

Abduljabbar, R., Dia, H., Liyanage, S., & Bagloee, S. A. (2019). Applications of artificial intelligence in transport: An overview. *Sustainability (Basel)*, 11(1), 189. 10.3390/su11010189

Agbehadji, I. E., Awuzie, B. O., Ngowi, A. B., & Millham, R. C. (2020). Review of big data analytics, artificial intelligence and nature-inspired computing models towards accurate detection of COVID-19 pandemic cases and contact tracing. *International Journal of Environmental Research and Public Health*, 17(15), 5330. 10.3390/ijerph1715533032722154

Alam, A. (2022). Employing adaptive learning and intelligent tutoring robots for virtual classrooms and smart campuses: Reforming education in the age of artificial intelligence. In *Advanced Computing and Intelligent Technologies* [Singapore: Springer Nature Singapore.]. *Proceedings of ICACIT*, 2022, 395–406.

Alam, A. (2023). Harnessing the Power of AI to Create Intelligent Tutoring Systems for Enhanced Classroom Experience and Improved Learning Outcomes. In *Intelligent Communication Technologies and Virtual Mobile Networks* (pp. 571–591). Springer Nature Singapore. 10.1007/978-981-99-1767-9_42

Alam, A., & Mohanty, A. (2022, December). Facial Analytics or Virtual Avatars: Competencies and Design Considerations for Student-Teacher Interaction in AI-Powered Online Education for Effective Classroom Engagement. In *International Conference on Communication, Networks and Computing* (pp. 252-265). Cham: Springer Nature Switzerland.

Ashri, R. (2019). *The AI-powered workplace: how artificial intelligence, data, and messaging platforms are defining the future of work.* Apress.

Baidoo-Anu, D., & Ansah, L. O. (2023). Education in the era of generative artificial intelligence (AI): Understanding the potential benefits of ChatGPT in promoting teaching and learning. *Journal of AI*, 7(1), 52–62. 10.61969/jai.1337500

Bennani, S., Maalel, A., & Ben Ghezala, H. (2022). Adaptive gamification in E-learning: A literature review and future challenges. *Computer Applications in Engineering Education*, 30(2), 628–642. 10.1002/cae.22477

Bernius, J. P., Krusche, S., & Bruegge, B. (2022). Machine learning based feedback on textual student answers in large courses. *Computers and Education: Artificial Intelligence*, 3, 100081. 10.1016/j.caeai.2022.100081

Brokensha, S., Kotzé, E., & Senekal, B. A. (2023). *AI in and for Africa: A Humanistic Perspective.* CRC Press. 10.1201/9781003276135

Chang, Z., Liu, S., Xiong, X., Cai, Z., & Tu, G. (2021). A survey of recent advances in edge-computing-powered artificial intelligence of things. *IEEE Internet of Things Journal*, 8(18), 13849–13875. 10.1109/JIOT.2021.3088875

Chen, L., Chen, P., & Lin, Z. (2020). Artificial intelligence in education: A review. *IEEE Access : Practical Innovations, Open Solutions*, 8, 75264–75278. 10.1109/ACCESS.2020.2988510

Chen, X., Zou, D., Xie, H., Cheng, G., & Liu, C. (2022). Two decades of artificial intelligence in education. *Journal of Educational Technology & Society*, 25(1), 28–47.

Cheng, X., Liu, S., Sun, X., Wang, Z., Zhou, H., Shao, Y., & Shen, H. (2021). Combating emerging financial risks in the big data era: A perspective review. *Fundamental Research*, 1(5), 595–606. 10.1016/j.fmre.2021.08.017

Cockburn, I. M., Henderson, R., & Stern, S. (2018). The impact of artificial intelligence on innovation: An exploratory analysis. In *The economics of artificial intelligence: An agenda* (pp. 115–146). University of Chicago Press.

Coviliac, A. (2022). *Improvement of AI-assessment systems in grading open questions based on the teaching assistant's view* [Bachelor's thesis, University of Twenty].

Delgado, H. O. K., de Azevedo Fay, A., Sebastiany, M. J., & Silva, A. D. C. (2020). Artificial intelligence adaptive learning tools: The teaching of English in focus. *BELT-Brazilian English Language Teaching Journal*, 11(2), e38749–e38749. 10.15448/2178-3640.2020.2.38749

Deng, L. (2018). Artificial intelligence in the rising wave of deep learning: The historical path and future outlook [perspectives]. *IEEE Signal Processing Magazine*, 35(1), 180–177. 10.1109/MSP.2017.2762725

Elliott, A. (2019). *The culture of AI: Everyday life and the digital revolution*. Routledge. 10.4324/9781315387185

Evans, C., Kandiko Howson, C., & Forsythe, A. (2018). Making sense of learning gain in higher education. *Higher Education Pedagogies*, 3(1), 1–45. 10.1080/23752696.2018.1508360

Feldman, J., Monteserin, A., & Amandi, A. (2015). Automatic detection of learning styles: State of the art. *Artificial Intelligence Review*, 44(2), 157–186. 10.1007/s10462-014-9422-6

George, B., & Wooden, O. (2023). Managing the strategic transformation of higher education through artificial intelligence. *Administrative Sciences*, 13(9), 196. 10.3390/admsci13090196

Grivokostopoulou, F., Perikos, I., & Hatzilygeroudis, I. (2017). An educational system for learning search algorithms and automatically assessing student performance. *International Journal of Artificial Intelligence in Education*, 27(1), 207–240. 10.1007/s40593-016-0116-x

Janiesch, C., Zschech, P., & Heinrich, K. (2021). Machine learning and deep learning. *Electronic Markets*, 31(3), 685–695. 10.1007/s12525-021-00475-2

Jin, C., Chen, W., Cao, Y., Xu, Z., Tan, Z., Zhang, X., Deng, L., Zheng, C., Zhou, J., Shi, H., & Feng, J. (2020). Development and evaluation of an artificial intelligence system for COVID-19 diagnosis. *Nature Communications*, 11(1), 5088. 10.1038/s41467-020-18685-133037212

Kaack, L. H., Donti, P. L., Strubell, E., Kamiya, G., Creutzig, F., & Rolnick, D. (2022). Aligning artificial intelligence with climate change mitigation. *Nature Climate Change*, 12(6), 518–527. 10.1038/s41558-022-01377-7

Kinshuk, C., Chen, N.-S., Cheng, I.-L., & Chew, S. W. (2016). Evolution is not enough: Revolutionizing current learning environments to smart learning environments. *International Journal of Artificial Intelligence in Education*, 26(2), 561–581. 10.1007/s40593-016-0108-x

Klimova, B., Pikhart, M., & Kacetl, J. (2023). Ethical issues of the use of AI-driven mobile apps for education. *Frontiers in Public Health*, 10, 1118116. 10.3389/fpubh.2022.111811636711343

Liang, W., Tadesse, G. A., Ho, D., Fei-Fei, L., Zaharia, M., Zhang, C., & Zou, J. (2022). Advances, challenges and opportunities in creating data for trustworthy AI. *Nature Machine Intelligence*, 4(8), 669–677. 10.1038/s42256-022-00516-1

Mahmud, M., Kaiser, M. S., Hussain, A., & Vassanelli, S. (2018). Applications of deep learning and reinforcement learning to biological data. *IEEE Transactions on Neural Networks and Learning Systems*, 29(6), 2063–2079. 10.1109/TNNLS.2018.279038829771663

Makransky, G., Borre-Gude, S., & Mayer, R. E. (2019). Motivational and cognitive benefits of training in immersive virtual reality based on multiple assessments. *Journal of Computer Assisted Learning*, 35(6), 691–707. 10.1111/jcal.12375

Miao, F., Holmes, W., Huang, R., & Zhang, H. (2021). *AI and education: A guidance for policymakers*. UNESCO Publishing.

Nishant, R., Kennedy, M., & Corbett, J. (2020). Artificial intelligence for sustainability: Challenges, opportunities, and a research agenda. *International Journal of Information Management*, 53, 102104. 10.1016/j.ijinfomgt.2020.102104

Nova, K. (2023). AI-enabled water management systems: An analysis of system components and interdependencies for water conservation. *Eigenpub Review of Science and Technology*, 7(1), 105–124.

Ouyang, F., Xu, W., & Cukurova, M. (2023). An artificial intelligence-driven learning analytics method to examine the collaborative problem-solving process from the complex adaptive systems perspective. *International Journal of Computer-Supported Collaborative Learning*, 18(1), 39–66. 10.1007/s11412-023-09387-z

Pokrivcakova, S. (2019). Preparing teachers for the application of AI-powered technologies in foreign language education. *Journal of Language and Cultural Education*, 7(3), 135–153. 10.2478/jolace-2019-0025

Rastrollo-Guerrero, J. L., Gómez-Pulido, J. A., & Durán-Domínguez, A. (2020). Analyzing and predicting students' performance by means of machine learning: A review. *Applied Sciences (Basel, Switzerland)*, 10(3), 1042. 10.3390/app10031042

Ravyse, W. S., Seugnet Blignaut, A., Leendertz, V., & Woolner, A. (2017). Success factors for serious games to enhance learning: A systematic review. *Virtual Reality (Waltham Cross)*, 21(1), 31–58. 10.1007/s10055-016-0298-4

Ritter, G. X., & Urcid, G. (2021). *Introduction to lattice algebra: With applications in ai, pattern recognition, image analysis, and biomimetic neural networks*. CRC Press. 10.1201/9781003154242

Rothman, D. (2020). *Artificial Intelligence By Example: Acquire advanced AI, machine learning, and deep learning design skills*. Packt Publishing Ltd.

Slack, D., Krishna, S., Lakkaraju, H., & Singh, S. (2023). Explaining machine learning models with interactive natural language conversations using TalkToModel. *Nature Machine Intelligence*, 5(8), 873–883. 10.1038/s42256-023-00692-8

Srinivasa, K. G., Kurni, M., & Saritha, K. (2022). Harnessing the Power of AI to Education. In *Learning, Teaching, and Assessment Methods for Contemporary Learners: Pedagogy for the Digital Generation* (pp. 311–342). Springer Nature Singapore. 10.1007/978-981-19-6734-4_13

Sun, L., Guo, Z., & Hu, L. (2023). Educational games promote the development of students' computational thinking: A meta-analytic review. *Interactive Learning Environments*, 31(6), 3476–3490. 10.1080/10494820.2021.1931891

Tan, S. (2023). Harnessing Artificial Intelligence for innovation in education. In *Learning intelligence: Innovative and digital transformative learning strategies: Cultural and social engineering perspectives* (pp. 335–363). Springer Nature Singapore. 10.1007/978-981-19-9201-8_8

Tang, J., Liu, G., & Pan, Q. (2021). A review on representative swarm intelligence algorithms for solving optimization problems: Applications and trends. *IEEE/CAA Journal of Automatica Sinica, 8*(10), 1627-1643.

Tapalova, O., & Zhiyenbayeva, N. (2022). Artificial Intelligence in Education: AIEd for Personalised Learning Pathways. *Electronic Journal of e-Learning*, 20(5), 639–653. 10.34190/ejel.20.5.2597

Touretzky, D., Gardner-McCune, C., & Seehorn, D. (2023). Machine learning and the five big ideas in AI. *International Journal of Artificial Intelligence in Education*, 33(2), 233–266. 10.1007/s40593-022-00314-1

Valle-Cruz, D., Fernandez-Cortez, V., & Gil-Garcia, J. R. (2022). From E-budgeting to smart budgeting: Exploring the potential of artificial intelligence in government decision-making for resource allocation. *Government Information Quarterly*, 39(2), 101644. 10.1016/j.giq.2021.101644

Wigderson, A. (2019). *Mathematics and computation: A theory revolutionizing technology and science*. Princeton University Press.

Williams, P. (2023). AI, Analytics and a New Assessment Model for Universities. *Education Sciences*, 13(10), 1040. 10.3390/educsci13101040

Yang, F. Y., Chang, C. Y., Chien, W. R., Chien, Y. T., & Tseng, Y. H. (2013). Tracking learners' visual attention during a multimedia presentation in a real classroom. *Computers & Education*, 62, 208–220. 10.1016/j.compedu.2012.10.009

Zawacki-Richter, O., Marín, V. I., Bond, M., & Gouverneur, F. (2019). Systematic review of research on artificial intelligence applications in higher education–where are the educators? *International Journal of Educational Technology in Higher Education*, 16(1), 1–27. 10.1186/s41239-019-0171-0

Zhou, L., Pan, S., Wang, J., & Vasilakos, A. V. (2017). Machine learning on big data: Opportunities and challenges. *Neurocomputing*, 237, 350–361. 10.1016/j.neucom.2017.01.026

Zreik, M. (2024a). The Transformation of Healthcare Marketing in the New Digital Era: A Comprehensive Analysis with Reference to China. In Djakeli, K. (Ed.), *Modern Healthcare Marketing in the Digital Era* (pp. 42–60). IGI Global. 10.4018/979-8-3693-0679-6.ch003

Zreik, M. (2024b). Cultivating ILCs in China: A Pathway to Culturally Sustaining and Transformative Education. In Adams, S., & Breidenstein, A. (Eds.), *Exploring Meaningful and Sustainable Intentional Learning Communities for P-20 Educators* (pp. 120–140). IGI Global. 10.4018/978-1-6684-7270-5.ch006

Zreik, M. (2024c). Artificial Intelligence and the Future of Chinese Language Pedagogy: An In-Depth Analysis. In Çetin Köroğlu, Z., & Çakır, A. (Eds.), *Fostering Foreign Language Teaching and Learning Environments with Contemporary Technologies* (pp. 1–20). IGI Global. 10.4018/979-8-3693-0353-5.ch001

KEY TERMS AND DEFINITIONS

Adaptive Learning: Adaptive learning is a subset of personalized learning that uses AI to adjust the difficulty and content of educational materials in real-time based on a student's performance and progress. It ensures that students receive content at their appropriate skill level.

Artificial Intelligence (AI): AI refers to the simulation of human intelligence in machines that are programmed to think and learn like humans. In education, AI is used to create intelligent systems and tools that can enhance learning and teaching processes.

Blended Learning: Blended learning combines traditional classroom instruction with online and digital learning experiences. It provides a flexible and personalized approach to education, often incorporating AI and digital resources.

Data Analytics: Data analytics involves the examination of data to uncover meaningful patterns, trends, and insights. In education, data analytics is used to analyse student performance, identify areas of improvement, and inform instructional decisions.

Digital Literacy: Digital literacy refers to the ability to use digital devices and technologies effectively and responsibly. It includes skills such as computer proficiency, internet navigation, and information literacy, all of which are essential in today's technology-driven education.

EdTech: Short for Educational Technology, EdTech encompasses the use of technology and digital tools in education. This includes hardware, software, apps, and online platforms designed to enhance teaching and learning.

Flipped Classroom: In a flipped classroom, traditional teaching methods are reversed. Students learn content online at their own pace (often through videos) before coming to class, where the teacher facilitates discussions, projects, and activities. EdTech plays a significant role in creating and delivering flipped classroom content.

Gamification: Gamification involves incorporating game elements and principles, such as competition and rewards, into non-game contexts, like education. Gamified educational platforms and apps make learning more engaging and motivating for students.

Learning Management System (LMS): An LMS is a software application or platform used to manage, deliver, and track educational content and activities. It's commonly used in schools and universities to facilitate online learning and course administration.

Machine Learning: Machine learning is a type of AI that enables systems to learn and improve from experience without being explicitly programmed. It's used in education for tasks like predicting student outcomes and recommending resources.

Personalized Learning: Personalized learning is an educational approach that tailors instruction, content, and pace to the individual needs and preferences of each student. AI is often used to implement personalized learning by analysing student data and providing customized learning experiences.

STEM Education: STEM stands for Science, Technology, Engineering, and Mathematics. STEM education focuses on teaching these disciplines in an interdisciplinary and applied approach. Technology, including AI and digital tools, plays a vital role in STEM education.

Chapter 3
Mathematics Teacher Education in a Globalized Age:
Heutagogy and Virtual Learning Environments in Context

Angel Mukuka
https://orcid.org/0000-0003-0292-1324
Walter Sisulu University, South Africa

Jogymol K. Alex
United Arab Emirates University, UAE

Benjamin Tatira
https://orcid.org/0000-0002-0118-3676
Walter Sisulu University, South Africa

ABSTRACT

Globalization has a significant bearing on mathematics teacher education and there is need for a comprehensive restructuring to meet the demands of the interconnected world. This study advocates for a paradigm shift that encompasses cross-cultural competencies, linguistic diversity, and a nuanced understanding of global educational trends. It further highlights the significant influence of virtual learning environments (VLEs) in responding to the challenges posed by globalization and emphasizes the synergy between VLEs and heutagogical practices. Drawing from the cited case studies and other existing literature, this chapter underscores the departure from conventional pedagogical methods towards a learner-cantered, self-determined approach. The chapter concludes by discussing the far-reaching

DOI: 10.4018/979-8-3693-2873-6.ch003

implications for policy, practice, and theory, urging stakeholders to recognize the benefits of heutagogy in preparing mathematics teachers for the complexities of the twenty-first century teaching and learning.

INTRODUCTION

Mathematics teacher education is continually evolving to meet the demands of a dynamic landscape shaped by globalization and digital technological innovation, particularly in the era of the fourth industrial revolution. This transformative period, marked by technologies such as the internet of things, virtual reality, robotics, artificial intelligence, and more (Traore, 2021), emphasizes the urgency of cultivating 21st-century skills among both mathematics learners and teachers (Luneta, 2022a). Mukuka and Alex (2024) emphasise the growing demand for these essential skills in mathematics education, highlighting the importance of equipping prospective teachers with the knowledge and competencies required for effective instruction.

As the educational landscape experienced a significant shift during the COVID-19 pandemic, online learning became a pivotal mode of instruction and assessment for mathematics education (Kamal et al., 2023; Mukuka et al., 2021; Mulenga & Marbán, 2020). This transition, while promising, stresses the critical need for rigorous research and effective practices to integrate and sustain these innovations within the policies, curricula, and pedagogies of mathematics teacher education. This chapter explores the potential of two global and innovative approaches—heutagogy and virtual learning environments (VLEs)—to enhance the quality and relevance of mathematics teacher education in the current era.

Heutagogy, a self-determined learning approach that empowers learners to lead their educational journey and develop indispensable 21st-century competencies (Blaschke & Hase, 2015), aligns with the evolving demands of mathematics teacher education. Similarly, VLEs, digital platforms leveraging technology to deliver instruction and foster interactive learning experiences (Korucu-Kış, 2021; Yildiz & Scharaldi, 2015), exemplify the transformative power of globalization on pedagogical landscapes. This perspective is consistent with LeTendre's (2022) view, wherein globalization is seen as a progression marked by heightened communication and information exchange, facilitated by the utilization of VLEs. The evolution towards an interconnected society necessitates digital communication and provides easy access to specialized knowledge and diverse models of practice. This is crucial in fostering teaching skills among both preservice and in-service teachers.

In this chapter, we contend that both heutagogy and VLEs epitomize responses to global trends and challenges within mathematics teacher education. Their innovative solutions offer unprecedented opportunities to augment mathematics teaching and learning methodologies. Halupa (2021) proposed that heutagogy could serve as a theoretical model or structure for curriculum design, thereby aiding students in learning and problem-solving. This approach is especially potent when combined with technology. Halupa further suggests that modern teaching technologies empower students to independently construct and investigate new ideas, leading to a more rewarding educational experience characterized by significant, applicable practice and evaluation. Additionally, this chapter presents evidence and practical illustrations (through case studies) demonstrating the successful implementation and evaluation of heutagogy and VLEs across various contexts and levels of mathematics teacher education.

By shedding light on the affordances of heutagogy and VLEs in mathematics teacher education, this chapter makes a substantial contribution to advancing the knowledge and insights that are cardinal for cultivating essential teaching skills among future mathematics teachers in the globalized era. It highlights the influence of these dynamic forces in reshaping pedagogical approaches, ensuring that mathematics teachers are equipped to address the evolving needs of learners in our fast-paced and ever-changing world. The chapter specifically focuses on achieving the following objectives:

i. To explore the impact of globalization on mathematics teacher education, particularly in the context of heutagogy and VLEs.
ii. To present empirical evidence and practical examples of heutagogy and VLEs in various contexts and levels of mathematics teacher education.
iii. To discuss the benefits, challenges, and implications of heutagogy and VLEs for mathematics teacher education.

GLOBALIZATION AND MATHEMATICS TEACHER EDUCATION

The role of globalization in shaping mathematics teacher education is profound and multifaceted, extending its influence across economic, cultural, and political domains. In acknowledging the transformative power of teachers in cultivating students equipped for a diverse and globally interconnected world, scholars like Keengwe (2018), and Mukuka and Alex (2024) emphasise the significance of teacher training programs. It has further been stressed that these programs are critical for imparting the essential skills imperative for navigating the complexities of the 21st century, especially in a digitalized and globally interconnected landscape.

Globalization, as posited by Burbules and Torres (2000), represents the intensification of worldwide social relations, intertwining distant localities to the extent that local events are shaped by occurrences happening thousands of miles away, and vice versa. Within education, this phenomenon translates into the universal acceptance and dissemination of local knowledge, values, principles, and educational frameworks, essentially permeating even the localized settings of classrooms. Clarkson (2009) emphasizes this pervasive influence, particularly evident in the language employed in the teaching and learning of mathematics. Ensuring effective communication of mathematical concepts in diverse linguistic environments requires teachers to undergo comprehensive training. Take, for instance, the scenario of a mathematics teacher in Zambia, where English serves as the official language, while various languages such as Bemba, Kaonde, Lozi, Lunda, Luvale, Nyanja, and Tonga are also prevalent. In this multilingual context, it is imperative for the teacher to adeptly convey mathematical ideas not only in English but also in the students' native languages to facilitate optimal learning.

Despite the significance of addressing language diversity in mathematics education, existing research, as highlighted by Clarkson (2009), indicates a deficiency in studies considering these distinct language contexts. Similarly, findings of a study by McLachlan and Essien (2022) emphasize the need for a stronger focus on language policy within the realm of multilingual education in South African mathematics education. Notably, Clarkson (2009) emphasizes the importance of employing informal or exploratory talk in students' first language before transitioning to formal mathematical language, a crucial insight that informs effective teaching practices. However, contemporary challenges, such as rapid immigration trends, as observed by Sharma and Sharma (2023), demand an additional layer of complexity in teacher training programs. The increasing prevalence of children learning mathematics in their second language highlight the urgency for teachers to be equipped with the skills to navigate these evolving multilingual environments. Therefore, teacher training programs should not only focus on utilizing the "global language" (English, in this instance) but also emphasize the incorporation of local languages in teaching. This approach ensures that all students, regardless of their first language, have equitable access to a high-quality mathematics education.

The globalized context also fosters comparative studies of mathematics education worldwide, facilitating the identification of best practices while learning from the challenges encountered across diverse regions. This dynamic interchange of ideas and methodologies accentuates the need for acknowledging global effects on education. As Kim (2007) notes, the evolution of the teaching field is interconnected with the process of adapting global tendencies into teacher training. Embracing diverse cultural contexts and languages within mathematics teacher training methodologies becomes not just beneficial but crucial for preparing teachers to navigate the complexities of today's interconnected world (Clarkson, 2009). The integration of diverse cultural perspectives and linguistic nuances in mathematics teacher education not only enriches the pedagogical approach but also equips teachers to cater to the needs of a diverse student population. Furthermore, it fosters an environment where teachers can draw upon a spectrum of teaching strategies and tools that resonate with students from varied cultural backgrounds, thereby fostering inclusive and effective learning environments (Olosunde, 2019).

In view of the above, it suffices to state that the impact of globalization on mathematics teacher education requires a shift in educational policies and practices. It calls for a reimagining of teacher training programs that encompass cross-cultural competencies, linguistic diversity, and an understanding of global educational trends (Keengwe, 2018). This paradigm shift not only enhances the proficiency of teachers but also empowers them to mould a generation of learners who are adept at navigating the complexities of a globalized world while acknowledging diverse perspectives.

In essence, globalization's influence on mathematics teacher education transcends mere integration—it demands a comprehensive restructuring that embraces cultural diversity, linguistic pluralism, and global perspectives. Only through such a holistic approach can teacher training programs effectively prepare teachers to foster inclusive learning environments and equip students with the requisite skills to thrive in a globally interconnected society.

HEUTAGOGY AND VIRTUAL LEARNING ENVIRONMENTS

Heutagogy

Heutagogy, also known as self-determined learning (Blaschke & Hase, 2016), is an educational approach centred on learners by granting them the freedom to choose their preferred learning methods to enhance their skills. Expanding on the self-directed learning principles of andragogy, which emphasizes students' development of their own learning, heutagogy asserts "no limit" to student learning. Though heutagogy has been characterized as a broader version of andragogy (Agonács & Matos, 2019), some scholars (*such as*, Ceylan, 2020; Hase & Kenyon, 2000, 2013) argue that heutagogy should not be perceived as a substitute for pedagogy and andragogy. Rather, it should be viewed as a facilitator for nurturing learners' individual capabilities, skills, values, and assumptions during the process of engaging in adaptable, critical, and meaningful learning experiences.

According to Hase and Kenyon (2000), heutagogy is rooted in humanistic and constructivist principles and is applicable to both formal and informal learning settings. Heutagogy highlights the skill of learning how to learn as pivotal in the 21st century (Blaschke & Hase, 2016). Additionally, heutagogy integrates various early learning theories into a unified perspective that shifts the focus from teachers to learners and learning itself (Hase and Kenyon, 2013). This contrasts with the predominant teacher-directed culture in most learning institutions, as observed by different scholars (Lankford, 2021; Mukuka et al., 2019, 2020; Stephan, 2020).

In essence, heutagogy empowers students to take charge of their learning, fostering the development of self-determined learners equipped to face 21st century challenges. Similarly, focusing on cultivating self-determined learners in mathematics teacher training could be beneficial. This approach enables teacher educators to encourage student teachers to establish their learning objectives and paths, leading to enhanced professional growth. Consequently, this method aids student teachers in acquiring critical thinking and problem-solving skills crucial for gaining relevant mathematical knowledge for effective teaching.

Virtual Learning Environments

In the context of this paper, the term Virtual Learning Environment (VLE) is used to describe a digital platform that supports online learning by offering various tools and resources. Caprara and Caprara (2022) state that a VLE enables learners to access content, communicate with instructors and peers, and engage in educational activities at any time and place. VLEs have gained more popularity in recent years when the COVID-19 pandemic forced many people to switch to online learning (Alex & Mukuka, 2024; Kamal et al., 2023; Mukuka et al., 2021; Mulenga & Marbán, 2020). The concept of VLE is based on several educational theories. One important theory is the Social Learning Theory (SLT), which provides a way to understand how human agency influences learning. According to Bandura (1977), SLT emphasizes the significance of learning by observation, social interaction, and imitating the behaviours, attitudes, and emotions of others. In the context of VLEs, SLT can help explain the underlying processes of learners' social learning behaviours, such as involvement and cooperation (Li et al., 2023).

Another relevant theory that underpins VLEs is Constructivism, which originates from theories of learning proposed by scholars like Piaget (*cited in*von Glasersfeld, 2014), Bruner (1966), and Vygotsky (1978) among others. Constructivism suggests that learners create their own understanding and knowledge through experiences and reflection. In the context of VLEs, learners are also seen as active participants in their learning process, taking part in activities that enable them to build knowledge, solve problems, and discover new ideas (Closs et al., 2022). In mathematics teacher training, VLEs can offer interactive playgrounds with a level of interactivity that may not be possible in the physical classroom environment. Therefore, mathematics teacher educators can use VLEs to create interactive and engaging learning experiences for preservice teachers. It is quite evident that these platforms can be used to deliver lectures, foster discussions, and provide resources for learning (Alex &

Mukuka, 2024). They can also be used to simulate real-world mathematical problems, allowing students to apply what they have learned in a controlled environment.

CONCEPTUAL MODEL

Drawing from the operationalisation of key concepts in the preceding sections, it is evident that globalization significantly influences the incorporation of VLEs within educational settings. Conversely, the rise of VLEs can amplify the effects of globalization on education. In fact, Sharma and Singh (2023) contend that achieving the United Nations' 2030 sustainable development goal number 4, which prioritizes enhancing access to high-quality education, calls for an extensive dependence on digital technology. According to data from the World Economic Forum (2022), the surge of digital technologies and the internet has enabled learners worldwide to access educational materials and participate in collaborative learning. As noted earlier, this trend was accelerated by the COVID-19 pandemic, which mandated a shift to remote learning, highlighting the significance of VLEs (Alex & Mukuka, 2024). Additionally, the adoption of VLEs has been shaped by diverse factors, including individual inclinations and national culture (Li et al., 2021). VLEs offer a platform for self-determined learning, allowing learners to choose methods that align with their preferences. This autonomy in learning aligns with heutagogy, a learner-centred approach focusing on developing students" capacity and capability (Blaschke, 2012).

In training mathematics teachers, employing heutagogical practices through VLEs is beneficial, as it emphasises self-determined learning and learner development. This approach empowers students to select learning methods that suit their preferences and appropriate assessments to gauge their comprehension (Blaschke, 2012, 2019). This shift towards a learner-centred education prepares future mathematics teachers for today's diverse classroom settings and the evolving global knowledge economy.

The model depicted in Figure 1 illustrates how globalization drives technological advancements, cultural exchange, and global interconnectedness. This leads to widespread VLE adoption characterized by accessible tools, collaborative learning, and adaptability to different cultural and geographical contexts (Maher et al., 1999). Consequently, VLEs create an environment conducive to self-determined learning (Heutagogy), characterized by learner autonomy, flexible learning paths, collaboration, and inquiry-based learning.

This suggests that heutagogical practices, facilitated by VLEs, offer opportunities for mathematics preservice teachers to engage in self-determined learning, emphasizing critical thinking, problem-solving, and continuous professional growth. It also resonates with the view that we live in an era where the economy, society, and technological advancements are interwoven in a globalized framework (Lacárcel, 2022). This demonstrates an inseparable linkage between the three concepts highlighted in Figure 1. Exposure to diverse teaching methods and cultural contexts enriches the training of mathematics teachers through VLEs.

Figure 1. Conceptual relationships among key variables

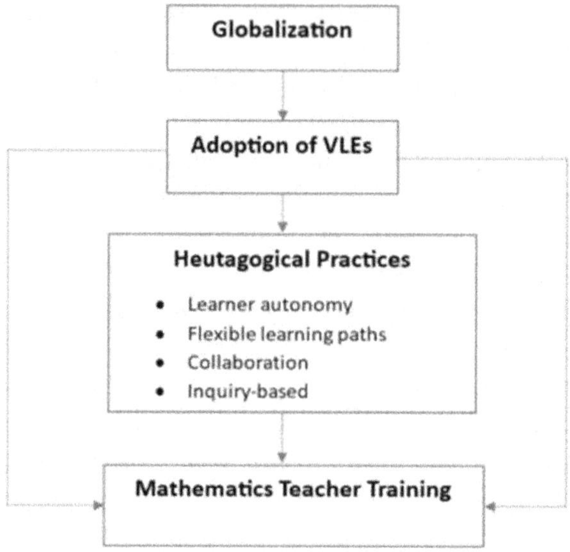

(Developed by authors)

EMPIRICAL EVIDENCE ON HEUTAGOGICAL PRACTICES IN TEACHER EDUCATION

As previously mentioned, heutagogy represents a student-centred instructional approach designed to foster lifelong learning and align learners with the evolving demands of the contemporary workplace. This approach emphasizes learner autonomy and self-determination, prioritizing the enhancement of capacity and capability. Given the dynamic nature of today's globalized world, where individuals must apply skills to navigate complexities, heutagogy proves particularly important. Nonetheless, it is worth noting that while existing literature on mathematics teacher training may not explicitly reference the application of heutagogy, its principles are universally applicable across educational settings. For instance, the integration of technology in training aligns with heutagogical principles, facilitating the creation of learner-generated content and promoting self-directed information discovery (Blaschke, 2012; Blaschke & Hase, 2016). Furthermore, heutagogical practices have successfully permeated diverse areas such as entrepreneurial education, executive coaching, journalism, graphics design, product design, public relations, and e-learning among others (Cochrane et al., 2012; Stoten, 2022) This suggests their adaptability for mathematics teacher training.

In the context of teacher training, VLEs have become very useful for the application of heutagogical practices. Despite limited research in this area, the case studies presented in Table 1 provide evidence on the successful integration of heutagogical methods in teacher training, thereby highlighting the efficacy of VLEs in this setting. Nonetheless, it is important to note that the articles listed in Table 1 encompass a broader scope beyond mathematics teacher education, which reflects the scarcity of empirical studies on heutagogy in this field.

Table 1. Samples of case studies on heutagogical implementation in teacher education context

Author(s)/Year	Article Title	Methodology	Country
Alex & Mukuka (2024)	Heutagogy in action: Unveiling the transformative power of virtual "air campus" experiences of mathematics trainee teachers	Qualitative case study through reflective journal writing	South Africa
Chimpololo (2020)	An analysis of heutagogical practices through mobile device usage in a teacher training programme in Malawi	Case study using questionnaires, semi-structured interviews, focus group discussions, and personal diaries	Malawi
Cochrane et al. (2012)	Heutagogy and mobile social media: Post Web 2.0 pedagogy	Participatory action research using surveys, e-portfolios, blogs, and collaborative writing.	New Zealand
Gerstein (2013)	Education 3.0 and the pedagogy (andragogy, heutagogy) of mobile learning.	This study evaluated the MOOC using qualitative data from participants' blogs, comments, and surveys	Not specific[1]
Khan & Thomas (2022)	Promoting positive education through constructivist digital learning heutagogy: An intervention outcome	Design-based research employing some elements of a mixed-methods approach	India
Wong et al. (2019)	Modelling the factor influencing the implementation of mobile-heutagogical practices among teachers: An application of invariance multi-group structural model	Quantitative approach involving a self-administered survey of 246 teachers and a structural equation modeling (SEM) analysis	Malaysia

A study by Alex and Mukuka (2024) applied heutagogical principles to explore the "school-based experiences" of preservice teachers through the utilization of a "virtual air campus." The focus was on first-year mathematics student teachers enrolled in a Bachelor of Education (B.Ed.) program at a rural university in South Africa's Eastern Cape province. The term "virtual air campus" in this context referred to the experience of observing freely available online school mathematics lessons. The implementation of this virtual initiative was necessitated by the constraints imposed by COVID-19 school closures, which prevented physical observation of lessons in schools. In contrast to traditional school-based experiences, Alex and Mukuka (2024) asserted that the virtual air campus initiative was rooted in the principles of the heutagogical approach. This approach mandated students to establish their own objectives, defining what they sought to learn from virtual mathematics lessons. Employing an exploratory qualitative case study design, the research aimed to address two fundamental questions: (1) what novel insights do student teachers gain from the observed lessons, and (2) how do these lessons influence their belief systems and professional identities? The study concluded that student teachers exhibited diverse experiences in their virtual "air campus" visits. The study established that the observed virtual mathematics classes provided insights into effective teaching methods and enhanced their understanding of how students learn. Additionally, student teachers indicated that they were able to rectify misconceptions they previously held regarding specific concepts in school mathematics. Furthermore, the study offered policy recommendations advocating for the integration of heutagogical practices and virtual learning environments into mathematics teacher education programs. These recommendations aimed to enhance the overall effectiveness and adaptability of teacher training in the face of challenges such as the constraints imposed by unforeseen circumstances like the COVID-19 pandemic.

Secondly, a study by Chimpololo (2020) implemented heutagogy by integrating mobile devices into a teacher training program in Malawi, focusing on three key heutagogical principles: interdependent learning, double- and triple-loop learning, and participation in communities of practice. Data collection involved questionnaires, semi-structured interviews, focus group discussions, and personal diaries. A noteworthy finding from this study was the near-universal ownership of mobile phones among student-teachers, with the majority possessing at least one Internet-enabled device. Despite limited laptop ownership and institutional access to computers, Chimpololo (2020) noted that a substantial number of student-teachers frequently utilized computers alongside their mobile phones for academic purposes. This underlines the adaptability and resourcefulness of learners in leveraging available technology for their studies.

In terms of heutagogical practices, student teachers reported engaging in interdependent learning, as well as double- and triple-loop learning. However, the study revealed that most participation in communities of practice occurred within student-exclusive groups. Notably, the intention to incorporate devices into teaching was expressed by only a minority of student-teachers. This indicates a potential gap between recognizing the value of heutagogical approaches and translating that recognition into practical implementation. The study's outcomes underscore the untapped potential of mobile devices in facilitating heutagogical learning approaches within teacher training programs. The widespread ownership of mobile phones among student-teachers, coupled with their resourcefulness in using technology, suggests an opportunity to further integrate and promote heutagogical practices. The findings emphasize the need for targeted interventions and support to bridge the intention-action gap, encouraging more pre-service and in-service teachers to incorporate mobile devices into their teaching practices for a more enriched educational experience.

Third, a study by Cochrane et al. (2012) investigated the impact of heutagogy on the development of professional learning networks and communities of practice among mobile web 2.0 learners. They used a design-based research approach to implement and evaluate a series of collaborative projects involving teacher education students and lecturers. Although this study was conducted by researchers from a university in New Zealand, it involved participants from six different countries, including New Zealand, United Kingdom, Spain, Germany, Australia, and Canada. They found that heutagogy supported the development of learner autonomy, social networking skills, and collaborative learning.

Fourth, a study by Gerstein (2013) examined the use of heutagogy as a lens for designing and implementing a massive open online course (MOOC) for educators. The author described the design principles and pedagogical strategies that were used to create a learner-centered, self-determined, and social learning environment. Gerstein also reported on the feedback and reflections of the participants, who were mostly teachers from the United States of America and other countries. Despite the absence of a specific location and departure from the standard research process, the study concluded that heutagogy proved to be a viable and effective approach for facilitating online professional development for teachers.

The fifth article among those illustrated in Table 1 was conducted by Khan and Thomas (2022). In this study, the heutagogical practices were applied through the Constructivist Digital Learning Heutagogy (CDLH) Model. A four-week training on how to teach with the use of the CDLH Model was given to mathematics teachers. These trained teachers then provided an intervention for six months to their students. The intervention aimed to assess the psychological indicators like positive emotions (such as, expression, emotional management, and gratitude expression), academic achievement (focus and persistence), learning engagement (love of learning) and positive relationship (peer relationships and empathy) taught by the CDLH Model. Khan and Thomas (2022) concluded that learning through the CDLH model supported academic achievement, learning engagement, and positive emotions. On the other hand, it was observed that the peer relationship was not supported by the intervention even though the CDLH model helped to break the monotony of the class. This study could inform educational policies that promote positive education and online learning, especially in the context of the COVID-19 pandemic and its aftermath. It could also provide guidelines for designing and implementing effective online curricula that incorporate CDLH principles and practices. Theoretically, the study offers insights into the mechanisms and processes that mediate the effects of CDLH on these outcomes. In terms of practice, this study could help teachers and learners to adopt CDLH as a pedagogical approach that fosters self-determined and autonomous learning, as well as positive well-being and development. It could also provide suggestions for creating and facilitating online learning environments that are engaging, interactive, and supportive of learners' needs and interests.

The last and sixth article presented in Table 1 was conducted by Wong et al. (2019). This study employed heutagogy to investigate the factors influencing teachers' behavioral intention to implement mobile-heutagogical practices. Conducted through a quantitative approach, the research encompassed 246 teachers, spanning both pre-service and practicing teachers. From the findings, four significant factors emerged as key influencers of teachers' behavioral intention: use expectancy, facilitating condition, social influence, and mobile teacher efficacy. Notably, the study revealed that mobile teacher efficacy had a more pronounced impact on the behavioral intention of practicing teachers compared to pre-service teachers.

The implications of this study (Wong et al., 2019) hold substantial importance for various stakeholders, particularly policymakers and those involved in teacher professional development. The identified factors provide valuable insights for tailoring strategies aimed at fostering the adoption of mobile-heutagogy among educators at different levels of education. The authors proposed a conceptual model derived from the study's outcomes, presenting it as a practical guideline to facilitate the effective implementation of mobile heutagogical practices. This model stands poised to enhance the integration of mobile-heutagogy in educational settings, aligning with the demands of Education 4.0, where technological integration in education is a focal point.

However, the study also brought to light a significant challenge: the current low level of mobile-heutagogy utilization among teachers in the Malaysian setting. This suggests a pressing need for concerted efforts to encourage and promote the adoption of mobile-heutagogy. The identified factors influencing behavioral intention serve as key touchpoints for intervention strategies, emphasizing the necessity for targeted initiatives and support systems to elevate the prevalence of mobile heutagogical practices among teachers. In essence, the study by Wong et al. (2019) not only provides insights into the determinants of teachers' behavioral intention but also sets the stage for proactive measures to elevate the adoption of mobile-heutagogy in the educational landscape.

Benefits Associated With Heutagogical Approach

Emerging from the analysis of the six case studies expounded in the preceding section, as well as drawing insights from other existing studies, it is evident that heutagogy stands out as notably effective within VLEs). The following are some of the many benefits associated with heutagogy in diverse virtual learning settings.

Supports Higher-Level Cognitive Function

According to Gregory et al. (2018), heutagogy serves as a catalyst for higher-level cognitive function and double-loop learning. This approach encourages learners to engage in reflective practices, enabling them to scrutinize their learning processes and make requisite adjustments. This, in turn, facilitates a profound comprehension of the subject matter and mastery thereof.

Facilitates Learner-Initiated Learning

In contrast to conventional learning paradigms wherein instructors orchestrate the learning journey, heutagogy directs the spotlight towards the learner, as posited by Gregory et al. (2018). This shift empowers learners to instigate and steer their own learning trajectories, instilling a sense of ownership and intrinsic motivation. Termed as *learner agency* by some scholars (Blaschke, 2018; Blaschke & Hase, 2015; Moore, 2020), heutagogy empowers learners to govern their learning endeavours, encompassing the determination of what and how they learn, as well as the way their understanding is assessed. Therefore, heutagogy is a promising student-initiated learning approach that differs from traditional, instructor-led learning in the sense that it gives learners more autonomy, responsibility, and motivation for their learning.

Promotes Lifelong Learning

The overarching objective of heutagogy lies in imparting the skills and mindset conducive to lifelong learning (Blaschke, 2018, 2019; Blaschke & Hase, 2015; Gregory et al., 2018). By cultivating a continuous and non-linear learning process, heutagogy equips learners with the acumen to adapt to intricate and dynamic work environments. This aligns with insights by Hase (2009) and Moore (2020) that heutagogy prepares learners to handle the complex nature of the workplaces. This means that, heutagogy is a learner-centered approach that enables learners to take charge of their learning and cope with the challenges of the modern world.

Complements Today's Technologies

Positioned as a "net-centric" theory (Blaschke, 2018, 2019; Gregory et al., 2018), heutagogy seamlessly aligns with contemporary technologies that enable learners to access, create, and share information in a networked and collaborative way. This inherent compatibility renders heutagogy particularly useful for application in online and virtual learning environments, where technology assumes a pivotal role in supporting learner autonomy, self-determination, and capability development. Heutagogy can also leverage the affordances of online learning technologies, such as social media, e-portfolios, and MOOCs, to facilitate learner-generated content, peer feedback, and personalized learning paths (Agonács & Matos, 2019; Ansari & Burns, 2023; Blaschke, 2012; Lou & Xu, 2022). This multifaceted utilization of technology in heutagogy transcends traditional boundaries, providing learners with a dynamic and interactive educational experience that extends beyond conventional classroom settings.

Builds and Develops Learner Capabilities

Heutagogy emerges as especially advantageous for tertiary-level students possessing industry-based competency, assisting them in augmenting and refining their capabilities, as indicated by Gregory et al. (2018). Insights from other studies also highlight that the transition from competency to capability, defined as the application of knowledge in novel contexts, is facilitated by heutagogy (Blaschke, 2012, 2014; Links, 2018). Within the realm of initial mathematics teacher training, Alex and Mukuka (2024) determined that implementing heutagogical practices empowers preservice teachers to cultivate professional identities and rectify misconceptions regarding the principles of effective mathematics instruction. This proves especially beneficial in virtual learning environments characterized by a diverse learner background and varying experience levels.

Provides a Learner-Centered Environment

According to Blaschke and Hase (2016), heutagogy establishes a learner-centered environment that empowers learners to delineate their unique learning paths based on their interests, goals, and preferences. This personalized approach accommodates individual learning styles and requirements, such as visual, auditory, or kinesthetic, and provides learners with flexibility and choice in how they access, create, and share information. By doing so, heutagogy amplifies engagement levels and enhances overall educational outcomes, as learners are more motivated, self-directed, and capable of applying their knowledge and skills in various contexts.

In view of the above stated benefits, it suffices to state that heutagogy is an approach laden with substantial advantages within virtual learning environments. By fostering self-directed learning, it not only empowers learners to assume command of their educational journey but also engenders improved learning outcomes and the cultivation of skills essential for lifelong learning.

Challenges Associated With Heutagogical Approach

It has already been noted that heutagogy offers tremendous opportunities in terms of access in a globalised age. However, it also presents several challenges and limitations, especially within virtual learning environments.

Understanding of Learning

Heutagogy challenges conventional notions of education, necessitating a paradigm shift in comprehending the learning process (Hase, 2009). This departure from traditional approaches may prove challenging for educators and learners accustomed to more structured, teacher-centered environments. Consequently, it is often recommended for mature students due to its demand for a heightened level of autonomy (Alex & Mukuka, 2024; Gregory et al., 2018). This shift in understanding learning dynamics is crucial for the successful implementation of heutagogy in virtual learning environments.

Learner Readiness

The promotion of self-determined learning in heutagogy may not align with the readiness of all learners, especially in virtual settings, introducing unique challenges for both students and instructors (Blackhe & Hase, 2016). For instance, the abrupt transition to online learning during the COVID-19 school closures highlighted disparities in preparedness, ranging from insufficient ICT skills to limited internet access and inadequate technology infrastructure, particularly in sub-Saharan Africa (Luneta, 2022b; Mukuka et al., 2021). Heutagogy places the onus on learners to take full responsibility for their education, a demand that can be particularly formidable for those lacking essential skills or motivation.

Designing Learning Experiences

Crafting heutagogical learning experiences presents a multifaceted challenge for educators (Moore, 2020). Successfully implementing this approach requires consideration of learners' needs and capabilities, coupled with the creation of an environment conducive to self-determined learning. Striking the right balance in designing such experiences is not easy but essential to foster an effective heutagogical learning environment.

Assessment Issues

The transition from instructor-led to learner-negotiated assessments, a fundamental aspect of heutagogy, poses implementation and evaluation challenges for both instructors and learners (Blaschke, 2019; Hase, 2009; Hase & Kenyon, 2013). Evaluating learner progress within a self-determined learning framework demands a focus on the learning process itself rather than solely on outcomes, adding complexity to the assessment process.

Support and Guidance

Despite heutagogy's emphasis on autonomy, learners still require support and guidance, creating a delicate balancing act for educators. Striking a balance that nurtures autonomy without undermining it is crucial for fostering a productive and supportive heutagogical learning environment.

Technology Integration

While technology can enhance heutagogical learning practices, its effective integration poses a significant challenge (Blaschke, 2018, 2019). Successful implementation demands a careful selection and utilization of technology that supports, rather than dictating the learning process. Ensuring technology complements the principles of heutagogy requires a nuanced and strategic approach.

Lack of Empirical Evidence

Heutagogy, not firmly established as a learning theory, warrants further empirical exploration to validate its effectiveness and outcomes (Agonács & Matos, 2019; Moore, 2020). Its development and research have been primarily concentrated within a limited community of scholars, predominantly from Australia and New Zealand, potentially limiting its generalizability and diversity. Addressing this gap involves investigating heutagogy in non-formal learning environments and with diverse learner populations to enhance its theoretical robustness.

Institutional Policies

Heutagogy encounters resistance within existing educational structures, such as accreditation, standardized curricula, and testing, which may impede learners' autonomy and self-determination (Blaschke, 2012; Blaschke & Hase, 2016; Hase, 2009). To fully unlock heutagogy's potential in virtual learning environments, a reassessment of institutional policies and a concerted effort to align them with the principles of heutagogy are essential. Overcoming these challenges necessitates a fundamental shift in mindset among educators and learners, coupled with thoughtful design and implementation of learning experiences. Hase (2009) emphasizes that these challenges, although formidable, can spur innovation and foster new perspectives on the learning process.

Implications for Mathematics Teacher Education

Drawing insights from the examined case studies, and other relevant literature, it is imperative to emphasise that the integration of heutagogy into mathematics teacher training demands a departure from conventional pedagogical methods towards a more self-directed and autonomous approach. This transformative shift involves trainees engaging in reflective practices concerning what and how they learn, with trainers assuming a role in equipping trainees with the skills to teach themselves (Blaschke & Hase, 2015).

The case studies not only affirm these pedagogical shifts but also resonate with the findings of Agonács and Matos (2019), whose systematic literature review delves into the empirical evidence supporting heutagogy's application while identifying research gaps. Agonács and Matos (2019) emphasize that heutagogy has predominantly been employed in conjunction with online or digital technological components. However, the specific implementation modes of heutagogy remain somewhat unclear. Notably, their study brings to light the underrepresentation of capability development and non-linearity in existing research, coupled with a prevalence of qualitative methodologies and a dearth of quantitative and longitudinal data collection methods.

Furthermore, Agonács and Matos (2019) shed light on the potential application of heutagogy in MOOCs, emphasizing that the learner-centric nature of heutagogy aligns with the autonomy and refined skills required in MOOCs. Expanding on this, we echo the call for further research, particularly into the viability of applying heutagogy within various realms of virtual learning environments. Such investigations have the potential to significantly contribute to the field, offering empirical evidence that fortifies the credibility of heutagogy, especially in the context of mathematics teacher training within the globalized era. These research endeavours can enhance our understanding of the practical implications and effectiveness of heutagogical approaches in the ever-evolving landscape of education.

Considering these insights, the ensuing implications for policy, theory, and practice are noteworthy.

Policy Implications

Heutagogy challenges the traditional policies and structures of mathematics teacher education, such as curriculum design, accreditation standards, and assessment methods. Heutagogy requires a more flexible and adaptive policy framework that supports learner autonomy, capability, and lifelong learning. Policy makers and stakeholders need to recognize and value the benefits of heutagogy for preparing mathematics teachers who can cope with the complexities and uncertainties of the twenty-first century. Policy makers and stakeholders also need to collaborate with mathematics teacher educators and researchers to develop and implement heutagogical policies and practices that are responsive to the needs and interests of learners and the society.

Practice Implications

Heutagogy transforms the roles and responsibilities of mathematics teacher educators and trainees. Mathematics teacher educators need to shift from being instructors and content experts to being facilitators and co-learners. They need to create heutagogical learning environments that foster learner agency, self-determination, and capability development. They need to provide learners with guidance, feedback, and support, but also respect their choices and preferences. Mathematics teacher trainees need to take ownership and control of their learning. They need to engage in self-directed and self-regulated learning activities, such as setting goals, planning strategies, monitoring progress, and reflecting on outcomes. They need to develop and demonstrate their mathematical knowledge, skills, and competencies in authentic and meaningful contexts.

Theoretical Implications

Heutagogy contributes to the advancement of the theoretical knowledge and understanding of mathematics teacher education. Heutagogy integrates and extends the existing theories and concepts of adult learning, such as andragogy, self-directed learning, self-efficacy, self-determination, and capability. Heutagogy also offers a new perspective and framework for exploring and explaining the phenomena and experiences of mathematics teacher learning in the digital age. Heutagogy invites further theoretical inquiry and debate on the nature, purpose, and process of mathematics teacher education.

CONCLUSION

This chapter has underscored the imperative paradigm shift required in mathematics teacher education due to the profound impact of globalization. Recognizing that globalization demands more than mere integration, the chapter emphasizes the need for a comprehensive restructuring that embraces cultural diversity, linguistic pluralism, and global perspectives. Central to this transformation is the integration of Virtual Learning Environments (VLEs) and heutagogical practices. The chapter contends that these approaches, when applied in tandem, not only respond to the challenges posed by a globalized world but also empower mathematics teachers to foster inclusive learning environments, equipping students with skills essential for thriving in a globally interconnected society.

In view of this, future research in the realm of mathematics teacher education must prioritize several key directions to elevate the value and significance of heutagogical practices within Virtual Learning Environments (VLEs) in the globalized era. Firstly, a thorough exploration of the specific implementation modes of heutagogy in various types of VLEs is imperative. Additionally, the research landscape should shift towards more quantitative and longitudinal studies, aiming to provide robust empirical evidence of the impact of heutagogical practices through VLEs on mathematics teacher education. A nuanced examination of capability development, emphasizing non-linearity and adaptability, is essential to comprehend the full extent of heutagogical benefits. Comparative studies across diverse cultural contexts will offer insights into the cultural influences on the success of heutagogical approaches. Furthermore, exploring the integration of heutagogy within MOOCs and investigating emerging technologies' incorporation will ensure scalability and relevance. Addressing the professional development of mathematics teacher educators and assessing the direct impact on student learning outcomes are vital for successful implementation. Collaborative efforts involving researchers, industry professionals, and educational institutions can contribute to innovative heutagogical practices aligned with the evolving global knowledge economy. Lastly, long-term impact assessment studies will provide valuable insights for sustained improvements in mathematics teacher training programs, ensuring their adaptability and effectiveness in preparing educators for the challenges of a globalized society.

REFERENCES

Agonács, N., & Matos, J. F. (2019). Heutagogy and self-determined learning: A review of the published literature on the application and implementation of the theory. *Open Learning*, 34(3), 223–240. 10.1080/02680513.2018.1562329

Alex, J. K., & Mukuka, A. (2024). Heutagogy in action: Unveiling the transformative power of virtual "air campus" experiences of mathematics trainee teachers. *Eurasia Journal of Mathematics, Science and Technology Education*, 20(3), em2416. 10.29333/ejmste/14320

Ansari, S., & Burns, E. (2023, November 23). *The evolution of online learning: heutagogy meets MOOCs in the digital age*. Jamk Arena Pro. https://urn.fi/urn:nbn:fi:jamk-issn-2984-0783-27

Bandura, A. (1977). Self-efficacy: Toward a unifying theory of behavioral change. *Psychological Review*, 84(2), 191–215. 10.1037/0033-295X.84.2.191847061

Blaschke, L. M. (2012). Heutagogy and lifelong learning: A review of heutagogical practice and self-determined learning. *International Review of Research in Open and Distance Learning*, 13(1), 56. 10.19173/irrodl.v13i1.1076

Blaschke, L. M. (2014). Moving students forward in the PAH continuum: Maximizing the power of the social web. In L. M. Blaschke, C. Kenyon, & S. Hase (Eds.), *Experiences in self-determined learning*. Amazon.

Blaschke, L. M. (2018). Self-determined learning (Heutagogy) and digital media creating integrated educational environments for developing lifelong learning skills. In Kergel, D., Heidkamp, B., Telléus, P., Rachwal, T., & Nowakowski, S. (Eds.), *The Digital Turn in Higher Education* (pp. 129–140). Springer. 10.1007/978-3-658-19925-8_10

Blaschke, L. M. (2019). The pedagogy–andragogy–heutagogy continuum and technology-supported personal learning environments. In Jung, I. (Ed.), *Open and Distance Education Theory Revisited* (pp. 75–84). Springer. 10.1007/978-981-13-7740-2_9

Blaschke, L. M., & Hase, S. (2015). Heutagogy, technology, and lifelong learning for professional and part-time learners. In Dailey-Hebert, A., & Dennis, K. (Eds.), *Transformative Perspectives and Processes in Higher Education* (Vol. 6, pp. 75–94). Springer International Publishing. 10.1007/978-3-319-09247-8_5

Blaschke, L. M., & Hase, S. (2016). Heutagogy: A holistic framework for creating twenty-first-century self-determined learners. In B. Gros, Kinshuk, & M. Maina (Eds.), *The Future of Ubiquitous Learning. Lecture Notes in Educational Technology* (pp. 25–40). Springer. https://doi.org/10.1007/978-3-662-47724-3_2

Bruner, J. (1966). *Towards a Theory of Instruction*. Norton.

Burbules, N. C., & Torres, C. A. (2000). An introduction to globalization and education. In Burbules, N. C., & Torres, C. A. (Eds.), *Globalization and education: Critical perspectives* (pp. 1–26). Routledge.

Caprara, L., & Caprara, C. (2022). Effects of virtual learning environments: A scoping review of literature. *Education and Information Technologies*, 27(3), 3683–3722. 10.1007/s10639-021-10768-w34629934

Ceylan, B. (2020). Ubiquitous learning and heutagogy in teacher education. In Durak, G., & Çankaya, S. (Eds.), *Managing and Designing Online Courses in Ubiquitous Learning Environments* (pp. 279–295). IGI Global. 10.4018/978-1-5225-9779-7.ch015

Chimpololo, A. (2020). An analysis of heutagogical practices through mobile device usage in a teacher training programme in Malawi. *Journal of Learning for Development*, 7(2), 190–203. 10.56059/jl4d.v7i2.391

Clarkson, P. C. (2009). Globalisation and mathematics teaching: The global importance of local language contexts. In Zajda, J., & Rust, V. (Eds.), *Globalisation, Policy and Comparative Research. Globalisation, Comparative Education and Policy Research* (Vol. 5, pp. 139–156). Springer. 10.1007/978-1-4020-9547-4_10

Closs, L., Mahat, M., & Imms, W. (2022). Learning environments' influence on students' learning experience in an Australian Faculty of Business and Economics. *Learning Environments Research*, 25(1), 271–285. 10.1007/s10984-021-09361-233814969

Cochrane, T., Antonczak, L., Gordon, A., Sissons, H., & Withell, A. (2012). Heutagogy and mobile social media: post Web 2.0 pedagogy. In M. Brown (Ed.), *ascilite 2012: Future challenges, sustainable futures* (pp. 204-214)4. Wellington, New Zealand: ascilite..https://bit.ly/3OHTnBM

Gerstein, J. (2013). *Education 3.0 and the pedagogy (andragogy, heutagogy) of mobile learning. [Blog post.] User generated education*. https://usergeneratededucation.wordpress.com/2013/05/13/

Gregory, S., Bannister-Tyrrell, M., Charteris, J., & Nye, A. (2018). Heutagogy in postgraduate education: Cognitive advantages for higher degree online students. In Erwee, R., Harmes, M., Harmes, M., & Danaher, P. (Eds.), *Postgraduate Education in Higher Education. University Development and Administration* (pp. 189–209). Springer. 10.1007/978-981-10-5249-1_32

Halupa, C. (2021). Reaching "creating" in Bloom's taxonomy: The merging of heutagogy and technology in online learning. In Information Resources Management Association (Ed.), *Research Anthology on Developing Critical Thinking Skills in Students* (pp. 15–35). IGI Global. https://doi.org/10.4018/978-1-7998-3022-1.ch002

Hase, S. (2009). Heutagogy and e-learning in the workplace: Some challenges and opportunities. *Impact: Journal of Applied Research in Workplace E-Learning*, 1(1), 43–52.

Hase, S., & Kenyon, C. (2000). From Andragogy to Heutagogy. In *ultiBASE In-Site*. RMIT. https://webarchive.nla.gov.au/awa/20010220130000/http://ultibase.rmit.edu.au/Articles/dec00/hase2.htm

Hase, S., & Kenyon, C. (2013). *Self-determined learning: Heutagogy in action* (1st ed.). Bloomsbury Publishing.

Kamal, T., Khan, M. A., Hamad, Z., Illiyan, A., Das, A. K., & Alkhuraydili, A. (2023). Perception and challenges of virtual classes with gender digital divide amidst and post-COVID-19 pandemic in Iraq: An empirical analysis. *Education Research International*, 2023, 1–14. 10.1155/2023/3665768

Keengwe, J. (2018). Globalization, digital technology, and teacher education in the United States. In *Oxford Research Encyclopedia of Education*. Oxford University Press. 10.1093/acrefore/9780190264093.013.366

Khan, T., & Thomas, S. (2022). Promoting positive education through constructivist digital learning heutagogy: An intervention outcome. *Journal of Learning for Development*, 9(2), 305–316. 10.56059/jl4d.v9i2.646

Kim, T. (2007). Old borrowings and new models of the university in East Asia. *Globalisation, Societies and Education*, 5(1), 39–52. 10.1080/14767720601133140

Korucu-Kış, S. (2021). Preparing student teachers for real classrooms through virtual vicarious experiences of critical incidents during remote practicum: A meaningful-experiential learning perspective. *Education and Information Technologies*, 26(6), 6949–6971. 10.1007/s10639-021-10555-733935573

Lacárcel, A. M. (2022). Artificial intelligence in education as a means to personalize learning. In Saura, J. R., & Debasa, F. (Eds.), *Handbook of Reearch on Artificial Intelligence in Government Practices and Processes* (pp. 285–308). IGI Global., 10.4018/978-1-7998-9609-8.ch016

Lankford, B. A. (2021). Multimedia in teacher-directed and student-centered mathematics classrooms. In Y. Wen, Y. Wu, G. Qi, S.-C. Guo, J. M. Spector, S. Chelliah, Kinshuku, & Y.-J. Lan (Eds.), *Expanding Global Horizons Through Technology Enhanced Language Learning. Lecture Notes in Educational Technology* (pp. 211–227). Springer. 10.1007/978-981-15-7579-2_13

LeTendre, G. (2022). Globalization and the impact of ICT on teacher's work and professional status. In *The Palgrave Handbook of Teacher Education Research* (pp. 1–22). Palgrave Macmillan. 10.1007/978-3-030-59533-3_83-1

Li, S., Hong, Y.-C., & Craig, S. D. (2023). A Systematic literature review of social learning theory in Online Learning Environments. *Educational Psychology Review*, 35(4), 108. 10.1007/s10648-023-09827-0

Links, M. J. (2018). Beyond competency-based continuing professional development. *Medical Teacher*, 40(3), 253–258. 10.1080/0142159X.2017.140121929228838

Lou, J., & Xu, Q. (2022). The development of positive education combined with online learning: Based on theories and practices. *Frontiers in Psychology*, 13, 1–10. 10.3389/fpsyg.2022.95278436118474

Luneta, K. (2022a). Can the Fourth Industrial Revolution Resolve Why the Teaching of Mathematics in the Current Paradigm Continues to Be Decontextualised and Ineffective. In Chirinda, B., Luneta, K., & Uworwabayeho, A. (Eds.), *Mathematics Education in Africa* (pp. 19–38). Springer International Publishing. 10.1007/978-3-031-13927-7_2

Luneta, K. (2022b). Special challenges in mathematics education in Sub Sahara Africa. *Current Opinion in Behavioral Sciences*, 48, 101211. 10.1016/j.cobeha.2022.101211

Maher, M., Skow, B., & Cicognani, A. (1999). Designing the virtual campus. *Design Studies*, 20(4), 319–342. 10.1016/S0142-694X(98)00043-X

McLachlan, K., & Essien, A. A. (2022). Language and multilingualism in the teaching and learning of mathematics in South Africa: A review of literature in Pythagoras from 1994 to 2021. *Pythagoras*, 43(1), a669. 10.4102/pythagoras.v43i1.669

Moore, R. L. (2020). Developing lifelong learning with heutagogy: Contexts, critiques, and challenges. *Distance Education*, 41(3), 381–401. 10.1080/01587919.2020.1766949

Mukuka, A., & Alex, J. K. (2024). Review of research on microteaching in mathematics teacher education: Promises and challenges. *Eurasia Journal of Mathematics, Science and Technology Education*, 20(1), em2381. 10.29333/ejmste/13941

Mukuka, A., Mutarutinya, V., & Balimuttajjo, S. (2019). Exploring the barriers to effective cooperative learning implementation in school mathematics classrooms. *Problems of Education in the 21st Century, 77*(6), 745–757. 10.33225/pec/19.77.745

Mukuka, A., Mutarutinya, V., & Balimuttajjo, S. (2020). Data on students' mathematical reasoning test scores: A quasi-experiment. *Data in Brief*, 30, 105546. 10.1016/j.dib.2020.10554632337327

Mukuka, A., Shumba, O., & Mulenga, H. M. (2021). Students' experiences with remote learning during the COVID-19 school closure: Implications for mathematics education. *Heliyon*, 7(7), e07523. 10.1016/j.heliyon.2021.e0752334307946

Mulenga, E. M., & Marbán, J. M. (2020). Is COVID-19 the gateway for digital learning in mathematics education? *Contemporary Educational Technology*, 12(2), ep269. 10.30935/cedtech/7949

Olosunde, G. (2019). Influence of globalization on mathematics content, teaching methods, and resource materials in colleges of education in Oyo State. In Tella, A. (Ed.), *Globalized Curriculum Methods for Modern Mathematics Education* (pp. 15–28). IGI Global. 10.4018/978-1-5225-6158-3.ch002

Sharma, R., & Singh, A. (2023). Use of digital technology in improving quality education. In Nadda, V., Tyagi, P. K., Vieira, R. M., & Tyagi, P. (Eds.), *Implementing Sustainable Development Goals in the Service Sector* (pp. 14–26). IGI Global. 10.4018/979-8-3693-2065-5.ch002

Sharma, S., & Sharma, S. (2023). Successful teaching practices for english language learners in multilingual mathematics classrooms: A meta-analysis. *Mathematics Education Research Journal*, 35(4), 821–848. 10.1007/s13394-022-00414-0

Stephan, M. (2020). Teacher-centered teaching in mathematics education. In Lerman, S. (Ed.), *Encyclopedia of Mathematics Education* (pp. 836–840). Springer International Publishing. 10.1007/978-3-030-15789-0_150

Stoten, D. W. (2022). Navigating heutagogic learning: Mapping the learning journey in management education through the OEPA model. *Journal of Research in Innovative Teaching & Learning*, 15(1), 83–97. 10.1108/JRIT-07-2020-0038

Traore, M. (2021). *Guidebook to a successful implementation of digital education*. World Bank. https://docs.edtechhub.org/lib/RTJ2Z8MX

von Glasersfeld, E. (2014). Piaget and the radical constructivist epistemology. *Costruttivismi, 1*, 94–107. https://www.aippc.it/wp-content/uploads/2019/04/2014.02.094.107.pdf

Vygotsky, L. S. (1978). *Mind in society: The development of higher psychological processes*. Harvard university press.

Wong, K.-T., Yeop, M. A., & Mazura, M. (2019). Modelling the factor influencing the implementation of mobile-heutagogical practices among teachers: An application of invariance multi-group structural model. *International Journal of Learning. Teaching and Educational Research*, 18(12), 1–16. 10.26803/ijlter.18.12.1

World Economic Forum. (2022, June 30). *Online learning: What next for higher education after COVID-19?* WEForum. https://www.weforum.org/agenda/2022/06/online-learning-higher-education-covid-19/

Yildiz, M., & Scharaldi, K. (2015). Cultivating transformative educator model among pre-service teachers while integrating global competencies, critical thinking and 21st century skills: A transdisciplinary approach. In D. Rutledge & D. Slykhuis (Eds.), *Proceedings of SITE 2015-Society for Information Technology & Teacher Education International Conference* (pp. 1760–1769). Association for the Advancement of Computing in Education (AACE). https://www.learntechlib.org/primary/p/150236/

ADDITIONAL READINGS

Durak, G., & Çankaya, S. (Eds.). (2019). *Managing and designing online courses in ubiquitous learning environments*. IGI Global. https://www.igi-global.com/book/managing designing-online-courses-ubiquitous/223463

Friedman, Z. L., & Nash-Luckenbach, D. (2024). Has the time come for Heutagogy? Supporting neurodivergent learners in higher education. *Higher Education*, 87(6), 1905–1920. 10.1007/s10734-023-01097-7

Kellam, H. (2024). Element 4: Inclusion and Differentiation—inclusive practices for online learning environments. In *Personalized and Inclusive Engagement for the Design, Delivery, and Evaluation of University eLearning. Monographs in the Psychology of Education*. Springer. 10.1007/978-3-031-49129-0_5

Yeigh, T., Handayani, S., Peddell, L., & Lee, M. (2023). Using Self-Determined Learning to Improve Teacher Professional Learning. In Lynch, D., Yeigh, T., & Boyd, W. (Eds.), *Re- imagining Teaching Improvement*. Springer. 10.1007/978-981-99-7746-8_14

KEY TERMS AND DEFINITIONS

Conventional pedagogical methods: Traditional teaching methods and practices.

Cross-cultural competencies: Skills and understanding necessary to interact effectively with people from different cultural backgrounds.

Global educational trends: Current patterns and shifts in education on a global scale.

Globalised Age: Refers to the interconnectedness and interdependence of the world due to advancements in technology, communication, transportation, and many more.

Heutagogy: A form of self-determined learning with practices and principles rooted in andragogy, a self-regulated learning approach.

Learner-centered approach: An educational approach that focuses on the needs and abilities of the learner, rather than the teacher or the curriculum.

Linguistic diversity: The variety and multilingualism of languages spoken within a specific area or around the world.

Mathematics Teacher Education: The field of study and practice related to the teaching and learning of mathematics.

Twenty-first century teaching and learning: Refer to a broad set of abilities, knowledge, habits, and traits that are considered essential for success in today's interconnected world. These skills are not limited to any specific academic subject or career but are applicable across various fields and throughout a person's life. Example of these skills include, but not limited to critical thinking, creativity, collaboration, communication, information literacy, technology literacy, and life skills.

Virtual Learning Environments (VLEs): Online platforms that replicate a physical learning environment, enabling teaching and learning to take place remotely.

ENDNOTE

1 The participants were not affiliated to a specific university, but rather they were part of a global community of learners who shared their experiences and reflections on heutagogy, and other topics related to education and technology.

Chapter 4
Innovative Pedagogical Practices in Mathematics Education

Omosa Elijah
Kisii University, Kenya

ABSTRACT

Mathematics is a very important human venture that finds its applications in all spheres of life. Despite this importance attached to it, the interest and performance in it is dismal. This is as a result of how it is taught. There is every need for teaching it better, thus the need for innovative and powerful approaches. This is this concept is "used to identify the knowledge and capabilities required to accommodate the mathematical demands of private and public life, and to participate in society as informed, reflective, and contributing citizens." Therefore, letting the students make sense of the mathematical content they learn can have a significant positive impact on multiple dimensions of their lives. Math enhances our capacity to deal with these kinds of day-to-day challenges. There is a need to develop innovative learning environments that benefit all learners. The second theme is centered on how both tasks and tools can be used to support powerful pedagogical practices. Finally, the third theme reviews the challenges of developing innovative mathematical learning environments.

INTRODUCTION

The study of mathematics enables learners to develop essential literacy, numeracy skills and comptences. Through the concepts of mathematics, learners can reason out of logic, think critically, and they can effectively solve problems. By solving

DOI: 10.4018/979-8-3693-2873-6.ch004

puzzles and logically thinking, the can apply matehmatical skills in solving real-life problems. Mathematics helps us learners in making connections to real life situations and the concepts that are learnt in it. The number patterns met early in life influence life at a later stage. Therefore a solid base at a lower level, starting with early years is very important.

Worldwide there is a growing focus on numeracy. Geiger et al. (2015, p. 531), posit this concept is "used to identify the knowledge and capabilities required to accommodate the mathematical demands of private and public life, and to participate in society as informed, reflective, and contributing citizens". it is therefore pertinent to guide learners to connect the mathematics they learn in class and its applications to real-life situations, for example, in analyzing information such as correlation or causality and put our mathematical thinking into practice. We could notice the importance of understanding diverse graphics related to epidemiology during a pandemic. Although many patients get lost in numbers, health numeracy is basic to understanding medical information and making informed decisions (Ancker & Kaufman, 2007).

In today's labour market, mathematics skills are basic to increasing employability to students of both STEM and social sciences. Persons from the creative backgrounds too can need knowledge of mathematics. Geometry for instance is useful for artistic and design projects. Mathecatics assists us in making our budgeting and planning. In our cooking, we need to measure proportions when cooking; we calculate and share costs; when shopping, estimating how likely our team is to win a competition. Math enhances our capacity to deal with these kinds of day-to-day challenges. Over a long time, mathematics has significantly influenced art and architecture worldover (Gamwell, 2015). Today, its influence can also be found in contemporary art and architecture. Additionally, we can appreciate the beauty of nature due to its underlying mathematical principles. As Stewart (2017) shows in the book "The Beauty of Numbers in Nature: Mathematical Patterns and Principles from the Natural World", there are fascinating patterns in flowers, dunes, zebras' stripes, snowflakes, and spiderwebs.

If students are to learn mathematics better, teachers must utilize innovative pedagogical practices in their teaching. This chapter, presents pedagogical practices that support creative and flexible mathematical thinkers for the 21st century. Innovation and technology have the potential to revolutionize the way mathematics is taught and learned. Here are some ways in which they can help:

INTERACTIVE LEARNING

Technology can be utilized as a tool to presnt mathematics content and experiences in an interactive way, so as engage learners and help them learn mathematics in a fun and engaging manner. For example, students can use online tools to visualize mathematical concepts and explore mathematical models.

These types include student-student interaction, stuent-teacher interaction, and interaction between students and leaning technology. Student-technology interactions are explained by computer programs that use problem-solving strategies and multiple representations. Teacher - students interactions are explained in two categories, classroom interaction and small group interaction. Teachers need to consider many factors in order to establish a classroom environment to enhance the mathematical understanding by their students. In small cooperative groups, factors that effect interaction are as follows: group composition, type of interaction, effect of teacher, interdependence of students and nature of the task. Some teaching implications then are: students are encouraged to use multiple representations to develop problem solving strategies; students' motivation to learn should be mastery goal oriented, teachers should try to create contexts for mathematical argumentation; teachers should encourage student participation in classroom discussions; students should be expected to provide mathematical reasoning rather than producing the right answer; and design of tasks should be suitable to promote skills such as mathematical reasoning and metacognition. There are indeed several interactive learning tools available for teaching and learning mathematics. Some of the popular ones include:

Brilliant: This website offers a wide range of courses in STEM. By going though the interactive quizzes, problems and games users learn and practice mathematical concepts. Through the community section they can discuss and share their knowledge with others.

Mathigon: The interactive textbooks, puzzles and games haboured in this website offer help to users learn mathematics. A wide range of topics ranging from basic arithmetic to advanced calculus are presented. Through the community section they can discuss and share their knowledge with others. *Developing eContent for Teaching and Learning of Mathematics (DETL):* DETL, an Indian government initiative provides free online resources for teaching and learning mathematics. It encompasses interactive e-books, videos, and quizzes to aid learners to learn mathematics. It targets all levels form primary school to undergraduate level. *GeoGebra:* Is a free platform which offers over 100 million users worldwide, a wide range of digital tools for class activities, graphing, geometry, collaborative whiteboard, and more. *Desmos:* This platform provides best-in-class calculators, digital math activities, and curriculum to help every student love math and love learning math.

Khan Academy: This platform offers free online courses in math, science, and humanities. It provides personalized learning experiences and has over 120 million registered users.

PERSONALIZED LEARNING AND MATHEMATICS TEACHING AND LEARNING

Technology can be utilized to create and present personalized learning experiences that cater to the individual needs of each learner. For example, learners can use adaptive learning software that adjusts the difficulty of the content based on their achievement. Personalized learning is one of the several instructional stratergies that support mathematics teaching and learning. However, educators do not have a consensus on the definition of personalized learning. Some see it as (*a*) customization, (*b*) student groupings, and (*c*) flexibility of instruction. Customization entails tailoring experiences and instruction for students' needs, interests, goals, and backgrounds. It suggests that some humanizing aspects should be considered in personalized learning-a knowledge and understanding on how students learn mathematics. One-to-one and whole-class approaches if used well, can complement each other leading to deep students' understanding of mathematics and for supporting them to develop a positive personal relationship with mathematics. Pacing of instruction and time and space for students to engage with mathematics is part of the flexible instruction stratergies that can lead to meaningful learning. Personalized learning is a teaching approach that caters for the needs of individual learners. Educationists mainly focus on how to improve achievement and test scores and achievement, but normally ignore the humanizing and social aspects of teaching and learning mathematics. Personalized learning can present an opportunity to learners to air their voice to the ways they think mathematically, represent and discuss their mathematical ideas, and use mathematics to make sense of their worlds. Personalized learning is based on constructivism, whereby learners see themselves as doers of mathematics through the facilitation by the teacher for developing understanding and perseverance, as well as for engaging in collaborations with peers to unpack mathematical thinking.

A desirable environment for personalized learning of mathematics is where each and every student is provided with a personalized teacher who understands his or her needs, knows interests and background, and possesses content knowlede and pedagogical content knowledge in mathematics. The ideal situation where where students could have broad access to their personalized teacher to ask questions and that their personalized teacher is able to extend their mathematical thinking. However, because of several issues, this may not be practical. Teachers who are presently teaching the classes need to be retooled so as to be able to handle the classes in a near personolized learning.

COLLABORATIVE LEARNING (CL)

Collaborative learning (CL) is a teaching strategy in which the teacher groups students into teams of students who s/he guides to learn through working together to share ideas, solve a problem, or accomplish a common goal. In mathematics education, CL's popularity gained popularity in the 1980s, but since then it has continusly evolved. Here are some examples of how CL can be used in teaching mathematics: a) **Group problem-solving**: Students work in groups to solve a mathematics problem. Each student in the group is responsible for contributing to the solution. This approach encourages students to not only learn from others but to share their ideas and thoughts.**b) Peer tutoring**: Students work in pairs, with one student acting as the tutor and the other as the tutee. The tutor helps the tutee understand a math concept or solve a problem. This approach can be particularly effective for students who are struggling with a particular topic. c) **Project-based learning**: Students work in groups to complete a project that involves math. For example, students might design a bridge or a building using math concepts. This approach encourages students to apply math concepts to real-world problems.

d) **Cooperative learning**: Students work in groups to complete a task. Each student in the group is responsible for a specific part of the task. This approach encourages students to work together and learn from each other. These are just a few examples of how CL can be used in teaching mathematics. By encouraging students to work together and learn from each other, CL can help students develop a deeper understanding of math concepts and improve their problem-solving skills. Collaborative learning is a teaching approach that involves students working together to explore, master, or apply course material and concepts. It can take many forms, including Think-Pair-Share, Number Talks, Jigsaw Activities, and Peer Review/Assessment. Here are some examples of how these techniques can be used in teaching mathematics:

Think-Pair-Share: This technique provides an opportunity and encourages individual participation and is applicable across all grade levels and class sizes. In a math class, the teacher can pose a question to the students, give them a minute to think about it, and then have them pair up to discuss their answers. Finally, the teacher can ask for volunteers to share their answers with the class.

Number Talks: This technique is a structured way to develop students' mental math skills. In a math class, the teacher can pose a problem to the students and ask them to solve it mentally. The students can then share their strategies with the class, and the teacher can facilitate a discussion about the different approaches.

Jigsaw Activities: This technique involves dividing students into groups and assigning each group a different part of a larger topic. In a math class, the teacher can divide the class into groups and assign each group a different concept or skill to master. The groups can then present their findings to the class, and the teacher can facilitate a discussion about the different concepts.

Peer Review/Assessment: This technique involves students evaluating each other's work. In a math class, the teacher can assign students to work in pairs or small groups to solve a problem. The students can then evaluate each other's work and provide feedback on how to improve their solutions.

Technology can be used to facilitate CL experiences that allow students to work together on math problems and projects. For example, students can use online tools to collaborate on math problems in real-time.

REAL-WORLD APPLICATIONS

Technology can be used to show students how mathematics is used in the real world. For example, students can use simulations and modeling tools to explore how math is used in fields such as finance, science, and engineering. Mathematics is often considered a difficult subject by students. One of the numerous reasons behind this is that topics such as algebra and calculus often seem abstract and disconnected from our daily lives, leaving many students questioning the practicality of studying them. However, if in our teaching we provide a connection between mathematics with real-life applications of the various branches or topics of the subject, for example through provision of a rationale, we can make the subject matter to look less abstract and more engaging. When learners see the relevance of given content, it can lead to creation and maintenance of interest, deeper comprehension, thus enabling them to apply their mathematical knowledge and skills to future scenarios. Certainly! Real-world applications of mathematics can make learning more engaging and relevant for students. Here are some creative ways teachers are connecting math to everyday life:

Real-World Math Wall

A fifth-grade teacher in Texas, Samantha Baumgardner, requests her learners to bring in common items and write three ways they relate to mathematics on notecards. These objects form the class's real-world math wall. Students contribute items like playing cards, cake pans, softball score sheets, and cookie recipes. For instance, one student brought in a medicine dropper and explained, "The medicine dropper helps you measure water, and it also helps you with cooking steak".

Water Audit

Krista Dunlop-Sheppard, a bilingual resource and home schoolteacher, guides her students in conducting a water audit at home and school. Students measure the flow rate of water from fountains, toilets, and urinals to calculate the school's daily water usage. They interview custodians and research water consumption by the cafeteria dishwasher. Based on their findings, students suggest conservation methods, such as collecting rainwater or using less water for cleaning.

Acting Out Restaurant Scenarios

Justin Ouellette, an educator in China, brings in menus for students to act out real-life restaurant scenarios. This exercise helps students understand negative and positive numbers, fractions, and decimals.

Cooking and Baking

Cooking and baking provide practical examples of math outside the classroom. Reading recipes involves understanding fractions, proportions, and measurements. Discussing cooking times and adjusting recipes reinforces mathematical concepts. By integrating these real-world applications, educators can make math more meaningful and applicable to students' lives. Problems from the learner's context make mathematics more accessible and enjoyable. This may help learners to develop their critical thinking, enhance their motivation, and boost active learning.

Several exercises presenting real-life problems have been uploaded to SOWISO's platform that needs mathematics to solve them. An extension has been created with applications for our best-selling course, Basic Math. Each chapter in Basic Math now includes an extra package with exercises demonstrating the practical applications of the theory covered. This extension is ideal for higher education Science and Engineering programs. Soon, we will add an extension for Social Sciences.

Since students' experiences are diverse, our platform is tailored to various learning backgrounds and needs. With features like randomized exercises and automated personalized feedback, we help students practice as much as they want while supporting them on their unique journeys.

As you can see below, SOWISO guides students towards the solution, providing hints and constructive feedback so they can master their math at their own pace. You can see examples of our new exercises and how the platform works here.

e) **Gamification**: Technology can be used to gamify the learning experience and make math more fun and engaging. For example, students can use online games and apps to practice math skills and compete with their peers.

These are just a few examples of how innovation and technology can help teach mathematics. By leveraging these tools, educators can create more engaging and effective learning experiences that help students develop a deeper understanding of math concepts.

Use of game elements in non-game contexts, such as education, to increase engagement and motivation is what is referred to as gamification. In the teaching and learning of mathematics, gamification is used to make learning more interactive and fun for students. It can help students develop problem-solving skills, improve their understanding of mathematical concepts, and increase their confidence in their abilities.

One study found that gamification techniques can be used to teach mathematical concepts in a way that is more engaging and effective than traditional teaching methods. The study found that students who were taught using gamification techniques had higher levels of engagement, motivation, and achievement than those who were taught using traditional methods.

Gamification can be used in a variety of ways to teach mathematics. For example, teachers can use games to teach the four basic arithmetic skills of addition, subtraction, multiplication, and division. They can also use games to teach more advanced mathematical concepts, such as algebra, geometry, and calculus. Games can be designed to be played individually or in groups and can be tailored to the needs and interests of individual students.

In conclusion, gamification is a promising technique for teaching mathematics that can help draw the attention, maintain interest increase student achievement. By using game elements in the classroom, teachers can make learning more interactive and fun, and help students develop the skills they need to succeed in mathematics and beyond.

INQUIRY-BASED LEARNING (IBL)

Inquiry-based learning (IBL) is a student-centered approach to learning that emphasizes the role of teacher as a facilitator and the learner as a constructor of his or her knowledge through exploration and discovery. It is an effective way to teach mathematics as it encourages students to ask questions, investigate and solve problems on their own. Here are some examples of how inquiry-based learning can be used in teaching mathematics:

a) *Exploring geometric shapes:* Students can be asked to investigate the properties of different geometric shapes such as triangles, circles, and squares. They can be encouraged to ask questions such as "What is the relationship between the area and perimeter of a square?" and "How can we calculate the circumference of a circle?".

b) *Data analysis:* Students can be given a set of data and asked to analyze it. They can be encouraged to ask questions such as "What is the mean, median, and mode of this data?" and "What is the range of this data?".

c) *Real-world problems:* Students can be given real-world problems that require mathematical thinking to solve. For example, they can be asked to calculate the amount of paint needed to paint a room or the amount of fabric needed to make a dress.

Inquiry-based learning is a great way to teach mathematics as it encourages students to think critically, ask questions, and develop problem-solving skills. By using this approach, students are more likely to retain the information they learn and apply it to real-world situations.

PROBLEM-BASED LEARNING (PBL)

Problem-based learning (PBL) is an innovative teaching methodology that can help students develop a deep understanding of mathematical concepts and utilize them to real-world scenarios. In a PBL classroom, learners are presented with a complex problem that requires them to apply their knowledge and skills to solve it. The teacher acts as a facilitator and provides limited guidance, allowing students to take ownership of their learning.

Here are some examples of how PBL can be used in teaching mathematics:

a) *Selling Geometry:* This project introduces students to a brief history of geometry, terms used in geometry, geometrical shapes, and transformation and manipulation of shapes through reflections, tessellations, and dilations.

b) *Mathematical Modeling:* Students are asked to create a mathematical model of a real-world problem, such as predicting the spread of a disease or analyzing the stock market.

c) *Designing a Playground:* Students are tasked with designing a playground that meets specific requirements, such as a budget, safety regulations, and accessibility guidelines.

These are just a few examples of how PBL can be used in teaching mathematics. By using PBL, students can develop critical thinking, problem-solving, and collaboration skills, which are critical for success in the 21st century.

FLIPPED CLASSROOM

In the flipped classroom model the traditional classroom approach is reversed. In this model, the learners gain the new content almost independently, often through homework, and then in class, the majority of the time is spent practicing, asking questions, and doing activities with the teacher there to support. Here are some examples of how to use the flipped classroom in teaching mathematics:

a) *Pre-recorded lectures:* Teachers can record their lectures and post them online for students to watch before class. This allows students to learn at their own pace and frees up class time for more interactive activities.
b) *Interactive activities:* In-class activities can be designed to help students apply the concepts they learned in the pre-recorded lectures. For example, teachers can use games, group work, or problem-solving activities to engage students and reinforce their understanding of the material.
c) *Collaborative learning:* The flipped classroom model encourages students to work together and learn from each other. Teachers can facilitate group discussions or assign group projects to help students develop their communication and teamwork skills.
d) *Real-world applications:* Teachers can use real-world examples to help students see the relevance of the concepts they are learning. For example, teachers can use data from current events or scientific research to illustrate how mathematical concepts are used in the real world.
e) *Immediate feedback:* The flipped classroom model allows teachers to provide immediate feedback to students. Teachers can use in-class activities to assess students' understanding of the material and provide feedback on their progress.
f) *The Khan Academy:* This is an online platform that provides free educational resources, including video tutorials, practice exercises, and quizzes. You can use this platform to supplement your classroom lectures and provide students with additional resources to help them understand mathematical concepts. For example, you can assign videos on a particular topic for students to watch before class, then use class time to work through practice exercises together.
g) *Mathantics:* This is another online resource that provides animated video tutorials on various mathematical concepts. You can use these videos to introduce new topics or reinforce concepts that students are struggling with. For example, you can use a video on fractions to introduce the concept to your class, then follow up with a group activity to reinforce the concept.
h) *Think-Pair-Share:* This is a collaborative learning strategy that presents a though-provoking situation, then the facilitator encourages learners to think independently, pair up and discuss with a classmate or in small groups and share their response with the class. This strategy can be to encourage students

to think critically about mathematical concepts and work together to solve problems. For example, you can ask an open-ended question and have students think quietly about it for a minute or two. Then, every student pairs up with another learner and they discuss the question for two to five minutes. Finally, the whole class engages in a discussion where students raise their hands and share all the thoughts and ideas they've gathered.

i) *Jigsaw Technique:* This is another collaborative learning strategy where the facilitator divides learners into small groups and assigning each group a specific topic to research. Each group member becomes an expert on their topic and then shares their knowledge with the rest of the group. This strategy can be utilized to teach complex mathematical concepts that require a lot of research and analysis. For example, when teaching a topic on graphs, the class can be divided into groups and assign each group a different type of graph to research. Each group member can then become an expert on their graph and share their knowledge with the rest of the group.

j) *Brainstorming:* This is a creative problem-solving technique that encourages students to generate as many ideas as possible in a short amount of time. You can use this technique to help students think outside the box and come up with creative solutions to mathematical problems. For example, you can ask your class to brainstorm different ways to solve a particular math problem, then have them share their ideas with the rest of the class.

By using the flipped classroom model, teachers can create a more engaging and interactive learning environment for their students, which can lead to better learning outcomes.

7. **Using virtual manipulatives** Virtual manipulatives are digital tools that can be used to teach mathematical concepts in a more interactive and engaging way. Here are some examples of virtual manipulatives and how they can be used in teaching mathematics: a) *Polypad:* Polypad is a free online tool that allows students to manipulate virtual tiles to explore mathematical concepts such as fractions, decimals, and algebraic equations. For example, students can use Polypad to model fractions by dividing a square into equal parts and coloring them in different colors to represent different fractions. They can also use Polypad to solve algebraic equations by dragging and dropping tiles to balance the equation. b) *Didax:* Didax is a website that offers a range of virtual manipulatives for teaching mathematics. Some of the manipulatives available on Didax include base-10 blocks, fraction tiles, and algebra tiles. For example, students can use base-10 blocks to model place value concepts or use fraction tiles to compare and order fractions. c) *K-5 Math Teaching Resources:* K-5 Math Teaching Resources is a website that provides free virtual manipulatives for teaching mathematics. Some of the manipulatives available on K-5 Math

Teaching Resources include polygons, number tiles, and fraction bars. For example, students can use polygons to explore geometric concepts such as symmetry and angles or use fraction bars to model fractions and compare them. d) *Toy Theater:* Toy Theater is a website that offers a range of interactive games and activities for teaching mathematics. Some of the virtual manipulatives available on Toy Theater include pentominoes, tangrams, and pattern blocks. For example, students can use pentominoes to explore geometric concepts such as area and perimeter or use pattern blocks to create and identify patterns. Using virtual manipulatives in teaching mathematics can help students to visualize abstract concepts and develop a deeper understanding of mathematical concepts. By providing students with interactive and engaging tools, teachers can help to make mathematics more accessible and enjoyable for all students.

Adaptive learning is a computer-based and/or online educational system that modifies the presentation of material in response to student performance. Here are some examples of how to use adaptive learning in teaching mathematics:

i. *DreamBox Learning:* DreamBox Learning is an adaptive learning platform that uses Intelligent Adaptive Learning™ technology to adapt in real-time to every interaction a student makes, both within and between lessons. DreamBox Learning provides a personalized learning experience for students, which is tailored to their individual needs and abilities. For example, DreamBox Learning can be used to teach students how to solve mathematical problems by providing them with interactive lessons that adapt to their learning style and pace.

ii. *ALEKS:* ALEKS is an adaptive learning platform that uses artificial intelligence to provide students with personalized learning experiences. ALEKS can be used to teach students a wide range of mathematical concepts, from basic arithmetic to advanced calculus. For example, ALEKS can be used to teach students how to solve algebraic equations by providing them with interactive lessons that adapt to their learning style and pace.

iii. *Knewton:* Knewton is an adaptive learning platform that uses machine learning algorithms to provide students with personalized learning experiences. Knewton can be used to teach students a wide range of mathematical concepts, from basic arithmetic to advanced calculus. For example, Knewton can be used to teach students how to solve mathematical word problems by providing them with interactive lessons that adapt to their learning style and pace. In conclusion, adaptive learning platforms such as DreamBox Learning, ALEKS, and Knewton can be used to teach mathematics by providing students with personalized learning experiences that adapt to their individual needs and abilities. These platforms can be used to teach a wide range of mathematical concepts, from basic arithmetic to advanced calculus, and can provide students with interactive lessons that adapt to their learning style and pace.

IMPROVISATION FOR TEACHING AND LEARNING MATHEMATICS

Improvisation techniques can be a fun and engaging way to teach mathematics. Here are some examples of how you can use improvisation techniques in your math lessons:

a) **Improv comedy**: Improv comedy can be used to teach math concepts such as probability, geometry, and algebra. For example, you can use improv games like "Yes, and…" to teach probability by having students create a probability tree based on a given scenario. You can also use improv games like "Freeze" to teach geometry by having students create shapes with their bodies.
b) **Scream circle**: Scream circle is a technique where students stand in a circle and scream out math facts. This can be a fun way to help students memorize math facts.
c) **Mathematical sets**: Mathematical sets can be used to teach students about logic and reasoning. For example, you can use Venn diagrams to teach students about set theory.
d) **Mathematical storytelling**: Mathematical storytelling can be used to teach students about problem-solving and critical thinking. For example, you can use stories to teach students about word problems.
e) **Manipulatives**: Physical manipulatives can be used to teach students about math concepts such as fractions, decimals, and geometry. For example, you can use blocks to teach students about fractions.
f) **Games**: Games can be used to teach math concepts such as addition, subtraction, multiplication, and division. For example, you can use card games to teach students about addition.
g) **ICT Technology**: ICT technology can be used to teach math concepts such as graphing, statistics, and geometry. For example, you can use graphing software to teach students about graphing.
h) **Real-world examples**: Real-world examples can be used to teach students about the practical applications of math. For example, you can use real-world examples to teach students about the importance of math in everyday life.

PROJECT-BASED LEARNING

Project-Based Learning (PBL) is a form of instruction that involves learners in challenging, real-life projects which give learners an opportunity to investigate and build knowledge via practical application. PBL involves learners in challenging, real-world projects which gives learners an opportunity to investigate and build knowledge via practical application. Learners actively participate in the project planning, execution, and evaluation processes, which promotes a sense of freedom and responsibility. Teams of learners utilize their varied skills and views to solve complicated challenges. Collaboration between the learners improves interpersonal skills, critical thinking, and communication abilities, reflecting the importance of teamwork in real-world workplaces (Thomas, 2000).

PBL tries to move away from emphasis on rote memorization in the classroom than traditional methods do. Learners are given the chance to carry out research issues from the real world, develop hypotheses, plan experiments, collect and analyze data and communicate their findings. PBL differs from traditional teaching techniques by incorporating a number of essential components. First, it places a focus on learner autonomy, empowering learners to take charge of their learning process. Learners actively participate in the project design, development, implementation and evaluation, thus promoting a sense of freedom and responsibility. Second, PBL is learner collaboration. Teams of learners use their varied views and skills to solve complicated challenges. Collaboration improves interpersonal skills, critical thinking, and communication abilities, reflecting the importance of teamwork in real-world workplaces.

Fielding et al. (2007) makes it clear that "rigor, engagement, lesson purpose and results are hallmarks of excellent instruction". It is possible to say that PBL has all these characteristics.

The merits of project-based learning for learners are numerous. As learners take on challenging, open ended problems, it improves their critical thinking and problem-solving abilities. PBL promotes critical thinking, creative problem-solving, and information analysis in learners. PBL also encourages knowledge retention and greater understanding. Learners gain a greater understanding of the significance and value of what they are studying by applying theoretical ideas to practical situations. This hands-on learning method encourages a deeper understanding of the material. PBL also fosters crucial 21st-century skills like collaboration, communication, and adaptability. Learners get the necessary interpersonal, leadership, and teamwork skills to succeed in the modern workplace.

According to Holm (2011), PBL is a teaching methodology that utilizes learner-centered projects to facilitate learner learning. More people are realizing that in order to succeed in the modern world, learners need to acquire a variety of abilities that go beyond the classroom. PBL is a method of instruction that successfully develops key 21st-century skills including teamwork, communication, and adaptability (Larmer et al., 2015; Hallerman et al., 2011). PBL is based fundamentally on collaboration. Learners participate in group projects to solve challenging issues or complete assignments. They acquire the ability to work well together through exchanging tasks, responsibilities, and resources. Learners gain vital interpersonal abilities through teamwork, including active listening, empathy, and respect for different points of view. Learners must communicate, reach an agreement and compromise as part of collaborative assignments; these are all essential skills for working in various and interdisciplinary circumstances.

Additionally, another key development through PBL is communication skills. Throughout the project, learners participate in conversations, presentations, and written reflections. They develop the ability to express their thoughts clearly, engage in active listening, and offer helpful criticism. In the current world, effective communication is crucial because it allows people to express their ideas, work together, and motivate others. Effective communication is vital in the present day because it allows people to clearly express their ideas, work with others in a variety of contexts, and inspire and motivate others to pursue similar objectives. In social, professional, and personal situations, it promotes relationships, fosters understanding, and leads to effective outcomes.

Another critical 21st-century potential that PBL fosters is adaptability. Learners work on actual, real-world challenges through project-based learning that may have several solutions or call for flexibility in approach. As new information becomes available, they learn to improvise, change their techniques, and revise their plans. This capacity for adaptation equips learners for the fast-paced environment of the contemporary workplace, where they will unavoidably face difficulties that call for adaptability and creative problem-solving. Moreover, PBL naturally fosters other crucial abilities like creativity, critical thinking, and self-direction. Learners develop their ability to think critically, examine data, and balance opposing viewpoints by working on open-ended projects. They are encouraged to think creatively, come up with original solutions, and take control of their education. Buck Institute for Education (2014) claims that "PBL is a rigorous, relevant, and engaging instructional model that supports authentic inquiry and autonomous learning for learners" (p. 6).

The eight significant features of PBL are (Philen, 2016, p.24):

Key Knowledge, Understandings and Success Skills

Authenticity

Sustained Inquiry

Challenging Problem or Question

Reflection

Voice and Choice

Critique & Revision

Project-based learning (PBL) has attracted an abundance of attention recently as a productive teaching strategy in a number of subject areas, including mathematics. PBL incorporates practical issues, cooperative cooperation, and hands-on activities to actively include learners in their learning process.

Learners are presented with challenges from the real world that provide opportunities for the use of their knowledge and abilities. They collaborate with their classmates and solve challenges by sharing ideas and resources through collaborative activities. The learners can actively participate and develop a greater understanding of the topic matter through practical application through hands-on activities, which offer experiential learning opportunities.

PBL offers an alteration from the typical classroom environment and gives learners the chance to use mathematical concepts in practical settings. Learners learn the value and meaning of mathematical concepts beyond abstract formulas and equations by working on real-world projects.

Learners must comprehend the concepts of measurement, geometry and structural stability, for instance in order to design and build a bridge. Learners are better able to understand mathematical ideas as they go through these problems because they can see how they are directly applied in real-world situations.

The capacity of PBL in mathematics to improve critical thinking and problem-solving abilities is one of its main benefits. Learners are encouraged to consider critically, reason, and connect various mathematical concepts rather than just memorizing formulas. Learners gain the ability to think creatively, recognize patterns, and develop original ways for finding solutions by working on complicated, open-ended problems. By involving learners in genuine, real-world problems, project-based learning in mathematics strengthens learners' critical thinking and problem-solving skills. It fosters an increased awareness of mathematical concepts and their practical application through the reinforcement of logical reasoning, pattern recognition, and the creation of multiple strategies.

Without placing sufficient focus on the underlying logic and problem-solving abilities, traditional approaches to mathematics education frequently highlight memorization of formulas and procedures.

PBL, on the other hand, offers a useful substitute by immersing learners in genuine, real-world mathematics issues. Learners in project-based learning are given challenging, open-ended challenges that call for the analysis, evaluation, and application of mathematical ideas. They are urged to exercise critical thought, divide the issue into components, and come up with numerous solutions. This procedure develops their capacity for logical thought, pattern recognition, and the ability to link various mathematical ideas.

Apart from that, project-based learning encourages learners to participate in problem-solving activities that correspond to real-world situations. Learners attain a deeper comprehension of how mathematics is used in many circumstances by solving real-world challenges. They improve understanding of the importance and applicability of mathematical ideas and develop more learning motivation. Plus, PBL promotes excellent teamwork and collaboration by encouraging learners to work cooperatively in groups. Learners who work together to solve problems can exchange ideas, discuss various strategies, and cultivate insight from one another's viewpoints. They build vital interpersonal skills and learn how to use group intelligence to generate the best answers through this process. Learners who participate in project-based mathematics learning demonstrate more strength in the face of difficulties. Trial and error is a common part of projects, and learners learn from their errors and modify their approaches as necessary. Through this iterative process, they develop stronger problem-solving abilities and a growth attitude, which encourages confidence in their own capacity for overcoming challenges.

Additionally, PBL develops learners' capacity for productive teamwork. Learners must share their ideas, consider the viewpoints of their peers, and come to a decision as a group when working on group projects in mathematics. Learners gain negotiation and compromise skills as well as a shared comprehension of the issue at hand through collaborative learning. These abilities are essential for success in math as well as other areas of life since they prepare learners for their future academic and professional pursuits.

Project-based learning goes beyond merely understanding the fundamentals of mathematical ideas. Learners gain a profound understanding of the underlying principles through working on projects that call for application and analysis. Learners can identify the connections between many mathematical concepts through projects, which offer a comprehensive view of mathematics.

PBL also encourages learners to look at issues from various perspectives, which results in a deeper comprehension of mathematical ideas. Learners are encouraged to look for alternative solutions, test hypotheses, and consider their methods as they come across difficulties and barriers while working on their projects. PBL in mathematics helps learners to investigate issues from various angles, which results in a deeper understanding of mathematical concepts. Learners work on projects and come across various methods, answers, and interpretations of mathematical ideas. This exposure to other experiences encourages a more thorough knowledge of the subject. Learners who take into account multiple points of view expand their understanding of mathematics and gain understanding of how various mathematical concepts are related to one another. Their capacity to think critically, connect ideas, and understand the complexity and beauty of mathematical ideas is improved by this multi-dimensional inquiry. With the help of this method, learners can develop strong problem-solving skills and a growth attitude.

CONCLUSION

An innovative teaching strategy that fosters teamwork, critical thinking, and problem-solving abilities is project-based learning. PBL encourages greater learning, knowledge retention, and the development of crucial skills needed in the twenty-first century by involving learners in real-world, practical projects. PBL is a potent tool for empowering and inspiring learners as education changes to meet the demands of a changing world. Math project-based learning offers a dynamic and efficient method for involving pupils and enhancing their mathematical abilities. PBL helps learners better understand mathematical ideas and develop critical thinking abilities by incorporating real-world situations, cooperative cooperation, and critical thinking.

Learners develop a greater understanding of the real-world uses of mathematics via hands-on projects, inspiring a passion for the subject that lasts a lifetime. Project-based learning is a formidable tool for educators to unleash learners' potential and revolutionize mathematics education as they work to design meaningful learning experiences.

MATHEMATICAL MODELING

Mathematical modeling is a process of creating mathematical models to represent real-life situations or phenomena. It is a powerful tool for teaching and learning mathematics as it helps learners to understand the relevance of mathematics in the real world. Mathematical modeling can be used to solve problems, make predictions, and analyze data. It involves identifying the key variables, constructing a model, and testing the model against real-world data. According to an article published in the *Teaching Mathematics and its Applications: An International Journal of the IMA*, teachers can use real data to teach mathematical modeling. The article suggests that teachers should approach mathematical modeling through mathematizing the context. It also provides a framework to support teachers' choice of modeling activities and demonstrates these using examples. Another article from Learning Sciences International explains that modeling with mathematics is simply the application of mathematics using the practices and habits of mind to solve real-life problems. This problem-solving process is key to the conceptual learning mathematics. Mathematical modeling is a versatile tool that can be applied to a wide range of fields, including engineering, mathematics, physics, biology and economics. It is a valuable skill for learners to learn as it helps them to develop critical thinking and problem-solving skills that are essential in today's world. By using real-world data and mathematizing the context, teachers can help learners to see the relevance of mathematics in their everyday lives and inspire them to pursue further studies in this field.

Mathematical modeling is a powerful method of teaching mathematics with several benefits. Here are some key advantages:

Real-World Relevance: Mathematical modeling connects classroom learning to real-life contexts. By using real data and scenarios, students see the practical applications of mathematical concepts.

Critical Thinking: Through modeling, students develop critical thinking skills. They learn to analyze situations, formulate mathematical structures, and solve complex problems.

Independence and Creativity: Modeling encourages students to explore mathematical ideas independently. They can generate their own theorems and solutions, fostering creativity and self-directed learning.

Metacognitive Skills: Students improve their metacognitive abilities by reflecting on their modeling process. They learn to evaluate their strategies, adjust approaches, and refine their mathematical thinking.

Communication Skills: Modeling requires students to communicate their mathematical reasoning effectively. They learn to express ideas clearly and justify their solutions.

In summary, mathematical modeling enhances students' understanding, motivation, and application of mathematics in real-world contexts. It's a valuable approach for teaching mathematical concepts.

ONLINE LEARNING

Online learning has become an integral part of education worldwide, and mathematics is no exception. However, teaching and learning mathematics online presents unique challenges that are not present in traditional classroom settings. These challenges include mathematics anxiety, conceptual understanding of mathematical ideas, communicating mathematically and the appropriate utilization of technology. To address these challenges, educators have developed a set of standard practices to improve the quality of online teaching and learning of mathematics. These practices include using interactive distance education technology, providing meaningful and practical solutions for teaching mathematics and statistics online, and sharing best practices with the professional community. There are several online resources available that mathematics educators may find useful in their effective instruction. For example, teachers can use the content knowledge and suggested teaching ideas available at the Department of Education and Training Additionally the Department hosts links to other resources, including sample schemes and lesson plan templates. Khan Academy is another free online resource that offers a library of trusted, standards-aligned practice and lessons covering math K-12 through early college, grammar, science, history, AP®, SAT®, and more. Again, James P. Howard II and John F. Beyers in their book, "Teaching and Learning Mathematics Online" provides meaningful and practical solutions for teaching mathematics and statistics online. It focuses on the problems observed by mathematics instructors currently working in the field who strive to hone their craft and share best practices with our professional community. Edutopia provides a comprehensive guide on how to better engage learners in online math classes. The guide suggests that teachers should shift their teaching methods to help learners thrive during distance learning. Some of the ideas include automated assessments for conceptual understanding, oral exams, and culminating projects.

ARTIFICIAL INTELLIGENCE (AI)

Artificial Intelligence (AI) has found its applications in various fields, mathematics education included. A systematic literature review conducted by Mohamed et al indicates that AI can assist teachers and learners to improve teaching and learning of mathematics. In this review, a complete overview of AI in mathematics teaching and learning for learners at all levels of education. The types of themes for AI in mathematics education were categorized into advantages and disadvantages, conceptual understanding, factors, role, idea suggestion, strategies, and effectiveness.

Edutopia: Use AI to generate math problems that are appropriate for a learner's level of understanding. Teachers can also use artificial intelligence tools like ChatGPT to generate rich math tasks and differentiate instruction. Artificial intelligence (AI) systems offer effective support for online learning and teaching, including personalizing learning for learners, automating instructors' routine tasks, and powering adaptive assessments.

Examples where artificial Intelligence (AI) has been applied in various ways to enhance learning mathematics:

Personalized Learning: AI-based platforms can collect and analyze learner data on interaction with educational materials, exercise completion time, test results, and overall performance to understand each learner's attitudes and needs. Drawing on this data, AI tools can design personalized training pathways and adapt them in real-time to the learner's progress.

Adaptive Learning: AI adaptive learning tools have been shown to quickly and dynamically make changes to the learning environment, content, and tasks to help individuals learn more and quickly improve. For instance, researchers at the Human-Computer Interaction Institute at Carnegie Mellon University taught a system how to solve a math problem.

Automated Grading: AI can be used to grade assignments and tests, freeing up teachers' time to focus on other aspects of teaching.

Virtual Tutors: AI-powered virtual tutors can provide learners with personalized feedback and guidance on their work.

Curriculum Design: AI can help educators design curriculums that are tailored to the needs of individual learners.

Assistance for Special Needs learners: AI can be used to help learners with special needs, such as those with dyscalculia, by providing them with additional support and resources.

Learning Outcomes Prediction: AI can be used to predict learner learning outcomes, allowing educators to intervene early and provide additional support where necessary.

Resource Planning: AI can help educational institutions plan and allocate resources more effectively.

These are just a few examples of how AI is being used to enhance learning mathematics. As AI technology continues to develop, it is becoming increasingly accessible for educators to deploy AI tools in their classrooms and provide personalized learning experiences.

In summary, AI has been used in mathematics education to improve teaching and learning performances by providing individualized support, enhancing cognitive skills, and developing effective strategies for mathematics education. AI does these by utilizing robots, tools, teachable agent, autonomous agent, and a comprehensive approach.

REFERENCES

Ancker, J. S., & Kaufman, D. (2007). Rethinking health numeracy: A multidisciplinary literature review. *Journal of the American Medical Informatics Association : JAMIA*, 14(6), 713–772. 10.1197/jamia.M246417712082

Anderson, J., & Kriesler, A. (2018). Making maths matter: Engaging students from low SES schools through social justice contexts. In E. Bergqvist, M. Österholm, C. Granberg, & L. Sumpter (Eds.), *Proceedings of the 42nd Conference of the International Group for the Psychology of Mathematics Education,* 2, 35–42). Umeå, Sweden: PME.

Anthony, G., Hunter, J., & Hunter, R. (2019). Working towards equity in mathematics education: Is differentiation the answer? In Hine, G., Blackley, S., & Cooke, A. (Eds.), *Proceedings of the 42nd Annual Conference of the Mathematics Education Research Group of Australasia.* MERGA.

Anthony, G., & Hunter, R. (2016). Grouping practices in New Zealand mathematics classrooms: Where are we at and where should we be? *New Zealand Journal of Educational Studies*, 52(1), 73–79. 10.1007/s40841-016-0054-z

Anthony, G., Hunter, R., & Hunter, J. (2018). Challenging teachers' perceptions of student capability through professional development: A telling case. *Professional Development in Education*, 44(5), 650–666. 10.1080/19415257.2017.1387868

Attard, C., Edwards-Groves, C., & Grootenboer, P. (2018). Dialogic practices in the mathematics classroom. In Hunter, J., Perger, P., & Darragh, L. (Eds.), *Proceedings of the 41st Annual Conference of the Mathematics Education Research Group of Australasia* (pp. 122–129). Auckland: MERGA.

Australian Curriculum, Assessment and Reporting Authority. (2015). *Australian curriculum: Mathematics, version 8.* Australia Curriculum. https://www.australiancurriculum.edu.au/Mathematics/Curriculum/F-10

Averill, R. (2018). Using rich investigative mathematics activities. In Hunter, R., Civil, M., Herbel-Eisenmann, B., Planas, N., & Wagner, D. (Eds.), *Mathematical discourse that breaks barriers and creates space for marginalized learners* (pp. 195–212). Sense Publishers. 10.1163/9789004378735_010

Bennison, A., Goos, M., & Bielinski, D. (2018). Identifying practices that promote engagement with mathematics among students from disadvantaged backgrounds. In J. Hunter, P. Perger, & L. Darragh (Eds.), *Proceedings of the 41st Annual Conference of the Mathematics Education Research Group of Australasia*, (pp. 154–16). Auckland: MERGA.

Bicknell, B., Young-Loveridge, J., & Nguyen, N. (2016). A design study to develop young children's understanding of multiplication and division. *Mathematics Education Research Journal*, 28(4), 567–58. 10.1007/s13394-016-0180-4

Blue, L. E., O'Brien, M., & Makar, K. (2018). Exploring the classroom practices that may enable a compassionate approach to financial literacy education. *Mathematics Education Research Journal*, 30(2), 143–16. 10.1007/s13394-017-0223-5

Bobis, J., & Tregoning, M. (2019). "The task is not the challenge": Changing teachers' practices to support student struggle in mathematics. In G. Hine, S. Blackley, & A. Cooke (Eds.), *Proceedings of the 42nd Annual Conference of the Mathematics Education Research Group of Australasia,* (pp. 133–139). Perth: MERGA.

Bobis, J., Way, J., Anderson, J., & Martin, A. J. (2016). Challenging teacher beliefs about student engagement in mathematics. *Journal of Mathematics Teacher Education*, 19(1), 33–55. 10.1007/s10857-015-9300-4

Bragg, L. A., Herbert, S., Loong, E., Vale, C., & Widjaja, W. (2016). Primary teachers notice the impact of language on children's mathematical reasoning. *Mathematics Education Research Journal, 28*(4). .10.1007/s13394-016-0178-y

Brown, R. (2017). Using collective argumentation to engage students in a primary mathematics classroom. *Mathematics Education Research Journal*, 29(2), 183–199. 10.1007/s13394-017-0198-2

Brown, R., & Redmond, T. (2017). Privileging a contextual approach to teaching mathematics: A secondary teacher's perspective. In A. Downton, S. Livy, & J. Hall (Eds.), *Proceedings of the 40th Annual Conference of the Mathematics Education Research Group of Australasia*, (pp. 109–116). Melbourne: MERGA.

Bryk, A. S. (2015). Accelerating how we learn to improve. *Educational Researcher*, 44(9), 467–477. 10.3102/0013189X15621543

Buck Institute for Education (BIE). (2014). *Introduction to project-based learning*. BIE. http://bie.org/images/uploads/general/20fa7d42c216e2ec171a212e97fd4a9e.pdf

Burdman, P. (2020). *Let's Make Math Education Relevant for Real Life*. EdSurge. https://www.edsurge.com/news/2020-10-26-let-s-make-math-education - relevant-for-real-life

Celedón-Pattichis, S., Peters, S. A., Borden, L. L., Males, J. R., Pape, S. J., & Chapman, O. (2018). Asset-based approaches to equitable mathematics education research and practice. *Journal for Research in Mathematics Education*, 49(4), 373–389. 10.5951/jresematheduc.49.4.0373

Chan, M. C. E., Clarke, D., & Cao, Y. (2018). Erratum to: The social essentials ofq learning: An experimental investigation of collaborative problem solving andq knowledge construction in mathematics classrooms in Australia and China. *Mathematics Education Research Journal*, 30(1), 51–55. 10.1007/s13394-017-0221-7

Cheeseman, J. (2018). Teachers' perceptions of obstacles to incorporating a problem solving style of mathematics into their teaching. In Hunter, J., Perger, P., & Darragh, L. (Eds.), *Proceedings of the 41st Annual Conference of the Mathematics Education Research Group of Australasia* (pp. 210–217). Auckland: MERGA.

Cheeseman, J., Downton, A., & Livy, S. (2017). Investigating teachers' perceptions of enabling and extending prompts. In A. Downton, S. Livy, & J. Hall (Eds.), *Proceedings of the 40th Annual Conference of the Mathematics Education Research Group of Australasia,* (pp. 141–148). Melbourne: MERGA.

Cheeseman, J., & McDonough, A. (2019). Coding young learners' pictorial responses to an open-ended assessment task. In Hine, G., Blackley, S., & Cooke, A. (Eds.), *Proceedings of the 42nd Annual Conference of the Mathematics Education Research Group of Australasia,* (pp. 112–116). Perth: MERGA.

Choy, B. H. (2016). Snapshots of mathematics teacher noticing during task design. *Mathematics Education Research Journal*, 28(3), 421–440. 10.1007/s13394-016-0173-3

Choy, B. H., & Dindyal, D. (2017a). Noticing affordances of a typical problem. In B. Kaur, W. K. Ho, T. L. Toh, & B. H. Choy (Eds.), *Proceedings of the 41st Conference of the International Group for the Psychology of Mathematics Education* (pp. 249–256). Singapore: PME.

Choy, B. H., & Dindyal, D. (2017b). Snapshots of productive noticing: Orchestrating learning experiences using typical problems. In A. Downton, S. Livy, & J. Hall (Eds.), *Proceedings of the 40th Annual Conference of the Mathematics Education Research Group of Australasia,* (pp. 157–16). Melbourne: MERGA.

Choy, B. H., & Dindyal, D. (2018). Orchestrating mathematics lessons: Beyond the use of a single rich task. In J. Hunter, P. Perger, & L. Darragh. (Eds.), *Making waves, opening spaces. Proceedings of the 41st annual conference of the Mathematics Education Research Group of Australasia.* Auckland: MERGA.

Choy, B. H., Thomas, M. O. J., & Yoon, C. (2017). The FOCUS framework: Characterising productive noticing during lesson planning, delivery and review. In Schack, E. O., Fisher, M. H., & Wilhelm, J. A. (Eds.), *Teacher noticing: Bridging and broadening perspectives, contexts, and frameworks* (pp. 445–466). Springer. 10.1007/978-3-319-46753-5_26

Colliver, Y. (2018). Fostering young children's interest in numeracy through demonstration of its value: The footsteps study. *Mathematics Education Research Journal*, 30(4), 407–428. 10.1007/s13394-017-0216-4

English, L. D., & King, D. (2019). STEM integration in sixth grade: Designing and constructing paper bridges. *International Journal of Science and Mathematics Education*, 17(5), 863–888. 10.1007/s10763-018-9912-0

English, L. D., & Kirshner, D. (Eds.). (2015). *Handbook of International Research in Mathematics Education* (3rd ed.). Routledge. 10.4324/9780203448946

Fielding, L., Kerr, N., & Rosier, P. (2007). *Annual growth for all students, catch-up growth for those who are behind.* The New Foundation Press, Inc.

Fielding-Wells, J., & Fry, K. (2019). Introducing guided mathematical inquiry in the classroom: Complexities of developing norm of evidence. In Hine, G., Blackley, S., & Cooke, A. (Eds.), *Proceedings of the 42nd Annual Conference of the Mathematics Education Research Group of Australasia,* (pp. 268–275). Perth: MERGA.

Gamwell, L. (2015). Mathematics: Geometries of beauty. *Nature*, 528(7583), 476–477. 10.1038/528476a

Geiger, V., Goos, M., & Forgasz, H. (2015). A rich interpretation of numeracy for the 21st century: A survey of the state of the field. *ZDM Mathematics Education*, 47(4), 531–548. 10.1007/s11858-015-0708-1

Geiger, V., Mulligan, J., Date-Huxtable, L., Ahlip, R., Jones, D. H., May, E. J., Rylands, L., & Wright, I. (2018). An interdisciplinary approach to designing online learning: Fostering pre-service mathematics teachers' capabilities in mathematical modelling. *ZDM Mathematics Education*, 50(1), 217–223. 10.1007/s11858-018-0920-x

Gibbs, B., & Hunter, B. (2018). Making mathematics accessible for all: A classroom inquiry. In J. Hunter, P. Perger, & L. Darragh (Eds.), *Proceedings of the 41st Annual Conference of the Mathematics Education Research Group of Australasia,* (pp. 330–336). Auckland: MERGA.

Ginns, P., Hu, F. T., Byrne, E., & Bobis, J. (2016). Learning by traced worked examples. *Applied Cognitive Psychology*, 30(2), 160–169. 10.1002/acp.3171

Gutstein, E., & Peterson, B. (Eds.). (2013). *Rethinking mathematics: Teaching social justice by the numbers*. Rethinking Schools Ltd.

Hallermann, S., Larmer, J., & Mergendoller, J. (2011). *PBL in the elementary grades: Step-by step guidance, tools, and tips for standards-focused K-5 projects.* Novato, CA: Buck Institute for Education.

Holm, M. (2011). Project-Based Instruction: A Review of the Literature on Effectiveness in Prekindergarten through 12th Grade Classrooms. *InSight: Riviera Academic Journal, 7*(2), 1-1. http://independent.academia.edu/MaggieHolm

Hunter, J. (2017). Developing interactive mathematical talk: Investigating student perceptions and accounts of mathematical reasoning in a changing classroom context. *Cambridge Journal of Education*, 47(4), 475–49. 10.1080/0305764X.2016.1195789

Hunter, R., & Hunter, J. (2017). Maintaining a cultural identity while constructing a mathematical disposition as a Pāsifika learner. In McKinley, E. A., & Tuhiwai Smith, L. (Eds.), *Handbook of Indigenous education*. Springer. 10.1007/978-981-10-1839-8_14-1

Hunter, R., & Hunter, J. (2018). Opening the space for all students to engage in mathematical talk within collaborative inquiry and argumentation. In R. Hunter, M. Civil, B. Herbel-Eisenmann, N. Planas, & D. Wagner (Eds.), *Mathematical discourse that breaks barriers and creates space for marginalized learners*. Rotterdam: Sense Publishers. 10.1163/9789463512121_001

Hunter, R., Hunter, J., Anthony, G., & McChesney, K. (2018 September). Developing mathematical inquiry communities: Enacting culturally responsive, culturally sustaining ambitious teaching. *SET, 2,* 25–3

Hunter, R., Hunter, J., & Bills, T. (2019). Enacting culturally responsive or socially response-able mathematics education. In C. Nicol, S. Dawson, J. Archibald, & F. Glanfield (Eds.), *Living culturally responsive mathematics curriculum and pedagogy: Making a difference with/in indigenous communities.* Springer: Netherlands. 10.1163/9789004415768_007

Hurst, C., & Hurrell, D. (2016). Investigating children's multiplicative thinking: Implications for teaching. *European Journal of STEM Education*, 1(3), 1–1. 10.20897/lectito.201656

Jazby, D., & Widjaja, W. (2019). Teacher noticing of primary students' mathematical reasoning in a problem-solving task. In Hine, G., Blackley, S., & Cooke, A. (Eds.), *Proceedings of the 42nd Annual Conference of the Mathematics Education Research Group of Australasia,* (pp. 380–387). Perth: MERGA.

Jorgensen, R. (2018). Language resources to scaffold mathematical learning. In Hunter, R., Civil, M., Herbel-Eisenmann, B., Planas, N., & Wagner, D. (Eds.), *Mathematical discourse that breaks barriers and creates space for marginalized learners* (pp. 235–256). Sense Publishers. 10.1163/9789004378735_012

Kalogeropoulos, P., Russo, J., Sullivan, P., Klooger, M., & Gunningham, S. (2019). Re-enfranchising mathematically-alienated Students: Teacher and tutor perceptions of the Getting Ready in Numeracy (G.R.I.N.) Program. *International Electronic Journal of Mathematics Education*, 14(1), 1–17. 10.29333/iejme/5881

Larmer, J., Mergendoller, J., & Boss, S. (2015). *Setting the standard for project-based learning*. ASCD.

Leach, G. (2019). Strength-based grouping: A call for change. In G. Hine, S. Blackley, & A. Cooke (Eds.), *Mathematics Education Research: Impacting Practice, Proceedings of the 42nd Annual Conference of the Mathematics Education Research Group of Australasia*. Perth: MERGA.

Lee, M. Y., & Choy, B. H. (2017). Mathematical teacher noticing: The key to learning from Lesson Study. In Schack, E. O., Fisher, M. H., & Wilhelm, J. A. (Eds.), *Teacher noticing: Bridging and broadening perspectives, contexts, and frameworks* (pp. 121–140). Springer. 10.1007/978-3-319-46753-5_8

Leong, Y. H., Cheng, L. P., Toh, W. T., Kaur, B., & Toh, T. L. (2019). Making things explicit using instructional materials: A case study of a Singapore teacher's practice. *Mathematics Education Research Journal*, 31(1), 47–66. 10.1007/s13394-018-0240-z

MacDonald, A., & Murphy, S. (2019). Using the drawing-telling approach to reveal young children's mathematical knowledge. In Hine, G., Blackley, S., & Cooke, A. (Eds.), *Proceedings of the 42nd Annual Conference of the Mathematics Education Research Group of Australasia,* (pp. 99–10). Perth: MERGA.

Marshman, M. (2018). Identifying the mathematics middle year students use as they address a community issue. *Mathematics Education Research Journal*, 30(4), 355–38. 10.1007/s13394-017-0195-5

Mason, J. (2002). *Researching your own practice: The discipline of noticing*. Routledge. 10.4324/9780203471876

McCluskey, C., Mulligan, J., & Mitchelmore, M. (2016). The role of reasoning in the Australian Curriculum: Mathematics. In B. White, M. Chinnappan, & S. Trenholm (Eds.), *Proceedings of the 39th Annual Conference of the Mathematics Education Research Group of Australasia,* (pp. 447–45). Adelaide: MERGA.

Mildenhall, P., & Sherriff, B. (2018). Using multiple metaphors and multimodalities as a semiotic resource when teaching year 2 students computational strategies. *Mathematics Education Research Journal, 30*(4), 383-406. .10.1007/s13394-017-0212-8

Miller, J., Warren, E., & Armour, D. (2018). Mathematical discourse when engaging in pattern generalisation. In R. Hunter, M. Civil, B. Herbel-Eisenmann, N. Planas, & D. Wagner (Eds.), *Mathematical discourse that breaks barriers and creates space for marginalized learners.* Rotterdam: Sense Publishers. https://doi.org/10.1163/9789004378735_011

Moala, J. G., & Hunter, R. (2019). Developing mathematical resilience among diverse learners: Preliminary progress and problematics. In Hine, G., Blackley, S., & Cooke, A. (Eds.). *Proceedings of the 42nd Annual Conference of the Mathematics Education Research Group of Australasia.* Perth: MERGA.

Muir, T., Beswick, K., Callingham, R., & Jade, K. (2016). Experiencing teaching and learning quantitative reasoning in a project-based context. *Mathematics Education Research Journal,* 28(4), 479–50. 10.1007/s13394-016-0176-0

Naimsamphao, P., & Katwibun, D. (2019). Investigation of 10th Grade students'Q agency and authority in a mathematics Problem-based learning classroom. In Hine, G., Blackley, S., & Cooke, A. (Eds.), *Proceedings of the 42nd Annual Conference of the Mathematics Education Research Group of Australasia,* (pp. 524–531). Perth: MERGA.

Parish, L. (2016). The power of creativity: A case-study of a mathematically highly capable Grade 5 student. In B. White, M. Chinnappan, & S. Trenholm (Eds.), *Proceedings of the 39th Annual Conference of the Mathematics Education Research Group of Australasia,* (pp. 447–45). Adelaide: MERGA.

Patahuddin, S. M., Usman, H. B., & Ramful, A. (2018). Affordances from number lines in fractions instruction: Students' interpretation of teacher's intentions. *International Journal of Science and Mathematics Education,* 16(5), 909–928. 10.1007/s10763-017-9800-z

Philen, J. (2016). *Implementing Meaningful and Sustainable Project Based Learning Pedagogy.* [Doctoral Dissertation, England].

Promsawan, S., & Katwibun, D. (2017). 11th grade students' self-regulated learning in a maths problem-based learning (PBL) classroom. In A. Downton, S. Livy, & J. Hall (Eds.), *Proceedings of the 40th Annual Conference of the Mathematics Education Research Group of Australasia,* (pp. 420–427). Melbourne: MERGA.

Quane, K., Chinnappan, M., & Trenholm, S. (2019). The nature of young children's attitudes towards mathematics. In Hine, G., Blackley, S., & Cooke, A. (Eds.), *Proceedings of the 42nd Annual Conference of the Mathematics Education Research Group of Australasia,* (pp. 108–11). Perth: MERGA.

Russo, J., & Hopkins, S. (2017a). Class challenging tasks: Using cognitive load theory to inform the design of challenging mathematical tasks. *Australian Primary Mathematics Classroom*, 22(1), 21–27.

Russo, J., & Hopkins, S. (2017b). Examining the impact of lesson structure when teaching with cognitively demanding tasks in the early primary years. In A. Downton, S. Livy, & J. Hall (Eds.), *Proceedings of the 40th Annual Conference of the Mathematics Education Research Group of Australasia,* (pp. 450–457). Melbourne: MERGA.

Russo, J., & Hopkins, S. (2017c). How does lesson structure impact teachers' willingness to teach with challenging tasks? *Mathematics Teacher Education and Development*, 19(1), 30–46. 10.1007/s13394-017-0197-3

Russo, J., & Hopkins, S. (2017d). Student reflections on learning with challenging tasks: "I think the worksheets were just for practice, and the challenges were for maths". *Mathematics Education Research Journal*, 29(3), 283–31. 10.1007/s13394-017-0197-3

Russo, J., & Hopkins, S. (2019). Teachers' perceptions of students when observing lessons involving challenging tasks. *International Journal of Science and Mathematics Education*, 17(4), 759–779. 10.1007/s10763-018-9888-9

Russo, J. A., & Russo, T. (2019). Teacher interest-led inquiry: Unlocking teacher passion to enhance student learning experiences in primary mathematics. *International Electronic Journal of Mathematics Education*, 14(3), 701–717. 10.29333/iejme/5843

Sangkaew, S., & Katwibun, D. (2019). 11th grade students' self-beliefs in a mathematics problem-based learning (PBL) classroom. In Hine, G., Blackley, S., & Cooke, A. (Eds.), *Proceedings of the 42nd Annual Conference of the Mathematics Education Research Group of Australasia* (pp. 620–627). Perth: MERGA.

Sawatzki, C. (2016). Insights from a financial literacy task designer: The curious case of problem context. In In B. White, M. Chinnappan, & S. Trenholm (Eds.), *Proceedings of the 39th Annual Conference of the Mathematics Education Research Group of Australasia,* (pp. 576–58). Adelaide: MERGA.

Siriwat, R., & Katwibun, D. (2017). Exploring critical thinking in a mathematics problem-based learning classroom. In A. Downton, S. Livy, & J. Hall (Eds.), *Proceedings of the 40th Annual Conference of the Mathematics Education Research Group of Australasia,* (pp. 474–48). Melbourne: MERGA.

Stewart, I. (2017). *The Beauty of Numbers in Nature: Mathematical Patterns and Principles from the Natural World.* MIT Press.

Sullivan, P., Boreck, C., Walker, N., & Rennie, M. (2016). Exploring a structure for mathematics lessons that initiate learning by activating cognition on challenging tasks. *Journal of Mathematical Behaviour, 41,* 159-170. https://doi.org/.10.1016/j.jmathb.2015.1002

Thomas, J. W. (2000). *A review of research on project-based learning*. Bobopearlman. http://www.bobpearlman.org/BestPractices/PBL_Research.pdf

Warren, E., & Miller, J. (2016). *Mathematics at the margins*. Springer. 10.1007/978-981-10-0703-3

Way, J., & Thom, J. (2019). Capturing the mathematical drawing process using a digital pen. In Hine, G., Blackley, S., & Cooke, A. (Eds.), *Proceedings of the 42nd Annual Conference of the Mathematics Education Research Group of Australasia,* (pp. 104–107). Perth: MERGA.

Wells, J., Geiger, V., Miller, J., Bruder, R., Roder, U., & Ratnayake, I. (2019). Designing challenging online mathematical tasks for initial teacher education: Motivational considerations. In M. Graven, H. Venkat, A. A. Essien, & P. Vale (Eds.), *Proceedings of the 43rd Conference of the International Group for the Psychology of Mathematics Education*. Pretoria, South Africa: PME.

Wilkie, K. J., & Ayalon, M. (2019). Learning through critiquing: Investigating students' responses to other graphs of a real-life functional situation. In Hine, G., Blackley, S., & Cooke, A. (Eds.), *Proceedings of the 42nd Annual Conference of the Mathematics Education Research Group of Australasia*. Perth: MERGA.

Wilkie, K. J., & Clarke, D. M. (2016). Developing students' functional thinking in algebra through different visualisations of a growing pattern's structure. *Mathematics Education Research Journal*, 28(2), 223–224. 10.1007/s13394-015-0146-y

Wilkie, K. J., & Sullivan, P. (2018). Exploring intrinsic and extrinsic motivational aspects of middle school students' aspirations for their mathematics learning. *Educational Studies in Mathematics*, 97(3), 235–25. 10.1007/s10649-017-9795-y

Yeager, D. S., & Dweck, C. S. (2012). Mindsets that promote resilience: When students believe that personal characteristics can be developed. *Educational Psychologist*, 47(4), 302–331. 10.1080/00461520.2012.722805

Chapter 5
Innovative Pedagogical Approaches and Practices in Mathematics Education in Public Universities of Burundi

Alexis Arakaza
Université du Burundi, Burundi

ABSTRACT

This chapter will give a thorough overview of the current state of mathematics education in the public universities in Burundi. The chapter examines how instructors in Burundi's universities are adjusting to the changing nature of education, with a particular focus on creative teaching approaches and practices. Understanding, identifying, and addressing the unique challenges and opportunities in the context of mathematics teaching in Burundi's higher education institutions are the main goals and concerns of this chapter.

DESCRIPTION

In Burundi, the University of Burundi is the topest university in the nation. Established in 1964, the University of Burundi is the very first public university in the country. The University of Burundi has already joined the BMD (three years Baccalaureate, two years Masters and three years Doctorate) system since its academic reopening in 2011–2012. The academic year is divided into two semesters and consists of sixty credits in total because there are thirty credits per semester.

DOI: 10.4018/979-8-3693-2873-6.ch005

Each credit is worth twenty five hours, consisting of ten hours of student work and fifteen hours of presentation.

Additionally, to study at University of Burundi, it is the weighted national examination score which determines a student's admission. The Ministry of Education similarly assigns a specialization to students depending on their performance in national examination. While rare, it is possible for students to ask for a different specialisation than what they were first assigned. The university has faculties of agriculture, engineering, education, science, and some languages. Applied pedagogy, applied economics, and statistic specialty institutes are also available. After being admitted into the national university, all students are granted free tuition as well as a loan to aid with living expenses in the city. The additional expenses of studying, however, continue to be prohibitive for many students from low-income rural backgrounds. However, a lot of students from low-income rural areas still find it too expensive to complete their education due to the additional expenses involved. After completing their secondary education, students matriculate to the national university for a full year and a half. Some students choose to continue their education at private universities as a result of this delay.

Therefore, this chapter will give a thorough overview of the current state of mathematics education in the public universities in Burundi. The chapter examines how instructors in Burundi's universities are adjusting to the changing nature of education, with a particular focus on creative teaching approaches and practices. Understanding, identifying, and addressing the unique challenges and opportunities in the context of mathematics teaching in Burundi's higher education institutions are the main goals and concerns of this chapter.

This chapter's main features are to examine and highlight cutting-edge instructional strategies being applied in mathematics classrooms since determining innovative practices is essential to comprehending how teachers are modifying and developing their pedagogical approaches to improve the educational experience for pupils and to examine the integration of technology and digital tools into mathematical instruction because recognizing how technology affects instructional strategies, student engagement, and learning results requires an understanding of its role.

A crucial component of this chapter will be an analysis of the difficulties that arise when putting innovative pedagogies into practice. It will provide well-considered answers and suggestions for getting beyond these barriers, guaranteeing a fair assessment of the conditions surrounding mathematics education in Burundi.

This chapter holds great importance for educators, scholars, and policymakers who aim to improve the quality and relevance of mathematics education in Burundi. By examining both successful practices and challenges, it will contribute to the current conversation on educational innovation and development within the unique context of Burundi universities.

Anybody interested in the dynamics of mathematics education in Burundi, including educators, academics, policymakers, and others, is the target audience. For anyone seeking to enhance the quality of mathematics education in Burundi's higher education institutions, as well as the techniques of teaching mathematics, this chapter provides insightful information.

EXAMINE AND HIGHLIGHT CUTTING-EDGE INSTRUCTIONAL STRATEGIES BEING APPLIED IN MATHEMATICS CLASSROOMS.

The main objective of instruction at all educational stages is to cause a necessary adjustment in the students' perceptions (Tebabal & Kahssay, 2011). To enhance the appropriate transmission of knowledge, teachers should apply teaching approaches suitable for specific goals to produce good results. The study conducted by Hightower, Lloyd, and Swanson (2011) revealed that questions concerning the effectiveness of student learning and teachers' strategies have consistently elevated significant interest in the field of educational research. In addition, educational research constantly endeavours to examine how different teachers' strategies enhance student learning growth.

In addition, for developing students' mathematical knowledge, formal academic lessons must be used by mathematics teachers as natural vehicles (Ünal, 2017). Mathematics teachers are also advised to provide a mathematically rich environment and use a chronological flow of ideas to introduce mathematical concepts (Ünal, 2017). In this regard, how mathematics teachers instruct is reflected on how they perceive what to be taught, what materials to be used, and which methodology to follow while teaching. For example, teachers who believe that students learn by creating their understanding during the problem-solving process spent more time developing counting plans for students before teaching numerical facts and applied more word problems during instruction than teachers who believed that they are the source of mathematics knowledge and learners are just recipients. Some teachers feel uncomfortable losing their role in the center stage, performing in front of students. Besides, some teachers believe that learners may want the teacher to explain and provide the correct answers rather than struggle with a problem by themselves.

Moreover, teachers have been encouraged to use teaching methods that involve and motivate the learners this is case of University of Burundi where for example a teacher have a course of forty five hours must divide that period into the guided work and student personnel work. The purpose of that is to help the students to do research. Teaching method is how the teacher manages and organizes the teaching-learning process, presents clear explanations, and checks the progress through questions, answers, and reactions (Shulman, 1987). A teaching method is a method to facilitate interaction between the learners and teachers to realize set goals. One of the teaching methods that support learners to develop their mathematical reasoning is a learner-centered teaching method that helps the learners make sense of mathematics by creating contexts that help learners develop meaning in mathematics (Yushau, Mji, & Wessels, 2005). Therefore, different ways of teaching methods and learning strategies have to be adopted.

Using various teaching methods in the teaching and learning mathematics process can produce the desired goal of making learners learn with understanding, particularly the constructivist ones (Ünal, 2017). The teaching method that promoted learners' involvement was found to hold students' interest (Mitchsell, 1993; Mitchell &Gilson, 1997). The main strategies which are being used in teaching and learning mathematics while delivering content are cooperative learning, collaborative learning, and problem-based learning. Collaboration is looked at as a philosophy of interaction and personal lifestyle where individuals are responsible for their actions, including learning and respecting their peers' contributions. In contrast, cooperative learning is looked at as an activity involving a small group of people who work together as a team to solve a problem, complete a task or accomplish a common goal (Arts & Newman, 1990). Also, Goodsell, Maher, and Tinto (1992) said that cooperative learning falls in the more general category of collaborative learning, which is described as working in groups of two or more, mutually searching for understanding, solutions, meaning, or creating a product. Bybee (2000) defined problem-based learning as a content understanding in which learners can construct concepts and patterns and create meaning about an idea to explain what they are experiencing.

Besides cooperative learning, collaborative learning, and problem-based learning, other strategies are being used in teaching and learning as inquiry discovery methods, games and simulations, inductive methods, and computer-assisted instruction (Michael &Jackie, 1996). However, the recent reforms in education of pedagogical shift from content to competency-based learning build on the exact nature that aims to produce students who can competently apply the knowledge gained in schools by using scientific and mathematical principles in solving different problems. To achieve this particular aspect, using those strategies is highly emphasized official educational documents such as syllabuses.

In the context of Burundi, competence-based approach is the basis of the curriculum. Teachers are required to use learner-centered instruction strategies which allow students to be engaged in active learning, cooperative and problem-based learning, to promote their abilities of creativity, innovation, critical thinking, and problem-solving. Ministere de l' Education de l' Enseignement Superieure et de la Recherche Scientifiques [MEESRS] (2015) shows how the teaching of mathematics should be conducted. It is indicated that the teaching of mathematics which aims to meet the country's needs, must be based on the students' experience from which the teacher leads them to build the knowledge themselves. Students must be active in the teaching - learning process, seek to give the meaning of the mathematics taught, and cultivate in students an interest for mathematics by introducing certain notions through a problem. It is up to the teacher to show students the value of learning how to solve problems.

In the search for solutions to the proposed activities, each student appropriates and masters his/her knowledge. When combining procedures, the new knowledge is identified and must be remembered. In this process of building knowledge, the student learns mathematics when he/she uses hypothetic deductive reasoning, structures, classifies orders, calculates, constructs models, or uses geometry instruments. This is done by solving problems encountered in everyday life or answering questions specific to mathematics submitted by the teacher. The teacher guides the student without giving a solution to the problem posed and only intervenes in the event of a student's blockage (MEESRS, 2015). The teacher must place the students at the center of the teaching and learning activities where they develop concepts and structure their knowledge with the teacher's help who facilitates the learning process.

Further, the teachers' practice classroom depends on the contents addressed in the curricular documents such as syllabuses and textbooks. The teacher's instructional practices involve the interaction of learners, teachers, and the environment in which teaching and learning improve learners' active participation (Kim & Chin, 2011). This kind of interaction takes into consideration the materials that are used to enhance effective learning in the process of generating new understandings. However, Marzano, Waters, and McNulty (2005) argue that many teachers have concentrated on the wrong classroom practices, which negatively impact learners' accomplishment. In other words, according to Marzano et al. (2005), teachers of Mathematics may be practicing teaching methods that do not provide an enabling environment that makes learners have direct interaction with each other on what they are learning. Moreover, contextual curricular materials bring significant improvement in learners' learning in terms of increased interaction, confidence, and active participation.

Furthemore, the sufficient materials in secondary school would help to administer different learning activities. This is the importance of curricular materials. The proper uses of the curricular materials in effective learning increase participation in the mathematics classroom. The curricula for state education are generally written from prescribing a particular practice, mainly due to the total dependence of curriculum services on the central administrative authority and lack of enough options for the designers. Curriculum designers essentially rely on personal intuition and teaching background to do their job.

Concerning the recommendations of curriculum on what the student must know, the skills he/she must acquire, the teacher of mathematics must first have built up disciplinary, didactic knowledge and education, particularly during initial and/or continuing training. McTighe and Wiggins (2012) contend that teachers are like coaches to students, who guarantee what is taught is expected to have been learned and understood. Teachers have to prepare a lesson and find what is essential for the student to know by using appropriate teaching strategies which help him/her to be engaged in active learning by developing their deep understanding of the concept. Consequently, as noted by Kazima, Pillay, and Adler (2008), a teacher must be able to select and clarify suitable mathematical aims for any lesson and link them with the strategies used to teach concepts or ideas. He/she must be able to arrange / organize mathematical tasks into suitable lesson designs; 'evaluate' on the spot the mathematical worth of a student's argument or explanation, and listen to what students are saying and also to pose questions that allow students to develop their mathematical thinking.

This is also the case in University of Burundi, where teachers are given the curriculum/content to teach and numbers hours to use. Teachers will be able to develop the content to what he/she will give to students. The mathematics curriculum knowledge involves understanding how mathematics topics are arranged. The mathematics curriculum also provides detailed information regarding textbooks' instructions, such as the topics and number of the lessons to be covered. The curriculum also gives the description and explanation of the required assessment for all content areas. When teachers understand the mathematics curriculum, they can construct a good lesson plan and schemes of work that follow the curriculum line and effectively conduct the teaching and learning process. It is in line with Wanyonyi (2013), who said that lesson plans serve as a guide to the teacher on the content to be taught to avoid vagueness and irrelevancies.

However, Marzano, Wters, and McNUlt (2005) argue that many teachers have concentrated on the wrong classroom practices, which negatively impact learners' accomplishment. In other words, according to Marzano et al. (2005), mathematics teachers may practice teaching methods that do not provide a favourable environment that makes learners have direct interaction with each other on what they are learning. Learners are more interested in learning activities when interacting with one another and participating actively in their learning. Nonetheless, a few professors at the University of Burundi do not allow time to their students to interact with them or to their classmates. They will come in class and provide syllabus to students. Additionally, they provide homework to the students without providing them any feedback. The outcome will not be revealed to the students until the end of the semester. All they will see are the marks, not the comments. If a student wishes to file a complaint or to see how the works were marked, he/she should meet the teacher or submit a letter to the department asking for marked work.

There is ample evidence to conclude that school assessment applies mainly to examinations, and grading referring to the research done by Bezuk, Cathcart, Vance, and Pothier (2001) and Lissitz and Schafer (2002). School leaders have come to a point where they conclude that without awarding grades, one cannot assess (Lissitz, & Schafer (2002).. While assessment seems to be common in all schools, many teachers tend to have different views and assessment abilities. However, McMillan (2001) analyzed the actual classroom assessment practice and grading activities concerning individual class and assessed whether there are a substantive connection between teacher assessment activities, grade level, content, and student skill rates. McMillan (2001) found no clear connection between the assessment activities of teachers, grade level, content, and level of skill and proposes that teachers require specialized preparation to learn to assess the behaviour of children by evaluating the language of the learners.

In a formal sense, assessment is process which encompasses testing, evaluation, and measurement. However, in assessment, both formal and informal methods of data collection can be employed. For example, in Burundi, the teachers must give the students homework in terms of formal assessment. In informal assessment, the teachers are required to use oral questions during the teaching, monitoring and listening to students' answers. However, the effectiveness of classroom assessment in schools has not been realized by many education stakeholders. The problems affecting this effectiveness have been attributed to such factors as large class sizes, lack of time, knowledge of assessment techniques, and workload on teachers (Okyere & Larbi, 2019). It is believed that the ineffectiveness of classroom assessment goes beyond these to include the kind of perception teachers hold on classroom assessment. Teachers who see assessment as a tool for improving learners' performance are likely to use it quite often in class. For the general goal of learning mathematics to be realized, teachers play a critical role in primary assessors of students' learning progress (Susuwele, 2005). Teachers' knowledge and expertise in assessment and practices are very crucial to job performance.

EXAMINE THE INTEGRATION OF TECHNOLOGY AND DIGITAL TOOLS INTO MATHEMATICAL INSTRUCTION.

According to certain studies, there are several reasons why implementing innovative teaching approaches in universities has proven difficult. For example, Bunoti (2011) observed that most African public universities lacked adequate ICT resources, such as computers, which limited students' opportunities for hands-on learning because most courses were theoretical in nature and students were unable to use ICT to access a variety of learning resources. ICT has been linked to enhanced student learning and more effective teaching strategies. The internet, computers, and new technology have had a significant impact on all facets of our society in recent years.

Furthermore, new occupations and challenges have emerged, and many old activities have become out dated. Technical drawing, for example, is no longer completed by hand. Instead, plotters and other technological instruments are used nowadays, and a lot of software for symbolic algebra is becoming readily accessible. Additionally, computers have made it feasible to create virtual realities and to produce amazing images or interactive animations. Furthermore, experiences that are either unattainable or only possible through laborious, time-consuming work can be had with the help of modern equipment. According to Ittigson and Zewe (2003), technology plays a crucial role in the instruction of mathematics.

Moreover, ICT boosts students' comprehension of foundational topics and improves the way mathematics should be taught. Numerous scholars have conducted studies to assess the advantages of ICT use in mathematics. Becta (2003) enumerated the main advantages: ICT fosters increased student collaboration, communication, and knowledge sharing. Students receive quick and accurate feedback from ICT, which helps to motivate them in a positive way. They can also concentrate on tactics and solution interpretations rather than wasting time on laborious computational calculations. ICT is also helpful for constructivist teaching, which encourages students to use technology to learn and explore mathematical ideas. As per the education minister's recommendations, this approach encourages higher order thinking and better problem-solving techniques (Ittigson & Zewe, 2003). Students would then use technology to focus on problem-solving procedures rather than problem-related computations.

Therefore, understanding the current software that mathematics teacher use is crucial for the effective integration of ICT into the mathematics curriculum. Since the educational process depends on the gathering, processing, and presenting of information, modern technologies more especially, information and communication technologies are seen as essential to the achievement of high-quality education systems (Eisenberg, 2008). Throughout the past one hundred years, the educational system has not undergone any significant changes, and instructors continue to impart knowledge to passive students through what are known as frontal instructions. Moreover, it's thought that effective education does not necessarily require the use of modern technologies.

According to Passey, Rogers, Machell, and McHugh (2004), students in the information age view intense technology use in the classroom as both necessary and motivating. As a result, technology-free instruction is considered antiquated and dull by students. It is important to stress that not all educational issues can be resolved by technology and that teaching can proceed quite successfully without its usage. Stated differently, technology cannot be utilized as a short cut to gaining knowledge and abilities (Toyama, 2011).

Even though there are still not enough resources, universities are using ICT considerably more frequently. Few students get to use the university network to browse the Internet because not all students have access to computers. The main cause of this lack of infrastructure access is a lack of funding. This has made it extremely challenging to incorporate vital online resources (such email and the internet) into higher education. The adoption and use of ICT for teaching and learning in higher education has been a significant part of the change in technological processes. According to Davis and Tearle (1999) and Lemke and Coughlin (1998), ICT has the potential to strengthen teaching, help schools transform, and accelerate, enrich, and deepen skills. It can also motivate and engage students, help link school experiences to work practices, and create economic viability for tomorrow's workers. Basic education is necessary in today's world of rapid change in order for someone to be able to obtain and apply information. ICT is one such expertise that is necessary in the global village.

Content has been prioritized in traditional education. Textbooks have shaped course design for many years. Lectures and presentations have been used by teachers. Today's environments are favouring curriculum that increase performance and proficiency. The emphasis in curriculum is shifting from content to capabilities, with a greater emphasis on how knowledge will be used. These demands can all be strongly supported by modern ICT, and there are already many excellent examples of top-notch environments for competency and performance-based curricula that effectively utilize the affordances of these technologies (Oliver, 2000). The revitalization of educators and learners can be facilitated by the incorporation of ICT.

Further, ICT can support curriculum in challenging subjects, which can enhance and advance the quality of education. Teachers must participate in cooperative projects and the creation of intervention change methods in order to meet these goals. The traditional method of teaching in the past has involved instructors creating lesson plans and guiding students through a sequence of lessons in order to meet predetermined learning objectives. These teaching methods have typically focused on the organized dissemination of a body of knowledge, which is then followed by some type of interaction with the material to help students solidify their understanding. However, some professors of University of Burundi are still using the traditional method contrary to the aspiration of the curriculum. Others use it but not in proper way, they are just projecting the notes and assign to the students the work to do in group without even construct the group. They let the students construct the group or put the students in group by following the alphabetically which will help the teacher to mark and record the result at the end of the semester.

Furthermore, teachers strategically use ICT to enhance learning by creating meaningful and engaging learning experiences for their students. Pupils take pleasure in their education and the independent research that is encouraged by creative and sensible ICT use. To stay up to date with the latest advancements, people must use ICT to access knowledge (Plomp, Pelgrum & Law, 2007). ICT can be used to break down barriers to communication, like time and space (Lim, & Chai, 2004). ICT also makes it possible to create digital resources, such as virtual libraries, where professionals, educators, and students can access course materials and research materials at any time and from any location (Bhattacharya, & Sharma, 2007; Cholin, 2005). ICT offers fresh methods for teaching (Sanyal, 2001). It can provide speedy dissemination of education to target disadvantaged groups.

Moreover, ICT can enhance the student's perspective and comprehension of the outside world. According to Plomp, Pelgrum, and Law, (2007), many early innovators in education have found that using ICT encourages both instructors and students to be motivated. The use of ICT can enhance performance, instruction, administration, and the development of pertinent skills in underprivileged areas, according to Bottino (2003) and Sharma (2003). Additionally, it enhances the quality of education by supporting problem-solving, critical thinking, information seeking and analysis, directed instruction, real-time and delayed-time conversation, learning by doing, and the capacity to communicate, cooperate, and learn. Numerous studies have demonstrated the advantages for educational quality.

Additionally, given how widely ICT is being used in modern societies, particularly by the younger generation known as the digital generation, it should be obvious that ICT will have an impact on the entire learning process both now and in the future. When creating learning settings, authenticity is a crucial factor that needs to be taken into consideration. To keep newly learned knowledge from becoming irrelevant, learning environments must take into account the possible applications of knowledge that students are expected to master. Teachers should also encourage students to actively develop their own knowledge. According to Collins, this necessitates the use of open-ended learning settings as opposed to those that only concentrate on imparting knowledge (Collins, 1996; Jonassen, Peck, & Wilson, 1999). ICT may contribute to creating powerful learning environments in numerous ways.

ICT promotes the authenticity of learning settings by offering possibilities to access a wealth of information through a variety of information resources and by viewing information from a variety of angles. ICT may also simplify complicated procedures by using simulations, which support real-world learning settings. ICT may therefore support higher-order thinking and active learning. ICT use has the potential to promote cooperative learning and content reflection (Susman, 1998). ICT can also be used as a tool for curriculum diversification, giving teachers the ability to modify assignments and content to fit the needs and skills of each student and to give customized feedback, ICT can be used in a variety of teaching strategies, from cutting-edge to conventional. Of course, having access to technology can also have an impact on how ICT is used. This is not only about how many computers there are; it's also about where they are located, like in a computer room or a classroom. According to Kennewell, Parkinson, and Tanner (2000), putting computers in the classroom is crucial to maximizing the chances for curriculum-related activities. The use of ICT in the classroom enhances both the learning experience for students and teachers and allows for more efficient use of class time. The ICT environment has been created through the use of various applications and extensive web-based and multimedia material development experience. ICTs are crucial to the modernization and transformation of educational practices and institutions.

ICT can improve education quality in a number of ways, including by improving teacher preparation, boosting learner motivation and engagement, and facilitating the acquisition of fundamental skills. ICTs are also transformative instruments that, in the right hands, can help facilitate the transition to a learner-centered environment. ICT, particularly computers and Internet technologies, allow educators and students to complete tasks more effectively rather than just improving upon previous methods of teaching- learning. ICT influences not just what students should learn, but also greatly influences how they should learn. Curricula have changed from being content-centered to being competence-based, and the way they are delivered has also changed from being teacher centered to being student centered. ICT gives people the will to learn. ICTs can be utilized to create authentic and challenging content that will include students in the learning process. Similarly, interactive radio uses songs, dramatizations, sound effects, comedic skits, and other performance traditions to urge students to listen and participate more actively in the concepts being taught.

An appropriate use of digital technologies in education can have significant positive effects on students' achievement and attitude toward the learning process, according to an evolving consensus in the analysis of the effects of methodological and technological innovations on students' performance and attitude toward the learning process. Studies have indicated that the right use of ICTs can accelerate the paradigm shift in pedagogy and content that is central to 21st-century education reform. According to Kulik's (1994) meta-analysis study, students who received education based on ICT typically performed better academically than those who did not have access to computers. ICT based instruction allowed the students to learn more in less time and made them enjoy their sessions more. When Attwell and Battle (1999) looked at the connection between home computers and academic achievement, they found that pupils who had access to a computer at home for learning reasons performed better in math and reading. ICT boosts student engagement, which results in more time spent working outside of class. Although Coates et al. (2004) demonstrated that students taking classes on campus typically do better than their online peers; this difference is not statistically significant in this case.

ICT, particularly computers and Internet technologies, not only make it possible for educators and students to carry out their previous tasks more effectively, but also open up new avenues for teaching and learning. ICT can enhance the development of sophisticated thinking abilities in students, deepen their understanding of the subject matter, and involve them in creating their own knowledge. Numerous constructivist learning practices, such as having students work in groups or produce products that represent what they are learning, have been found to be effective in altering how students engage with the material (Windschitl, 2002). ICT have the power to improve education's relevance and quality while also expanding access to it. The use of ICT in educational settings, by itself acts as a catalyst for change in this domain. When students utilize ICT for learning, they become fully engaged in the process of learning, and as more and more students use computers as information sources and cognitive tools, technology will continue to play a larger role in supporting students' learning.

CONCLUSION

In this chapter, we conclude that it is useful for public universities to collaborate with the government in ensuring that teaching and learning facilities are sufficiently available. Technology significantly affects everyday life and by extension the teaching process. In this 21st century, teachers are encouraged to use the practices that put the student in the instruction center. In Burundi, through the minister of national education, the government encourages the teacher to use learner-centered instruction. Therefore, ICT can improve education quality in a number of ways, including by improving teacher preparation, boosting learner motivation and engagement, and facilitating the acquisition of fundamental skills. ICTs are also transformative instruments that, in the right hands, can help facilitate the transition to a learner-centered environment. ICTs, particularly computers and Internet technologies, allow educators and students to complete tasks more effectively rather than just improving upon previous methods of teaching and learning. Further, the integration and application of ICT in education enhance research, teaching, and learning. ICT can influence how education is delivered and provide more access to it. Furthermore, it will boost adaptability so that students can receive the instruction despite schedule and location constraints. It may have an impact on how pupils learn and are taught. It would give a stimulating environment and motivation for the teaching and learning process, which appears to have a significant influence on the educational process by presenting both teachers and students with new opportunities. These options may affect the success and performance of the students. Comparably, more accessibility to best practices and high-quality course materials in education that can be disseminated via ICT can promote better instruction and raise students' academic performance. The body of research points to effective ICT integration in the classroom.

REFERENCES

Artzt, A., & Newman, C. (1990). *How to Use Cooperative Learning in the Mathematics Class*. National Council of Teachers of Mathematics.

Attwell, P; Battle, J. (1999). Home Computers and School Performance. *The Information Society.*

Becta. (2003). What the Research Says about Using ICT in Maths. *UK:Becta ICT Research.*

Bezuk, N. S., Cathcart, W. G., Vance, J. H., & Pothier, Y. M. (2001). Learning Mathematics in Elementary and Middle Schools. In *Merrill Prentice Hall* (6th ed.). Pearson Education.

Bhattacharya, I., & Sharma, K. (2007). India in the knowledge economy-an electronic paradigm. *International Journal of Educational Management*, 21(6), 543–568. 10.1108/09513540710780055

Bunoti, S. (2011). *The quality of higher education in developing countries needs professional support.* A paper presented in 22nd International Conference on Higher Education: Quality in Higher Education, Ankara, Turkey.

Bybee, R. (2000). Achieving technological literacy: A national perspective. *Technology Teacher*, 60(1), 23–28.

Cholin, V. S. (2005). Study of the application of information technology for effective access to resources in Indian university libraries. *The International Information & Library Review, 37*(3).

Coates, D., & Humphreys, B. R. (2004). No Significant Distance' between Face-to-face and Online Instruction: Evidence from Principles of Economics. *Economics of Education Review., 23*(6), 533-546.

Collins, A. (1996). Design issues for learning environments. In Vosniadou, S. (Ed.), *International perspectives on the design of technology-supported learning environments* (pp. 347–361). Lawrence Erlbaum.

Davis, N. E., & Tearle, P. (Eds.). (1999). *A core curriculum for telematics in teacher training*. Telemetics. www.ex.ac.uk/telematics.T3/corecurr/tteach98.htm

Eisenberg, M. B. (2008). Information Literacy: Essential Skills for the Information Age. *DESIDOC Journal of Library and Information Technology*, 28(2), 39–47. 10.14429/djlit.28.2.166

Hightower, A. M., Lloyd, S. C., & Swanson, C. B. (2011). *Improving student learning by supporting quality teaching : Key Issues, Effective Strategies*. Editorial Projects in Education, Inc.

Ittigson, R. J., & Zewe, J. G. (2003). Technology in the mathematics classroom. In L. A. Tomei (Ed.), Challenges of teaching with technology across the curriculum: Issues and Solutions. Hershey: Information Science Publishing. 10.4018/978-1-59140-109-4.ch004

Jonassen, D., & Reeves, T. (1996). Learning with technology: Using computers as cognitive tools. In Jonassen, D. (Ed.), *Handbook of Research Educational on Educational Communications and Technology* (pp. 693–719). Macmillan.

Kazima, M., Pillay, V., & Adler, J. (2008). Mathematics for teaching: Observations from two case studies. *South African Journal of Education, 28*(2), 283–299.

Kennewell, S., Parkinson, J., & Tanner, H. (2000). *Developing the ICT capable school*. Routledge Falmer.

Kulik, J. (2003). *Effects of using instructional technology in elementary and secondary schools: What controlled evaluation studies say (Final Report No. P10446.001)*. Arlington, VA: SRI International.

Lemke, C., & Coughlin, E. C. (1998). *Technology in American schools*. PNBS. www.mff.org/pnbs/ME158.pdf

Lim, C. P. & Chai, C.S. (2004), An activity-theoretical approach to research of ICT integration in Singapore schools: Orienting activities and learner autonomy. *Computers & Education, 43*(3).

Lissitz, R., & Schafer, W. (2002). *Assessment in Educational Reform: Both Means and Ends*. Allyn and Bacon.

Marzano, R. J., Waters, J. T., & McNulty, B. A. (2005). *School leadership that works: From research to results*. Association for Supervision and Curriculum Development.

McMillan, J. (2001). *Essential Assessment Concepts for Teachers and Administrators*. Corwin Press, Inc.

McTighe, J., & Wiggins, G. (2012). *Understanding by Design* (2nd ed.). Association for Supervision of Curriculum Development.

Michael, T., & Jackie, M. (1996). Using Computers in the Nathematics Classroom: The Role of the Teacher. *Mathematics Education Research Journal*, 6(1), 38–57.

Ministere de l' Education de l' Enseignement Superieure et de la Recherche Scientifiques (MEESRS). (2015). *Curriculum de l'Enseignement fondamental*. MEERS.

Mitchell, M. (1993). Situational Interest. *Its Multifaceted Structure in the Secondary School Mathematics Classroom.*, 85(3), 424–436.

Mitchell, M., & Gilson, J. (1997). *Interest and Anxiety in Mathematics*. American Educational Research Association.

Okyere, M., & Larbi, E. (2019). *Senior high school mathematics teachers ' perception and use of assessment in the classroom, 15*(2), 43–54.

Oliver, R. (2000). *Creating Meaningful Contexts for Learning in Web-based Settings. Proceedings of Open Learning 2000. (Pp.* Learning Network, Queensland.

Passey, D., Rogers, C. G., Machell, J., & McHugh, G. (2004). *The Motivational Effect of ICT on pupils*. Department for Education and Skills Research Project.

Plomp, T.; Pelgrum, W. J. & Law, N. (2007). SITES2006 International comparative survey of pedagogical practices and ICT in education. *Education and Information Technologies, 12*(2), 83- 92.

Sanyal, B. C. (2001). New functions of higher education and ICT to achieve education for all. *Paper prepared for the Expert Roundtable on University and Technology-for-Literacy and Education*. Research Gate.

Sharma, R. (2003). *Barriers in Using Technology for Education in Developing Countries*. IEEE.

Susman, E. B. (1998). Co-operative learning: a review of factors that increase the effectiveness of computer-based instruction. *Journal of Educational Computing Research, 18*(4), 303-322.

Susuwele, B. W. J. (2005). *Classroom Assessment in Malawi : Teachers ' Perceptions and Practices in Mathematics by Classroom Assessment in Malawi : Teachers ' Perceptions and Practices in Mathematics*. Virginia Polytechnic Institute and State University.

Tebabal, A., & Kahssay, G. (2011). The effects of student-centered approach in improving students' graphical interpretation skills and conceptual understanding of kinematical motion. *Lat. Am. J. Phy. Edu*, 5(2), 374–381.

Toyama, K. (2011). *There are no technology shortcuts to good education*. EduTech Debate. http://edutechdebate.org/ict-in-schools/there-are-no-technology-shortcuts-to-good-education/#sthash.nOJ2LCNe.d

Ünal, M. (2017). Preferences of Teaching Methods and Techniques in Mathematics with Reasons. *Universal Journal of Educational Research*, 5(2), 194–202. 10.13189/ujer.2017.050204

Wanyonyi, P. W. (2013). *An Assement of Factors Affecting Students' Perfomonce in Mathematics at K.C.S.E Level in Secondary Schools in Country*. Kenyatta Universty.

Windschitl, M. (2002). Framing constructivism in practice as the negotiation of dilemmas: An analysis of the conceptual, pedagogical, cultural, and political challenges facing teachers. *Review of Educational Research, 72*(2), 131–175.

Yushau, B., Mji, A., & Wessels, D. C. J. (2005). The role of technology in fostering creativity in the teaching and learning of mathematics. *Pythagoras*, 62(62), 12–22. 10.4102/pythagoras.v0i62.110

Section 2
Assessment

Chapter 6
Assessment in Mathematics Education:
An Inquiry Into Methods and Skills Assessed in the Era of Globalization

Vedaste Mutarutinya
http://orcid.org/0000-0001-6116-8219
University of Rwanda, Rwanda

Sudi Balimuttajjo
https://orcid.org/0000-0002-8044-6162
Mbarara University of Science and Technology, Uganda

Marcellina Andrea Mjenda
University of Rwanda, Rwanda & The University of Dodoma, Tanzania

Robert Wakhata
University of Rwanda, Rwanda

Angel Mukuka
https://orcid.org/0000-0003-0292-1324
Walter Sisulu University, South Africa

Joseph Njiku
Dar es Salaam University College of Education, University of Dar es Salaam, Tanzania

ABSTRACT

This chapter is based on the systematic review that investigates assessment methods in mathematics education in the global era where pertinent skills to address specific problems are at the centre of assessment purposes. Of 442 records obtained from Google Scholar, 50 studies published between 2013 and 2023 were reviewed. These studies, sourced from journals, conference proceedings, and book chapters indexed by Scopus, were analysed using document and thematic analysis. The review aimed to identify the assessment methods applied, the specific skills measured, and

DOI: 10.4018/979-8-3693-2873-6.ch006

how these methods are implemented. Findings revealed that formative assessment, including performance-based, peer, oral, observation, written tests, portfolios, self, and open-ended methods, is central to classroom practices. These methods assessed mathematical and generic skills such as problem-solving, reasoning, thinking, representations, conceptual understanding, and feedback generation. The review provides insights into these methods and their implications for mathematics education in the globalized era.

CONCEPT OF GLOBALIZATION IN MATHEMATICS LEARNING

Globalization is a multifaceted phenomenon that is characterized by the interconnectedness and interdependence of various aspects such as cultures, economies, and perceptions on a global scale, individuals need a specific set of skills to navigate and thrive in this interconnected world (Burbules & Torres, 2013; Morrow & Torres, 2013). Competing in the era of globalisation requires a combination of skills such as technology literacy, multicultural adaptation, communication, and critical thinking in understanding and collaborating with people from diverse backgrounds (Hamdi, 2013). These skills can be developed through various interfaces including the teaching and learning process (Mulenga-Hagane et al., 2019). Mathematics plays a pivotal role in developing critical thinking and problem-solving abilities (Firdaus et al., 2015). The logical and analytical nature of mathematical concepts equips individuals with the skills to approach complex issues systematically. Mathematical literacy is essential for interpreting and analysing data, a skill highly valued in the data-driven global economy (Ojose, 2023). Learning mathematics enhances quantitative literacy, which is vital for making informed decisions in various fields. In business, for example, mathematical skills are essential for financial analysis and strategic planning (Bradley, 2013; Ojose, 2023). In science and technology, mathematical proficiency is fundamental for research and development (Maass et al., 2019). Moreover, Lusyana and Wangge (2016) confirmed that mathematical reasoning fosters a structured approach to problem-solving, a skill applicable across diverse professional and personal contexts. People who possess mathematical literacy are better equipped to make judgments in a world where data-driven insights drive development (Mulenga-Hagane et al., 2019). Mathematical skills are highly valued in the global workforce because they improve analytical thinking, logical reasoning, and quantitative skills, which in turn foster adaptation and resilience in the face of changing difficulties. Consequently, having a solid mathematical foundation is essential for personal and professional development as well as for handling the many problems that happen in the modern linked, and contemporary society.

Assessment Methods in Mathematics Education in a Globalized World

Currently, the world is experiencing significant changes in the field of education, particularly in assessment practices (Selwyn, 2012). There is a need to re-evaluate traditional methods of classroom assessment to meet the current educational demands. Education is vital in and to any global change, one of the cornerstones of implementing a curriculum that attempts to close the knowledge gap between theory and practice is investing in innovative assessment practices (Mjenda et al., 2023; Ndibalema, 2021; Paine, 2017). When relevant assessment methods are carried out well, they facilitate the recognition and exchange of acquired competencies among students. Suurtamm et al. (2016) state that measures and techniques intended to gather data and facilitate the learning process should be integrated through assessment. Therefore, assessment serves as an inquiry into what is being taught, how it is learned, and the motivations behind the process of learning. Griffin and Care (2014) explained that effective classroom assessment methods integrated during teaching and learning have a great impact on acquiring the mathematical skills necessary in this global era. Incorporating real-world applications and projects into assessments can deepen understanding and showcase the practical relevance of mathematical skills (Suurtamm et al., 2016). Collaborative assessments, where students work together to solve complex problems, mirror the teamwork required in the global workforce (Care et al., 2016). Additionally, assessments that emphasise the communication of mathematical ideas help develop effective cross-cultural communication skills.

Current Study

The study examines the usefulness of classroom assessments in the modern era and how well they foster the acquisition of necessary mathematical skills necessary in adapting to the changing demands of the labour market. The study is anticipated to provide suggestions for the integration of relevant assessment methods to ensure learners acquire the skills needed in a world (Boitshwarelo et al., 2017; Mulenga-Hagane et al., 2019) that promote creativity, critical thinking, and problem-solving by studying how mathematical abilities are currently tested in educational settings. Given that success in the modern economy requires more than just the ability to calculate, abilities necessary for success in a variety of professional domains are imperative. Through the identification and refinement of assessment procedures that are in line with these changing demands, teachers may enhance students' readiness to negotiate the intricacies of the modern economy. Debates persist regarding the effectiveness of various assessment methods that are used in mathematics classrooms. While traditional assessment methods like paper and pencil tests have stood the test of time (Ukobizaba et al., 2021), the changing landscape of globalization demands a more expansive evaluation of competencies that transcend these conventional measures. Clarkson (2009) notes that globalization has not only altered the skills required but also the methods necessary for assessing these skills within mathematics education. Despite the wealth of innovative assessment methods proposed, such as performance-based evaluations, project-based assessments, portfolios, peer assessments, and open-ended or problem-based approaches (Suurtamm et al., 2016; Ukobizaba et al., 2021), their effectiveness within mathematics education remains underexplored. Moreover, the potential variance in meaning and usage of these methods across disciplines, especially concerning subject-specific skills, requires further scrutiny. Thus, the goal of this chapter is to provide important new understandings of how mathematical skills, classroom assessment particularly formative assessment, and the dynamic environment of global developments in the modern period interact. Specifically, this chapter responds to the following questions:

1. What assessment methods are applied in mathematics education in the globalized era?
2. What specific skills are measured through the assessment methods in mathematics education?
3. In what ways are assessment methods applied in mathematics education to evaluate specific learning skills?

METHODOLOGY

Overview of Literature Search and Article Selection Process

This study is based on a comprehensive examination of existing literature, focusing on assessment methods utilized in mathematics education and the corresponding mathematical skills assessed in the era of globalisation. Adhering to established systematic review guidelines (Atkinson & Cipriani, 2018; Page et al., 2021; Turney, 2023), our approach encompassed a series of steps. The search was conducted in January 2024 to identify potentially relevant literature in the domain of mathematics education. The initial phase involved formulating a working title, the study's aims, and crafting research questions to guide our investigation. Subsequently, we developed a detailed plan outlining the study's purpose, objectives, and review criteria. The latter encompassed the specifications of inclusion and exclusion criteria as outlined in Table 1.

Another crucial component of our systematic review was the extensive literature search conducted from relevant sources, ensuring a comprehensive coverage of studies published within the past decade (2013 to 2023). Following this, a rigorous evaluation was undertaken to assess the quality and reliability of the selected articles, adhering to the predefined inclusion and exclusion criteria outlined in Table 1.

The subsequent development of a data extraction protocol was based on our research questions, which facilitated a structured approach to glean relevant information from the identified studies. This was followed by analysis and synthesis of the extracted data to answer the research questions. This process not only allowed for a comprehensive understanding of the existing landscape of assessment methods in mathematics classrooms but also provided insights into the evaluated mathematical skills within the contemporary globalized context.

Table 1. Inclusion and exclusion criteria

Inclusion Criteria	Brief Description
Research field	Only papers that are based on mathematics education were considered
Study focus	Papers that focus on the assessment methods or tools used to measure students' mathematical skills or competencies in the context of globalization.
Methodology	Papers that provide sufficient details about the design, implementation, and evaluation of the assessment methods or tools.
Language	Papers that are written in English
Indexing database	Papers that are peer-reviewed and published in a journal, conference proceedings, or edited book indexed by Scopus
Publication year	Papers published in the last 10 years, from 2013 to 2023
Exclusion Criteria	
Non-empirical studies	Theoretical papers, opinion pieces, and editorials that do not present original empirical findings
Non-relevant research focus	Studies that do not focus on assessment methods in mathematics education
Non-global perspectives	Studies that are not of international standard or do not consider the era of globalization or its impact on mathematics education
Non-English language studies	Studies published in languages other than English
Poor methodological quality	Studies with significant methodological flaws that could bias the results.
Outdated studies	Studies conducted before the year 2013
Non-peer reviewed sources	Studies that have not undergone a peer-review process for quality assurance.

Article Selection and Data Extraction

The methodology for this literature review followed careful sequential steps and procedures to identify relevant articles on assessment methods in mathematics education literature from January 2013 to December 2023. Google Scholar was used to search and identify articles in mathematics education with their titles containing "assessment" and "mathematics education" as well as those in abstracts, and contents that were used to elaborate the search terms. However, we were mindful of the fact that Google Scholar incorporates into its index any material resembling an academic article, research report, thesis, working paper, or book chapter based on its internal algorithm. Additionally, it is noteworthy that Google Scholar does not utilize stable document identifiers, such as the digital object identifier (DOI). Consequently, there exists no assurance that the outcomes observed in one search can be precisely reproduced in a subsequent search. Given these constraints, we restricted our focus to articles sourced from the Google Scholar search engine that have been formally published in journals, conference proceedings, and edited books or book chapters listed in the Scopus indexing database. This collective decision by all authors of this paper was motivated by the recognition that Scopus is widely acknowledged as one of the most dependable and recommended sources for academic research information (Baas et al., 2020; Mukuka & Alex, 2024; Singh et al., 2021).

Utilizing the above highlighted criteria, a total of 442 accessible records to the authors were initially gathered by two of the authors involved in this study, with 67 identified as duplicates. Subsequently, the titles and abstracts of 375 articles underwent further screening, focusing on their relevance and scope. During this stage, 244 articles were excluded for either or both of the following reasons: (1) the article lacked a focus on assessment methods in mathematics education, or (2) the study did not meet international standards or somehow failed to consider the impact of globalization on mathematics education. This screening process left us with 131 articles for a thorough review.

To facilitate the review, a data extraction tool was developed, incorporating predetermined codes related to research question 1, which focused on assessment methods in mathematics education. Various assessment methods, such as self-assessment, peer assessment, task-based assessment, portfolio/e-assessment, project-based assessment, and game-based assessment, were considered, with flexibility for potential variations in terminology used across studies. Concerning the second research question, all skills listed in Figure 3 were considered during the data extraction process. For research question 3, no predetermined codes were generated, as the application of the implemented assessment methods was going to be derived from each research article.

Following these criteria, a data extraction tool with eight columns was agreed upon and distributed to four authors who were involved in the extraction process. The columns included the name of the author(s) and publication year, research title, assessment method(s) employed/ explored, skills/competences assessed, practices or ways used to implement the identified assessment method(s), the country where the study was conducted or the country of the first author, research design employed, and the target group or participants. All 131 articles were distributed among the four authors and underwent full-text review, resulting in the exclusion of 81 articles from the final analysis. The primary reason for exclusion was the non-empirical nature of these articles, comprising theoretical papers, opinion pieces, or editorials lacking original empirical findings and detailed information on design and assessment methods. The exclusion of these 81 articles led to a final set of 50 articles, from which relevant information was extracted for further analysis. Table 2 shows the included studies.

Table 2. Summary of included studies on assessment methods in mathematics education

	Author/year	Research title
1.	Adelabu & Alex (2022)	Online Baseline Assessment in Mathematics: Initial Teacher Education Entry-Level Student Performance
2.	Bjørkli (2014)	The impact on learning outcomes in mathematics of mobile-enhanced, combined formative and summative assessment
3.	Buchholtz et al. (2018)	Combining and integrating formative and summative assessment in mathematics teacher education
4.	Burgos & Godino (2022)	Assessing the epistemic analysis competence of prospective primary school teachers on proportionality tasks
5.	Chen et al. (2020)	Formative assessment with interactive whiteboards: A one-year longitudinal study of primary students' mathematical performance
6.	Connor et al. (2017)	Using assessment to individualize early mathematics instruction
7.	Cross & Palese (2015)	Increasing Learning: Classroom Assessment Techniques in the Online Classroom
8.	Dewantara (2015)	Assessing Seventh Graders' Mathematical Literacy in Solving PISA-Like Tasks
9.	Fan & Zhu (2023)	Using performance assessment in secondary school mathematics: An empirical study in a Singapore classroom
	Hsu et al. (2023)	Examination of Taiwanese Mathematics Teacher Questioning
	Jones & Alcock (2014)	Peer assessment without assessment criteria
	Jones et al. (2015)	Assessing mathematical problem solving using comparative judgment
	Kim (2019)	Teacher learning opportunities provided by implementing formative assessment lessons: Becoming responsive to student mathematical thinking
	Kunwar et al. (2023)	Construction and Validation of Self-Assessment Instrument for Students' Mathematics Classroom Learning Behaviour
	Kyaruzi et al. (2019)	Students' formative assessment perceptions, feedback use, and mathematics performance in secondary schools in Tanzania
	Leenknecht et al. (2020)	Formative assessment as practice: the role of students' motivation
	Leung et al. (2014)	A study of the alignment of learning targets and assessment to generic skills in the new senior secondary mathematics curriculum in Hong Kong
	Logan (2015)	The influence of test mode and visuospatial ability on mathematics assessment performance
	Luz & Yerushalmy (2018)	Students' conceptions through the lens of a dynamic online geometry assessment platform
	Mahlambi et al. (2021)	Exploring the use of assessment for learning in the mathematics classroom
	Murchan & Oldham (2017)	Exploring the role of computer-based assessment in diagnosing children's mathematical errors in primary education in Ireland
	Mushin et al. (2013)	Language matters in demonstrations of understanding in early years mathematics assessment

continued on following page

Table 2. Continued

Author/year	Research title
Mutodi & Mugonga (2016)	Concept Map as an Assessment Tool in Secondary School Mathematics: An Analysis of Teachers' Perspectives
Nama & Ayalon (2023)	Exploring change in secondary mathematics teachers' noticing of argumentation through experiencing peer-assessment strategies
Nortvedt et al. (2015)	Assessment for learning in Norway and Portugal: the case of primary school mathematics teaching
Oktiningrum & Hartono (2016)	Developing PISA-" Like" Mathematics Task with Indonesia Natural and Cultural Heritage as Context to Assess Students Mathematical Literacy
Olsher et al. (2016)	How might the use of technology in formative assessment support changes in mathematics teaching?
Papic (2015)	An early mathematical patterning assessment: Identifying young Australian indigenous children's patterning skills
Rakoczy et al. (2017)	Formative Assessment in Mathematics Instruction: Theoretical Considerations and Empirical Results of the Co2 CA Project
Riyanto & Putri (2019)	Senior High School Mathematics Learning through Mathematics Modelling Approach
Robinson et al. (2015)	Student perceptions of screencast feedback on mathematics assessment
Rubin & Rajakaruna (2015)	Teaching and assessing higher-order thinking in the mathematics classroom with clickers
Şahin & Öztürk (2018)	How Classroom Assessment Affects Science and Mathematics Achievement? Findings from TIMSS 2015
Savard & Freiman (2016)	Investigating complexity to assess student learning from a robotics-based task
Schoenfeld (2015)	Summative and Formative Assessments in Mathematics Supporting the Goals of the Common Core Standards
Schroeder & Dorn (2016)	Enabling and Integrating Online Formative Assessment in a Flipped Calculus Course
Serow et al. (2016)	Assessment of Mathematics Learning: What Are We Doing?
Soto & Ambrose (2016)	Screencasts: Formative Assessment for Mathematical Thinking
Syaifuddin (2020)	Implementation of authentic assessment on mathematics teaching: Study on junior high school teachers
Szeibert et al. (2022)	A Case Study of Using Test-Enhanced Learning as a Formative Assessment in High School Mathematics
Tanudjaya & Doorman (2020)	Examining Higher Order Thinking in Indonesian Lower Secondary Mathematics Classrooms
Thanheiser (2018)	Brief report: The effects of preservice elementary school teachers' accurate self-assessments in the context of whole number
Thomas et al. (2023)	Assessing students' understanding of time concepts in Years 3 and 4: insights from the development and use of a one-to-one task-based interview
Urrutia & Araya (2022)	Do Written Responses to Open-Ended Questions on Fourth-Grade Online Formative Assessments in Mathematics Help Predict Scores on End-of-Year Standardized Tests?
van den Berg et al. (2018)	Testing the effectiveness of classroom formative assessment in Dutch primary mathematics education
Vankúš (2021)	Influence of game-based learning in mathematics education on students' affective domain: A systematic review
Veldhuis & Van den Heuvel-Panhuizen (2020)	Supporting primary school teachers' classroom assessment in mathematics education: Effects on student achievement
Walick & Burns (2017)	A Proposed Algebra Assessment for Use in a Problem-Analysis Framework
Weiland et al. (2013)	Preservice formative assessment interviews: the development of competent questioning
Yang & Lin (2015)	Assessing 10-to 11-year-old children's performance and misconceptions in number sense using a four-tier diagnostic test

Data Analysis

The data analysis methods employed in this study encompass two primary types: document analysis and thematic analysis. These methodologies were systematically applied to review and assess the selected articles, aligning with the predefined criteria. Document analysis entailed an examination of relevant articles to extract themes or patterns relevant to the research topic (Hassan, 2024; Kitchenham, 2004). The documents scrutinized included articles formally published in journals, conference proceedings, and edited books or book chapters indexed in Scopus (n=50). This process involved several steps, including sourcing, contextualizing, interrogating, making inferences, and ensuring reliability and validity. The data extraction tool, developed for the review, integrated predetermined codes aligned with the research questions.

On the other hand, thematic analysis involved a phased approach, including familiarization with the data, coding, generation of themes, reviewing themes, defining, and naming themes, and summarizing findings (Braun & Clarke, 2006). This method facilitated an understanding of diverse assessment methods in mathematics education, such as self-assessment, peer assessment, task-based assessment, portfolio/e-assessment, project-based assessment, and game-based assessment, among others. Additionally, it aided in identifying the specific mathematical skills assessed in each reviewed article. Figure 3 outlines the mathematical skills that were assessed in all the reviewed articles. The combination of thematic and document analysis methods provided a comprehensive approach to scrutinizing the selected articles, ensuring a thorough review by capturing all relevant information. Thus, the chosen analytical methods enhanced the depth and rigor of the review, contributing valuable insights to the understanding of assessment methods in mathematics education within the globalized era.

FINDINGS AND DISCUSSION

Characteristics of the Reviewed Studies

The analysed studies were published between 2013 and 2023 were distributed unevenly across the years. Most of the studies assessed in this study were published in 2015 followed by 2016. The distribution does not provide an obvious prediction as to whether studies in mathematics assessment are increasing or decreasing (see Figure 1).

Figure 1. Number of reviewed studies across years from 2013 to 2023

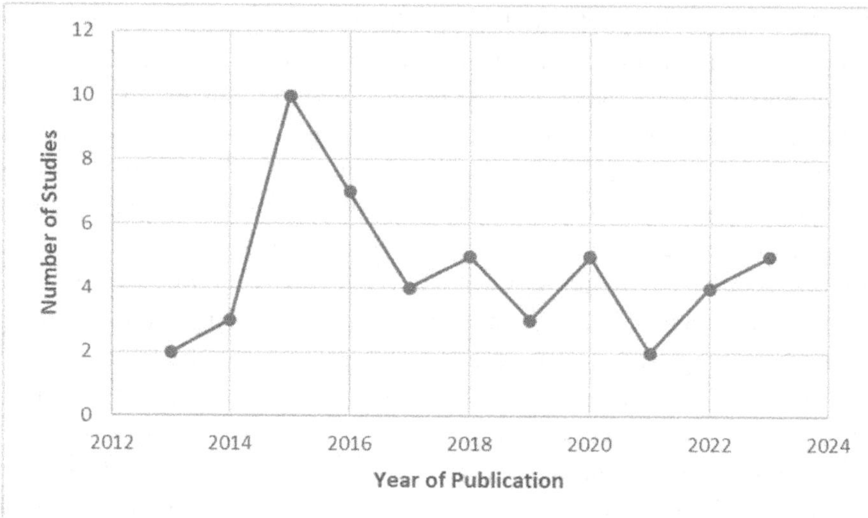

The reviewed studies came from different countries. When papers did not explicitly mention where the study was done, we used the first author's country as the country from which the paper came. Most papers come from the USA, and then Indonesia, and the UK (see Table 3). The analysis shows that only three of the reviewed studies come from African countries, suggesting a dearth of research on assessment in mathematics from the region.

Table 3. Distribution of the reviewed studies by country

Country																					
No. of Studies	8	5	5	4	3	3	3	3	2	2	1	1	1	1	1	1	1	1	1	1	1	1

We also analysed the nature of the participants sampled in the reviewed studies. We categorised the participants as preschool if they had not begun their grade one of formal education, primary school if they were in grade six or below, secondary school students if they were above grade six, and undergraduate students if they were university or university college students. In some studies, participants were university students but precisely named as preservice teachers. Some other studies were done on in-service teachers. Most studies were dedicated to primary school students followed by secondary school students as presented in Table 4.

Table 4. Nature of participants in the reviewed studies

Nature of Participants	Number of Studies
Primary School Students	17
Secondary School Students	15
Undergraduate Students	7
Pre-service teachers	5
Inservice Teachers	4
Preschool Children	1

Assessment Methods

The learning of mathematics is diverse and assessment methods and strategies may differ widely. In this chapter, we explored common assessment methods as identified and reported in the existing and accessible literature to the authors. Identification of these methods and strategies was important to guide mathematics teachers in applying them to effectively assess students for mathematics learning. With this aim in mind, the methods and strategies we explored were those used in the classroom or learning contexts. We focused our work on formative assessment where the information obtained is to be used either to plan or improve mathematics instruction. Therefore, in this section, we focus our attention on assessment as learning and assessment for learning, both of which make up a big portion of formative assessment where the learning process is the focus. We therefore discuss the assessment methods and strategies that are reported in the mathematics education literature that may be used to inform mathematics learning practices (See figure 2).

To inform the learning process, many of the reviewed studies adopted classroom-based assessments to deliberate on the learning activities. Many of the existing studies (see for example Kim, 2019; Kyaruzi et al., 2019; Leenknecht, et al., 2020; and Leung et al., 2014) do not categorically mention the assessment methods or strategies adopted by teachers or educators in the assessment of mathematics competencies. These studies tend to use the term formative assessment indicating the type of assessment they address in their discussion. It becomes unclear from such studies what methods or activities happened in the classroom as a way of assessing the learning of mathematics. Although the use of the term formative assessment suggests the process of obtaining information to form the teaching and learning process, it is too generic to guide practices.

Figure 2. Assessment methods

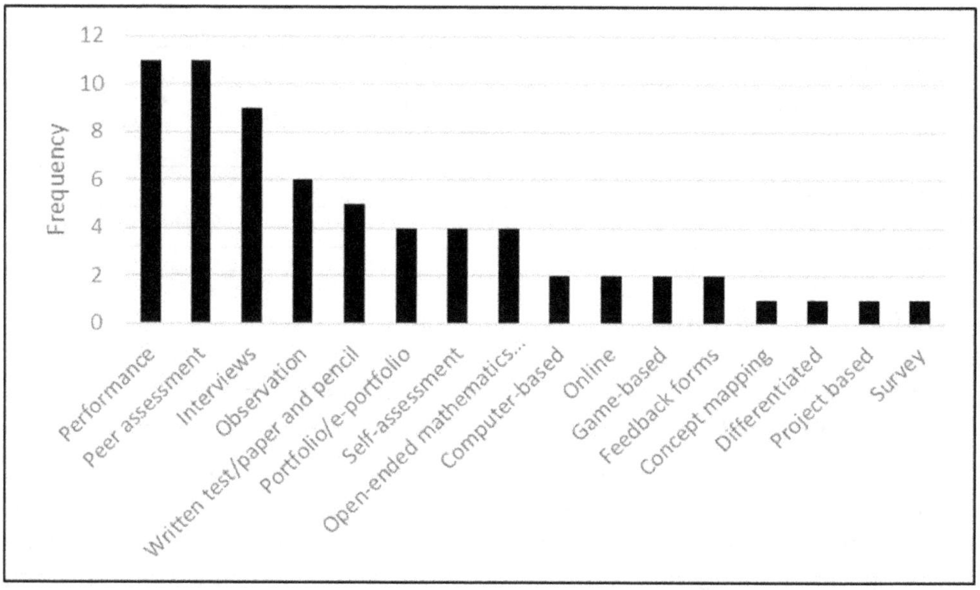

This study was able to identify some specific methods or strategies used in various mathematics classrooms as reported in the current literature. The common assessment methods include performance or task-based assessment (Burgos & Godino, 2022; Connor, et al., 2018; Cross & Palese, 2015; Fan & Zhu, 2023; Murchan & Oldham, 2017; Mutodi & Chigonga, 2016; Syaifuddin, 2020), peer assessment (Jones & Alcock, 2014; Jones et al., 2015; Nama & Ayalon, 2023; Savard & Freiman, 2016; Schoenfeld, 2015; Schroeder & Dorn, 2016; Syaifuddin, 2020), interviews (Mushin et al., 2013; Papic, 2015; Serow et al., 2016; Thomas et al., 2023; Veldhuis & van den Heuvel-Panhuizen, 2020; Weiland et al., 2014), and observations (Nortvedt et al., 2016; Papic, 2015; Schoenfeld, 2015; Soto & Ambrose, 2016; Syaifuddin, 2020). Other common methods are written assessment (Syaifuddin, 2020; Szeibert et al., 2023; Urrutia & Araya, 2022; van den Berg et al., 2018), open-ended problems (Olsher et al., 2016; Robinson et al., 2015; Tanudjaya & Doorman, 2020; Urrutia & Araya, 2022), self-assessment (Chen et al., 2020; Mahlambi et al., 2022; Syaifuddin, 2020), and portfolio assessment (Adelabu & Alex, 2022; Buchholtz et al., 2018; Syaifuddin, 2020). These can also be extended to include diagnostic (Schoenfeld, 2015; Yang & Lin, 2015), computer-based (Logan, 2015; Murchan & Oldham, 2017), online assessment (Luz & Yerushalmy, 2019; Urrutia & Araya, 2022), game based (Chen et al., 2020; Vankúš, 2021), feedback forms (Robinson et al., 2015; Veldhuis & van den Heuvel-Panhuizen, 2020), concept mapping (Mutodi & Chigonga, 2016), differentiated assessment (van den Berg et al, 2018), project-based assessment (Syaifuddin, 2020), and survey (Thanheiser, 2018).

The list is long suggesting that there is a range of methods at the disposal of mathematics teachers, although assessment activities may cut across methods. Also, researchers might have used different terminologies for the same activities. For example, performance assessment might also be referred to as task assessment, peer assessment may be named group assessment. Interviews are referred to in some other studies as verbal and oral assessments. Written assessment was also termed as paper and pencil or classroom tests. However, there might be some differences between similar methods. For example, open-ended mathematics problems may be written, oral, task-based, or project-based assessments. To discern between methods, it is imperative to look into the activities inherent in each method. We explore specific practices in the next section.

Mathematics lessons are mainly delivered through talk chalk (Iannone & Miller, 2019) where paper and pencil tests may be the norm. Nevertheless, in addition to active assessment methods mentioned earlier, studies have also shown the possibility of integrating technology into assessment. In most cases, technology is used as a tool to afford various methods rather than being a method in itself. Terms such as computer-based assessment (Logan, 2015; Murchan & Oldham, 2017) may be perceived as methods by themselves but technology is being used to facilitate the implementation of assessment methods such as online verbal assessment, electronic-based tests, compilation of learning evidence like an electronic portfolio, and observation. As an innovation, technology enables effective observation, data collection, marking or scoring, and feedback provision.

The assessment affordances of technology allow the transformation of classroom assessment practices. This may not necessarily mean the adoption of new methods. In this study, we identified assessment methods that adopted technology-based assessment including electronic portfolios (Adelabu & Alex, 2022; Buchholtz et al., 2018), observation through screencasts (Soto & Ambrose, 2016), online assessment, and provision of feedback through screencast feedback forms (Robinson et al, 2015).

These findings indicate that although mathematics classrooms differ greatly, the assessment methods reported are not very much different. One reason for this observation may be that the teaching and learning of mathematics is dominated by one method – chalk talk (Iannone & Miller, 2019) which in turn influences the assessment methods. Another reason for this observation may be the tendency of many researchers to report the assessment type such as formative assessment as we observed and reported previously instead of explaining methods of formative assessment adopted in mathematics classrooms. Additionally, when technology is integrated, it is adopted as a tool that teachers must innovatively use to facilitate assessment procedures effectively.

Skills Assessed in Mathematics

The learning of mathematics is expected to develop various mathematical and non-discipline-based skills. Although such skills may be numerous (see Figure 3), in this chapter we explore common mathematics and other skills assessed in mathematics classrooms. Understanding such skills may help teachers and educators assess their importance and plan to develop and access them. The existing literature is rich in enumerating various skills that are assessed in mathematics classrooms as presented here next.

Studying mathematics involves more mental processes but must be evidenced through various representations (Jones & Alcock, 2014; Mutodi & Chigonga, 2016; Olsher et al., 2016) including writing skills. Following the nature of mathematics, many studies assessed cognitive skills, especially thinking (Kim, 2019; Leung et al., 2014; Tanudjaya & Doorman, 2020), reasoning (Adelabu & Alex, 2022; BjÁ¸rkli, 2014; Fan & Zhu, 2023), and problem-solving (Leung et al., 2014; Murchan & Oldham, 2017; Riyanto et al., 2019).

Other studies assessed topic or content-specific skills such as algebraic skills (Burgos & Godino, 2022; Walick & Burns, 2017), estimation and approximation skills (Logan, 2015), measurement skills (Dewantara & Zulkardi, 2015; Thomas et al., 2023), operation with numbers (Veldhuis & van den Heuvel-Panhuizen, 2020). This assessment focus suggests that in mathematics, learning topics carry diverse and specific mathematical skills. These skills may be assessed in numerous ways. Teachers may need to consider effective ways of determining the development of such skills.

In addition to mathematics-allied skills such as reasoning and content-specific skills measurement skills, the literature indicates that assessment in mathematics goes beyond cognitive skills. Various studies have assessed skills such as collaboration (Schroeder & Dorn, 2016)), organisation (Mutodi & Chigonga, 2016), and generic attitude competencies (Syaifuddin, 2020). This suggests that assessment in mathematics includes 21st-century skills.

the learning of mathematics requires students to have positive attitudes toward the subject. Unfortunately, this is not always the case. The situation is more challenging because mathematics is compulsory for most school-going children. Determining students' attitudes may provide important information for addressing attitude-related competencies. Studies show that assessment in mathematics includes motivation, engagement and enjoyment, affirmation, and confidence (Leenknecht, et al., 2020; Rubin & Rajakaruna, 2015; Vankúš, 2021).

Figure 3. Mathematical skills

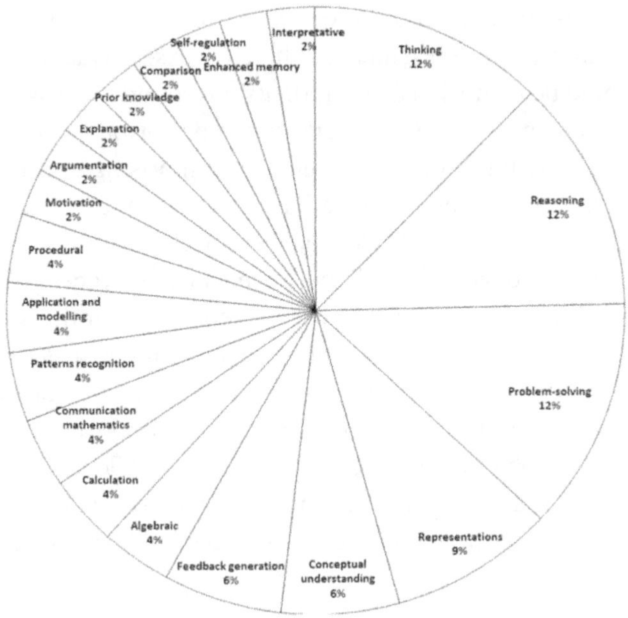

Application of Assessment Methods in Mathematics Classrooms

Knowing what assessment methods are used in the mathematics classrooms, the work that we have done in the previous section is one thing, and knowing how to enact them in the classroom is another. Many mathematics teachers may be aware of the methods mentioned earlier. They also need to know how to implement them for learning. In this section, we discuss ways of implementing commonly reported methods for the assessment of mathematics learning.

Performance Assessment

There is no direct definition of performance assessment as used in mathematics. Although reviewed studies mention performance assessment as one of the methods to be used in assessing learning of mathematics, many forgo explain what this means or how it is done. Only one study (Fan & Zhu, 2023) tried to categorically define what performance-based assessment means. We use ideas from Fan and Zhu (2023) and Amsami et al. (2015) to explain the method. Fan and Zhu (2023) argue that performance-based assessment uses students' work or tasks as evidence of students learning. This means it is not based on teacher-made tests, multiple-choice items, or standardised tests. It is based on students' learning tasks. As such, performance assessment is an alternative assessment method that assesses actual problem-solving abilities that should be identified from students' tasks and processes (Amsami et al., 2015). Whereas Fan and Zhu (2023) focus on the performance of tasks, Amsami et al., (2015) consider performance assessment as comprising portfolio and self-peer assessment. Also, whereas performance-based assessment is used interchangeably with task-based assessment, this performance may either need to be reflected upon by learners themselves during the collaborative learning process – peer assessment or through a collection of prices of learning evidence – portfolio. Syaifuddin (2020) considers performance-based assessment as an authentic assessment together with self and peer assessment, observation, paper and pencil test, portfolio, and project assessment. As an authentic assessment, the performance-based assessment focuses on mathematics problems that are open-ended and address important real-life problems (Fan & Zhu, 2023) which may not have only one solution. Teachers need to assess the learning and thinking process as evidenced in students' learning tasks.

Peer Assessment

Learning in mathematics classrooms involves several students. A class is therefore an opportunity for students to work together and learn from each other. Teachers must use this setting for instructional purposes including assessment. Studies have shown preference by teachers for peer assessment (Syaifuddin, 2020) which is also referred to as group assessment or participative assessment. It is a participatory assessment method that involves students assessing the work of their peers under the guidance of the teacher (Syaifuddin, 2020) allowing collaborative knowledge building (Schroeder & Dorn, 2016). Schoenfeld (2015) explains peer assessment as involving students displaying and sharing posters, and making comparisons and contrasts collaboratively. Nama and Ayalon (2023) present six stages of peer assessment: analysing written arguments presented by peers in the classroom (they may also be verbal arguments), collaboratively assessing peers' work under the guidance of the teacher, giving feedback to peers, receiving feedback, individually (it may be collaboratively) refining the initial points or arguments, and making a reflection of the whole experience. The last two stages provide room to retreat and for this reason, Amsami et al. (2015) and Syaifuddin (2020) term this method as self and peer assessment suggesting that peer assessment integrates self-assessment. Although peer assessment can be open, teachers tend to establish clear and detailed common understanding between teachers and students about what is being assessed (Jones & Alcock, 2014).

Verbal Assessment

Classroom interaction involves communication between students and teachers and among students. This interaction includes verbal exchange of communication including mathematical messages. Studies have shown that verbal mathematical messages can be collected as evidence for learning (see (Buchholtz et al., 2018; Mushin et al., 2013; Papic, 2015; Rakoczy et al., 2017; Serow et al., 2016). Mushin et al. (2013) report the use of oral mathematics assessment especially with young children where mathematical language was the skill in focus. To elicit mathematical language competencies among children, the oral assessment was seen as relevant (Mushin et al., 2013; Papic, 2015). Interview assessments in mathematics tend to be on a one-to-one basis (Thomas et al, 2023), and many focus on young children (see Mushin et al., 2013; Papic, 2015; Serow et al., 2016; Thomas et al., 2023). When the interview is focused on performance, students may be informed before the interview about the key areas of focus whose competencies they need to demonstrate (Thomas et al., 2023). Studies adopting this method tend to use interviews, verbal, and oral assessment interchangeably.

Observation

Observation of the learning process was also a common assessment method in the literature. The observation assessment method is common in lower levels of schooling and is done in an unstructured way (Nortvedt et al., 2016; Papic, 2015). However, Nortvedt et al. (2016) are critical to this modality pointing out that unstructured observation does not capture specific levels of competencies. Similarly, Papic (2015) adds that observation should not focus on engagement only but also and more importantly on what students think and can do mathematically. To obtain evidence about students thinking, teachers need to observe students as they work and reason. The thinking process much be observed in students' tasks such as their writing or drawing. At an advanced level, concept mapping activities by students may provide observational data regarding their thinking process (Mutodi & Chigonga, 2016). Furthermore, to aid in collecting observational data, especially about task performance and thinking processes, screencasts afforded by technology have been integrated (Robinson et al., 2015; Soto & Ambrose, 2016).

Other Practices

Despite paper and pencil assessments being considered traditional, it was common in the reviewed studies. Different terminologies are used to refer to paper and pencil including tests and written assessment. Written assessment may include open-ended mathematical problems that are not limited to one correct answer since they tend to model real-life scenarios (Tanudjaya & Doorman, 2020). They may also extend to more complex project-based assessments. This means written assessments are not entirely limited to multiple choice and standardised tests. They are also referred to as essay questions (Urrutia & Araya, 2022) where reasoning and communication skills may be assessed. Another method mentioned earlier is the portfolio, which may also be done electronically as hinted under performance-based assessment. The electronic portfolio includes a collection of electronic materials as evidence of students learning. Portfolios also involve reflection which acts as self-assessment (Buchholtz et al., 2018). Considering the rich information in portfolios, teachers need to have a well-designed assessment rubric to collect learning evidence from students' portfolios.

This section has presented a few strategies for adopting various assessment methods that go beyond paper and pencil, that we believe readers may find useful in guiding their practices. The list has become short because of the observed overlap between the methods especially when the activities involved are analysed. Also, this is contributed by the limited details that existing studies provided when explaining these assessment methods.

Summative Assessment in Mathematics

Preceding sections were dedicated to formative assessments. However, equally important is the summative assessment. Summative assessments are essential in gauging students' understanding and competency with mathematical ideas and techniques in the field of mathematics education (Buchholtz et al., 2018; Marinho et al., 2017). It functions as an in-depth evaluation after a unit, course, or academic term, in contrast to formative assessment, which focuses on continuing feedback and improvement during the learning process (Gallardo, 2021). Since the aim of education is to achieve the broad education goal, formative assessment should be smoothly incorporated into the lesson plan to make sure that students are learning the material and meeting the course's wider objectives, which are evaluated through summative assessment (Kibble, 2017). The idea and consequences of summative assessment in mathematics education take on greater significance in the context of today's globalized world, where technological breakthroughs and increased interconnectedness affect the educational viewpoint (Gallardo, 2021). In this view, summative assessments in mathematics are primarily used to assess students' grasp of the subject matter, their capacity for problem-solving, and their application of the concepts in practical settings. Standardized tests, term exams, or capstone projects are frequently used for this evaluation. They evaluate a variety of mathematical talents, from fundamental arithmetic abilities to sophisticated problem-solving techniques. Summative assessment provides a picture of students' mathematical competency and provides information to educators, policymakers, and stakeholders on the efficacy of curriculum design, instructional strategies, and overall learning outcomes (Chytry & Kubiatko, 2021).

The effects of summative assessment on mathematics education are complex in the globalized age when knowledge and abilities are valued more and more on a worldwide basis. First and foremost, summative assessments function as standards for gauging students' preparedness for participation and teamwork in a globalized job market powered by STEM (science, technology, engineering, and mathematics) disciplines (Marinho et al., 2017). Furthermore, summative assessment in mathematics offers a way to test students' performance globally in an interconnected world where educational standards and practices are compared across borders. Assessments like the Programme for International Student Assessment (PISA) reveal areas where national education systems need to improve and provide insights into how students perform in mathematics across national boundaries. These comparative analyses support cooperation and knowledge sharing across countries looking to improve their mathematical education in addition to informing educational policy.

Moreover, summative assessment in mathematics can use technology to improve its efficacy and accessibility in the era of digital transformation and data-driven decision-making (Marinho et al., 2017). Online assessment tools and adaptive testing platforms can offer customized feedback and adaptable learning experiences based on the requirements of each individual student. Furthermore, large-scale assessment data can be analyzed by data analytics and machine learning algorithms to find patterns, trends, and places that need instructional intervention, which will optimize teaching and learning strategies (Premkumar, 2016). Summative assessment in mathematics, however, has drawbacks and restrictions even if it has the potential to provide information about students' mathematical competency. Also, students of different origins may perform differently on final examinations due to linguistic and cultural variations, underscoring the need for culturally sensitive assessment procedures. Thus, in the globalized era, mathematics education requires a diverse array of summative assessment strategies to effectively enhance learning outcomes (Premkumar, 2016). One method involves traditional written examinations, assessing students' understanding of mathematical concepts, problem-solving abilities, and application of formulas. In most of educational settings, summative assessment practices are based on traditional ways and leave behind performance-based assessments such as projects or presentations which could enable students to demonstrate their comprehension in real-world contexts. In addition, the use of computer-based assessments offers interactive platforms for students to engage with mathematical problems, utilizing simulations and interactive exercises to enhance understanding. Furthermore, portfolio assessments allow students to compile a collection of their work, showcasing their progress and development over time. By integrating these diverse assessment methods, mathematics education can adapt to the demands of the globalized era, while preparing students for success in an interconnected world.

LIMITATIONS OF THE STUDY

This systematic review may not provide sufficient information on the overall assessment methods due to the limited number of studies searched and that was accessible to the authors with their heterogeneity in the study design and their method of analysis. In the review of 50 studies, only articles written in English and published in peer-reviewed journals in at least the last decade were included, which may cause a biased representation and analysis of our literature review. We also conducted a handsearching which was limited to Google Scholar electronic literature search with specific search terms on assessment methods. The findings of the reviewed studies neither propose comprehensive explanations nor provide sufficient background information on assessment challenges in different educational settings and contexts in the current globalization era. Yet, the assessment challenges faced by students and educators need to be addressed to identify suitable skills for planning and implementing assessment strategies in mathematics education. In addition, our review may have produced limited evidence of the role of ICT in assessment and the challenges faced by educators with variations in their experiences of technology-supported assessment in mathematics instruction. Overall, this systematic review may guide in understanding the complexity and dynamics of assessment methods in mathematics education, as a foundation for future research studies, and for developing effective teacher professional development programs.

CONCLUSION AND RECOMMENDATIONS

Mathematics classroom assessment involves a range of activities for both teachers and students for the learning process. This Systematic literature review was about ways of determining and informing the teaching and learning of mathematics. As such, studies on assessment of mathematics learning were reviewed where a range of methods and skills analysed were identified and discussed.

The reviewed studies show advocacy for mathematics formative assessment rather than summative assessment. This advocacy is geared towards informing classroom practices, teaching methods, students' needs, students' difficulties, and required resources. Formative assessment using different methods may help teachers to identify students' needs and difficulties and plan ways of addressing them. Also, teachers may need to use formative assessment to inform their teaching methods, teaching material, and the progress at which students are learning. This calls for the need to ensure assessment is for learning, especially skills development rather than assessment of learning.

The reviewed studies show that there are diverse assessment methods that may be used to inform the teaching and learning process. The number of methods provides an opportunity for teachers to adopt suitable methods based on classroom contexts. Teachers are expected to possess the skills necessary to implement these methods as the context may require. From these comes the role of teacher education both pre-service and in-service especially continuing professional development for teachers.

The development of diverse mathematical and generic skills is central in the era of globalization. Teachers need to be aware of the skills needed to be developed through mathematics instruction. This awareness may help teachers to plan for relevant assessment methods. Therefore, teachers should not use assessment methods for their sake, but to assess specific skills. For this, more research is needed to match between skills and assessment methods.

REFERENCES

Adelabu, F. M., & Alex, J. (2022). Online baseline assessment in mathematics: Initial teacher education entry-level student performance. *Academic Journal of Interdisciplinary Studies*, 11(1), 68–68. 10.36941/ajis-2022-0006

Amsami, B. U., Mohammed, Y., & Mazila, E. A. (2015). Visual art teachers and performance assessment methods in Nigerian senior secondary schools. Mgbakoigba. *Journal of African Studies*, 4, 1–18. https://www.ajol.info/index.php/mjas/article/view/118501

Atkinson, L. Z., & Cipriani, A. (2018). How to carry out a literature search for a systematic review: A practical guide. *BJPsych Advances*, 24(2), 74–82. 10.1192/bja.2017.3

Baas, J., Schotten, M., Plume, A., Côté, G., & Karimi, R. (2020). Scopus as a curated, high-quality bibliometric data source for academic research in quantitative science studies. *Quantitative Science Studies*, 1(1), 377–386. 10.1162/qss_a_00019

Bjørkli, K.BjÅ¸rkli. (2014). The impact on learning outcomes in mathematics of mobile-enhanced, combined formative and summative assessment. *International Journal of Technology Enhanced Learning*, 6(4), 343–360. 10.1504/IJTEL.2014.069025

Boitshwarelo, B., Reedy, A. K., & Billany, T. (2017). Envisioning the use of online tests in assessing twenty-first century learning: A literature review. *Research and Practice in Technology Enhanced Learning*, 12(1), 1–16. 10.1186/s41039-017-0055-730595721

Braun, V., & Clarke, V. (2006). Using thematic analysis in psychology. *Qualitative Research in Psychology*, 3(2), 77–101. 10.1191/1478088706qp063oa

Buchholtz, N. F., Krosanke, N., Orschulik, A. B., & Vorhölter, K. (2018). Combining and integrating formative and summative assessment in mathematics teacher education. *ZDM Mathematics Education*, 50(4), 715–728. 10.1007/s11858-018-0948-y

Burbules, N. C., & Torres, C. A. (2013). Globalization and education: An introduction. In *Globalization and education* (pp. 1–26). Routledge. 10.4324/9781315022642-5

Burgos, M., & Godino, J. D. (2022). Assessing the epistemic analysis competence of prospective primary school teachers on proportionality tasks. *International Journal of Science and Mathematics Education*, 20(2), 367–389. 10.1007/s10763-020-10143-0

Care, E., Scoular, C., & Griffin, P. (2016). Assessment of collaborative problem solving in education environments. *Applied Measurement in Education*, 29(4), 250–264. 10.1080/08957347.2016.1209204

Chen, I.-H., Gamble, J. H., Lee, Z.-H., & Fu, Q.-L. (2020). Formative assessment with interactive whiteboards: A one-year longitudinal study of primary students' mathematical performance. *Computers & Education*, 150, 103833. 10.1016/j.compedu.2020.103833

Chytry, V., & Kubiatko, M. (2021). Pupils' Summative Assessments in Mathematics as Dependent on Selected Factors. *Eurasia Journal of Mathematics, Science and Technology Education*, 17(8). https://eric.ed.gov/?id=EJ1309268

Connor, C. M., Mazzocco, M. M., Kurz, T., Crowe, E. C., Tighe, E. L., Wood, T. S., & Morrison, F. J. (2018). Using assessment to individualize early mathematics instruction. *Journal of School Psychology*, 66, 97–113. 10.1016/j.jsp.2017.04.00529499792

Cross, T., & Palese, K. (2015). Increasing learning: Classroom assessment techniques in the online classroom. *American Journal of Distance Education*, 29(2), 98–108. 10.1080/08923647.2015.1023594

Dewantara, A. H., & Zulkardi, D. (2015). Assessing seventh graders' Mathematical literacy in solving PISA-like tasks. *Indonesian Mathematical Society Journal on Mathematics Education, 6*(2), 39-49. https://eric.ed.gov/?id=EJ1079604

Fan, L., & Zhu, Y. (2023). Using performance assessment in secondary school mathematics: An empirical study in a Singapore classroom. *Journal of Mathematics Education*, 1(1), 132–152. https://journalofmathed.scholasticahq.com/article/90517.pdf

Firdaus, F., Kailani, I., Bakar, M. N. B., & Bakry, B. (2015). Developing critical thinking skills of students in mathematics learning. [EduLearn]. *Journal of Education and Learning*, 9(3), 226–236.

Gallardo, K. (2021). The importance of assessment literacy: Formative and summative assessment instruments and techniques. *Workgroups eAssessment: Planning, Implementing and Analysing Frameworks*, 3-25. 10.1007/978-981-15-9908-8_1

Griffin, P., & Care, E. (Eds.). (2014). *Assessment and teaching of 21st-century skills: Methods and approach*. Springer.

Hamdi, F. M. (2013). The impact of globalization in the developing countries. *Developing Country Studies*, 3(11), 142–144.

Hassan, M. (2024, January 1). *Documentary analysis: A research method for analysing documents*. Scribbr. https://www.scribbr.com/methodology/documentary-analysis/

Iannone, P., & Miller, D. (2019). Guided notes for university mathematics and their impact on students' note-taking behaviour. *Educational Studies in Mathematics*, 101(3), 387–404. 10.1007/s10649-018-9872-x

Jones, I., & Alcock, L. (2014). Peer assessment without assessment criteria. *Studies in Higher Education*, 39(10), 1774–1787. 10.1080/03075079.2013.821974

Jones, I., Swan, M., & Pollitt, A. (2015). Assessing mathematical problem solving using comparative judgement. *International Journal of Science and Mathematics Education*, 13(1), 151–177. 10.1007/s10763-013-9497-6

Kibble, J. D. (2017). Best practices in summative assessment. *Advances in Physiology Education*, 41(1), 110–119. 10.1152/advan.00116.201628188198

Kim, H. (2019). Teacher learning opportunities provided by implementing formative assessment lessons: Becoming responsive to student mathematical thinking. *International Journal of Science and Mathematics Education*, 17(2), 341–363. 10.1007/s10763-017-9866-7

Kitchenham, B. (2004). *Procedures for Performing Systematic Reviews. Keele University Technical Report.* TR/SE-0401. http://artemisa.unicauca.edu.co/~ecaldon/docs/spi/kitchenham_2004.pdf

Kyaruzi, F., Strijbos, J.-W., Ufer, S., & Brown, G. T. (2019). Students' formative assessment perceptions, feedback use and mathematics performance in secondary schools in Tanzania. *Assessment in Education: Principles, Policy & Practice*, 26(3), 278–302. 10.1080/0969594X.2019.1593103

Leenknecht, M., Wijnia, L., Köhlen, M., Fryer, L., Rikers, R., & Loyens, S. (2020). Formative assessment as practice: The role of students' motivation. *Assessment & Evaluation in Higher Education*, 46(2), 236–255. 10.1080/02602938.2020.1765228

Leung, K.-C., Leung, F. K., & Zuo, H. (2014). A study of the alignment of learning targets and assessment to generic skills in the new senior secondary mathematics curriculum in Hong Kong. *Studies in Educational Evaluation*, 43, 115–132. 10.1016/j.stueduc.2014.09.002

Logan, T. (2015). The influence of test mode and visuospatial ability on mathematics assessment performance. *Mathematics Education Research Journal*, 27(4), 423–441. 10.1007/s13394-015-0143-1

Lusyana, E., & Wangge, M. (2016). Increasing Higher Order Thinking Skill to Build Students" s Character by Using Mathematical Reasoning. In *Proceeding of 3rd International Conference On Research, Implementation And Education Of Mathematics And Science* (pp. 119-126). IEEE.

Luz, Y., & Yerushalmy, M. (2019). Students' conceptions through the lens of a dynamic online geometry assessment platform. *The Journal of Mathematical Behavior*, 54, 100682. 10.1016/j.jmathb.2018.12.001

Maass, K., Geiger, V., Ariza, M. R., & Goos, M. (2019). The role of mathematics in interdisciplinary STEM education. *ZDM Mathematics Education*, 51(6), 869–884. 10.1007/s11858-019-01100-5

Mahlambi, S. B., van den Berg, G., & Mawela, A. S. (2022). Exploring the use of assessment for learning in the mathematics classroom. *Journal of Education (University of KwaZulu-Natal), 89*, 23-46. 10.17159/2520-9868/i89a02

Marinho, P., Leite, C., & Fernandes, P. (2017). Mathematics summative assessment practices in schools at opposite ends of performance rankings in Portugal. *Research in Mathematics Education*, 19(2), 184–198. 10.1080/14794802.2017.1318085

Mjenda, M., Mutarutinya, V., & Owiti, D. (2023). Investigating secondary school mathematics teachers' knowledge, preparation and use of assessment rubrics in teaching and learning 3D-geometry in Tanzania. *International Journal of Education and Practice*, 11(3), 557–571. 10.18488/61.v11i3.3434

Morrow, R. A., & Torres, C. A. (2013). The state, globalization, and educational policy. In *Globalization and education* (pp. 27–56). Routledge.

Mukuka, A., & Alex, J. K. (2024). Review of research on microteaching in mathematics teacher education: Promises and challenges. *Eurasia Journal of Mathematics, Science and Technology Education*, 20(1), em2381. Advance online publication. 10.29333/ejmste/13941

Mulenga-Hagane, M., Daka, H., Musango, H., Mwelwa, K., & Kakupa, P. (2019). Formative Assessment as a Means of Improving Learner Achievement: Lessons from Selected Primary Schools of Lusaka, Zambia. *Journal of Lexicography and Terminology, 3*(1), 33-54.

Murchan, D., & Oldham, E. (2017). Exploring the role of computer-based assessment in diagnosing children's mathematical errors in primary education in Ireland. *Irish Educational Studies*, 36(4), 489–510. 10.1080/03323315.2017.1393765

Mushin, I., Gardner, R., & Munro, J. M. (2013). Language matters in demonstrations of understanding in early years mathematics assessment. *Mathematics Education Research Journal*, 25(3), 415–433. 10.1007/s13394-013-0077-4

Mutodi, P., & Chigonga, B. (2016). Concept map as an assessment tool in secondary school mathematics: An analysis of teachers' perspectives. *Eurasia Journal of Mathematics, Science and Technology Education*, 12(10), 2685–2696. 10.12973/eurasia.2016.2301a

Nama, S., & Ayalon, M. (2023). Exploring change in secondary mathematics teachers' noticing of argumentation through experiencing peer-assessment strategies. *Journal of Mathematics Teacher Education*, 1–46. 10.1007/s10857-023-09586-3

Ndibalema, P. (2021). Online Assessment in the Era of Digital Natives in Higher Education Institutions. *International Journal of Technology in Education, 4*(3).10.1080/0969594X.2015.1108900

Ojose, B. (2023). Mathematics literacy: Are we able to put the mathematics we learn into everyday use? *Journal of Mathematics Education*, 4(1).

Olsher, S., Yerushalmy, M., & Chazan, D. (2016). How might the use of technology in formative assessment support changes in mathematics teaching? *For the Learning of Mathematics*, 36(3), 11–18. https://www.jstor.org/stable/44382716

Page, M. J., McKenzie, J. E., Bossuyt, P. M., Boutron, I., Hoffmann, T. C., Mulrow, C. D., Shamseer, L., Tetzlaff, J. M., Akl, E. A., Brennan, S. E., Chou, R., Glanville, J., Grimshaw, J. M., Hróbjartsson, A., Lalu, M. M., Li, T., Loder, E. W., Mayo-Wilson, E., McDonald, S., & Moher, D. (2021). The PRISMA 2020 statement: An updated guideline for reporting systematic reviews. *BMJ (Clinical Research Ed.)*, (71), n71. Advance online publication. 10.1136/bmj.n7133782057

Paine, L., Aydarova, E., & Syahril, I. (2017). Globalization and teacher education. *The SAGE handbook of research on teacher education, 2*, 1133-1148.

Papic, M. (2015). An early mathematical patterning assessment: Identifying young Australian indigenous children's patterning skills. *Mathematics Education Research Journal*, 27(4), 519–534. 10.1007/s13394-015-0149-8

Premkumar, K. (2016). Use of student response systems for summative assessments. *Creative Education*, 7(13), 1851–1860. 10.4236/ce.2016.713187

Rakoczy, K., Klieme, E., Leiß, D., & Blum, W. (2017). Formative assessment in mathematics instruction: Theoretical considerations and empirical results of the Co 2 CA project. In Leutner, D., Fleischer, J., Grünkorn, J., & Klieme, E. (Eds.), *Competence assessment in education: Research, models and instruments* (pp. 447–467). Springer. 10.1007/978-3-319-50030-0_26

Riyanto, B., Zulkardi, P., & Ratu Ilma Indra, D. (2019). Senior high school mathematics learning through mathematics modeling approach. *Journal on Mathematics Education*, 10(3), 425–444. https://eric.ed.gov/?id=EJ1229897. 10.22342/jme.10.3.8746.425-444

Robinson, M., Loch, B., & Croft, T. (2015). Student perceptions of screencast feedback on mathematics assessment. *International Journal of Research in Undergraduate Mathematics Education*, 1(3), 363–385. 10.1007/s40753-015-0018-6

Rubin, J., & Rajakaruna, M. (2015). Teaching and assessing higher order thinking in the mathematics classroom with clickers. *International Electronic Journal of Mathematics Education*, 10(1), 37–51. 10.29333/iejme/290

Savard, A., & Freiman, V. (2016). Investigating complexity to assess student learning from a robotics-based task. *Digital Experiences in Mathematics Education*, 2(2), 93–114. 10.1007/s40751-016-0016-6

Schoenfeld, A. H. (2015). Summative and formative assessments in mathematics supporting the goals of the common core standards. *Theory into Practice*, 54(3), 183–194. 10.1080/00405841.2015.1044346

Schroeder, L. B., & Dorn, B. (2016). Enabling and integrating online formative assessment in a flipped calculus course. *PRIMUS (Terre Haute, Ind.)*, 26(6), 585–602. 10.1080/10511970.2015.1050619

Selwyn, N. (2012). *Education in a digital world: Global perspectives on technology and education*. Routledge. 10.4324/9780203108178

Serow, P., Callingham, R., & Tout, D. (2016). Assessment of mathematics learning: What are we doing? In Makar, K., Dole, S., Visnovska, J., Goos, M., Bennison, A., & Fry, K. (Eds.), *Research in Mathematics Education in Australasia 2012-2015* (pp. 235–254). Springer. 10.1007/978-981-10-1419-2_12

Singh, V. K., Singh, P., Karmakar, M., Leta, J., & Mayr, P. (2021). The journal coverage of Web of Science, Scopus, and Dimensions: A comparative analysis. *Scientometrics*, 126(6), 5113–5142. 10.1007/s11192-021-03948-5

Soto, M., & Ambrose, R. (2016). Screencasts: Formative assessment for mathematical thinking. *Technology. Knowledge and Learning*, 21(2), 277–283. 10.1007/s10758-015-9272-6

Suurtamm, C., Thompson, D. R., Kim, R. Y., Moreno, L. D., Sayac, N., Silver, S. S. E., & Vos, U. S. P. (2016). Assessment in mathematics education. In Kaiser, G. (Ed.), *Assessment in Mathematics Education. ICME-13 Topical Surveys*. Springer. 10.1007/978-3-319-32394-7_1

Syaifuddin, M. (2020). Implementation of authentic assessment on mathematics teaching: Study on junior high school teachers. *European Journal of Educational Research*, 9(4), 1491–1502. 10.12973/eu-jer.9.4.1491

Szeibert, J., Muzsnay, A., Szabó, C., & Bereczky-Zámbó, C. G. (2023). A case study of using test-enhanced learning as a formative assessment in high school mathematics. *International Journal of Science and Mathematics Education*, 21(2), 623–643. 10.1007/s10763-022-10264-8

Tanudjaya, C. P., & Doorman, M. (2020). Examining Higher Order Thinking in Indonesian Lower Secondary Mathematics Classrooms. *Journal on Mathematics Education*, 11(2), 277–300. https://eric.ed.gov/?id=EJ1252003. 10.22342/jme.11.2.11000.277-300

Thanheiser, E. (2018). Brief report: The effects of preservice elementary school teachers' accurate self-assessments in the context of whole number. *Journal for Research in Mathematics Education*, 49(1), 39–56. 10.5951/jresematheduc.49.1.0039

Thomas, M., Clarke, D. M., McDonough, A., & Clarkson, P. C. (2023). Assessing students' understanding of time concepts in Years 3 and 4: Insights from the development and use of a one-to-one task-based interview. *Mathematics Education Research Journal*, 35(S1), 1–22. 10.1007/s13394-023-00451-3

Turney, S. (2023, November 20). *Systematic review: Definition, example and guide*. Scribbr. https://www.scribbr.com/methodology/systematic-review/

Ukobizaba, F., Nizeyimana, G., & Mukuka, A. (2021). Assessment strategies for enhancing students' mathematical problem-solving skills: A review of literature. *Eurasia Journal of Mathematics, Science and Technology Education*, 17(3), em1945. 10.29333/ejmste/9728

Urrutia, F., & Araya, R. (2022). Do written responses to open-ended questions on fourth-grade online formative assessments in mathematics help predict scores on end-of-year standardized tests? *Journal of Intelligence*, 10(4), 82. 10.3390/jintelligence1004008236278604

van den Berg, M., Bosker, R. J., & Suhre, C. J. (2018). Testing the effectiveness of classroom formative assessment in Dutch primary mathematics education. *School Effectiveness and School Improvement*, 29(3), 339–361. 10.1080/09243453.2017.1406376

Vankúš, P. (2021). Influence of game-based learning in mathematics education on students' affective domain: A systematic review. *Mathematics*, 9(9), 986. 10.3390/math9090986

Veldhuis, M., & van den Heuvel-Panhuizen, M. (2020). Supporting primary school teachers' classroom assessment in mathematics education: Effects on student achievement. *Mathematics Education Research Journal*, 32(3), 449–471. 10.1007/s13394-019-00270-5

Walick, C. M., & Burns, M. K. (2017). A Proposed Algebra Assessment for Use in a Problem-Analysis Framework. *Assessment for Effective Intervention*, 42(3), 150–159. 10.1177/1534508417690000

Weiland, I. S., Hudson, R. A., & Amador, J. M. (2014). Preservice formative assessment interviews: The development of competent questioning. *International Journal of Science and Mathematics Education*, 12(2), 329–352. 10.1007/s10763-013-9402-3

Yang, D.-C., & Lin, Y.-C. (2015). Assessing 10-to 11-year-old children's performance and misconceptions in number sense using a four-tier diagnostic test. *Educational Research*, 57(4), 368–388. 10.1080/00131881.2015.1085235

KEY TERMS AND DEFINITIONS

Assessment Methods: Ways of collecting data during the teaching and learning process.

Assessment Strategy: A plan for implementing a specific assessment method.

Formative Assessment: Assessment done to inform or improve the teaching and learning process. It is done before and during the teaching and learning process.

Generic Skills: The ability to demonstrate or do activities that require general skills including soft skills.

Globalised Era: A period in human history where knowledge and skills need to be transferable across borders.

Mathematical Skills: The ability to demonstrate or do activities that require the application of mathematical concepts

Mathematics Education: A discipline involving the teaching and learning of mathematics

Summative Assessment: Assessment done to determine the extent to which learning objectives have been achieved, commonly done at the end of an educational programme.

Section 3
Gender

Chapter 7
Globalisation Gender Sensitivity in the Teaching of Mathematics in Higher Education

William Chakabwata
https://orcid.org/0000-0002-4224-5239
University of South Africa, South Africa

ABSTRACT

The teaching of mathematics globally at university level, has created an environment that favours the performance of men over women. Men tend to perform better than women at university level. The gender gap has been explained by using biological and social cultural theories. It is also noted that at doctoral level the number of women who earn degrees is very minimal and the women graduates tend to be viewed as less competent when compared with men. The study employs poststructural feminism, socialcultural, and situated learning theories to explore the gender gap in mathematics. Gender stereotypes seem to have a major impact on performance of women in mathematics.

BACKGROUND

Research reflects that male students tend to surpass female students in performance in mathematics at university level. The number of females who engage in mathematics in higher education also declines as one moves up the academic spectrum. The chapter presented a background to the study, gender sensitivity, gender identity, globalization of mathematics among other headings. Theories that are

DOI: 10.4018/979-8-3693-2873-6.ch007

used to undergird this chapter are poststructural, socialcultural and situated learning theories. The chapter also presents a discussion, conclusion and recommendations. The study is based on a desk study. The inclusion criteria will be based on peer reviewed publications which included doctoral thesis, peer reviewed journal articles, conference papers on global praxis on the teaching of mathematics and gender studies and Google Scholar articles on globalization and teaching of mathematics. The exclusion criteria are articles that are not peer reviewed.

There is a growing global concern regarding non-participation of women in mathematics at advanced level and tertiary levels (Mandina et al.,2013;Lubienski and Pinheiro, 2020; Marcus and Joakim, 2016). Herzig (2010) noted that women's engagement in post -graduate mathematics still falls behind that of men. Statistics gleaned from the United States of America (USA), showed that women's resilience in mathematics, tended to decline as they moved up the academic spectrum, from primary to doctorate level. At doctoral level women tend to grapple with severe issues of retention, raising questions about their capacity to perform at the same level as men after graduation.

The contemporary society has to seek to meet the Sustainable Development Goals (SDGs), as enunciated by the United Nations (UN), which included the provision of quality education, SDG 4,and gender equality SDG 5. Global challenges have accentuated the demands for social justice in many arenas including in the teaching of mathematics. The instruction in mathematics is also assumed to be able to elevate the Global Citizenship Education among students (Carlana, 2019). The teaching of mathematics is anticipated to promote the competencies which include identifying the relationship between mathematics and social formations and also cultural plurality. In addition, students are expected to acquire skills to solve challenging societal issues and develop democratic values and tolerance. Mathematics curriculum is also expected to promote sustainable attitudes among students.

Explicatory work on disparity between men and women in studying mathematics has tended to focalize on diverse factors which included teacher and peer support, enhancing student attainment, classroom factors, student motivation, and level of student anxiety among many other factors. Mathematics is a prerequisite that opens opportunities for students in areas such as science, engineering and technology. Even in developed countries such as the United States (US), gender disparities which had narrowed over time are beginning to become noticeable in mathematics again. Gender differences have been identified among men and women who are located within the same learning environment, highlighting the need for educators and policy makers to engage with this critical issue (Organization for Economic Cooperation and Development (OECD, 2014). In spite of the statistics which reflects that in the US, more women were likely to attend college than men, the enrolment figures in mathematics showed that in 1997, the percentage of bachelor's degree earned by women stood at 46%. The figures of women who earned a doctoral degree in mathematics by 2017, was 29% and only 19% of bachelor's degree in computer science (National Centre for Science and Engineering Statistics, 2019).

OECDiLibrary (2024) noted that women tend to be in the minority in given areas at doctoral level, although they are in the majority at master's degree level. The number of women who graduated at doctoral level in 2018 constituted 54%, in natural sciences, mathematics and statistics and they epitomise only 46% of the doctoral graduates in OECD countries. A major decline in the number of women who earned doctoral degrees in the domain of natural science, statistics and mathematics was noted in Denmark, India, Saudi Arabia, Slovenia, Korea, Poland and Czech Repubic in 2018.

Lubienski and Pinheiro (2020) unequivocally asserted that progress towards gender equality in mathematics in the United States has stalled. Using USA as an example, it is apparent that gender imbalances in mathematics at K-12 is skewed in favour of boys who occupy a disproportionate position in mathematics performance, at the peak of the pyramid (Ganley and Lubienski, 2016; Cimpian et al., 2016).

In general, the pattern tends to assume a global proportion of males preponderances in performances in mathematics. The gap of male dominance is still visible in mathematics although, it appears minuscle in countries such as US and OECD. Major imbalances are visible beyond high school, up to bachelor, master and doctoral levels. Dickerson et al., (2015) noted that in 19 Africa countries, as in other countries gender gap in mathematics persist. The same study noted that the gender gap varied with the region in which the learners lived. Numerous studies have been undertaken in order to determine the attitudes of women towards mathematics as a factor that contribute to their marginalization and underrepresentation in the field. It is also further argued that attitudes develop over time, and that they are likely to impact negatively on learning (Sarmah and Puri, 2014; Syyeda, 2016). Attitude is associated with a tendency to view mathematics as either useful or not useful, or appropriate for one's gender or inappropriate (Kibrislioglu, 2015).

Numerous people habour sentiments that mathematics is a very sophisticated and demanding subject. Such misconceptions have a negative effect on the enrolment of females in mathematics including countries such as Nepal. The implication is that women tend to struggle with proficiency in mathematics not because it is unassailable, but due to the stigma that accompanies the subject particularly in higher education. Consequently, very few women would want to enrol for mathematics at university level (Bhandary, 2017; Chipman, 2005).

Ruhle (2022) averred that in South Africa the rate of attrition of boys or class repetition is higher compared to girls. In grade 5 and 9 girls performance supersede that of boys. In South Africa girls do better than boys in literacy assessments and and the gender gaps in mathematics has tended to become insignificant with time.

Mkhize (2022) averred that at doctoral level South African women are rendered nearly inconspicuous. Females are underepresented and marginalised, at doctoral level, at universities in South Africa, due to historical background. The same author opines that the field of mathematics in South Africa, remains largely racialised, gendered and politicised.Mkhize (2022) expounds that African women remains on the periphery in a field that is rooted in an ideology of racialism, gender, cultural and political motivation to inferiorise Africans.

In South Africa perceptions of mathematics being a preserve of the white race are still being upheld by some institutions. Gender gaps in mathematics in South Africa are attributed to colonial policies which has created a racial stratification in terms of mathematical performance.

GENDER SENSITIVITY

Vasanthan & Ancymol (2021) explains that gender sensitivity refers to the way people apprehend the role of sex, in line with the society that they are a part of. Ozumba (2005) averred that gender sensitivity refers to the level of consciousness and assessment that society has to sustain gender differences between men and women. Colineauty et al., (2022,p2) defined gender as the "observed,experienced, prescribed, or favored social differences based on sex attributed at the time of birth". The concept of gender transcends this binary classification, to hide numerous nuances that are multidimensional and that are experienced by people and which are ascribed by society. Gender results in the organisation of people into a hierarchy which places other categories at the bottom (Poudel, 2017). Esen (2013) averred that gender inequality do not emanate from biological disparities, but from the social and cultural environment, which are subject to transformation. It is added that one of the primary tools that may be used to transform society is gender education, which has a primal goal of raising awareness, knowledge and competence in the field of gender.

Esen (2013) identifies two strategies that are designed to transform gender stereotypes as training in gender sensitivity and gender awareness. Gender sensitivity is conceptualised as the ability to embrace the differences between genders, as well the challenges and issues that emerge due to these disparities. Esen (2013,p 2547) defined gender awareness as " the more critical and exacting understanding of the fact that differences between sexes-which affect individuals' skills for accessing and controlling resources and services –are dependent on acquired behaviour." Gender awareness entails the capacity to identify, even if they are not prominent, challenges that arise due to gender inequality and discrimination.Gender. Sensitivities in the teaching of mathematics alludes to the fact that teachers in the fields of mathematics, should be able to isolate attitudes and behaviours in the learning spaces that may be categorised as stereotypical and sexist. UNESCO (2004) indicated that the concept of gender sensitivity owes its origin to the efforts to eliminate obstacles to any individual's economic or social progress rooted in conceptions of gender roles. Developing gender awareness demands a mind that is open to inquiry of new ideas and that is also sensitive. Valez and Ochoas (2020) expounded that gender sensitivity integrates a cross-cultural perspective to increase knowledge on how inequalities transpires on the basis of gender.

Gender sensitivity may be viewed as a spectrum, ranging from gender negativity to gender transformative. Within this spectrum gender sensitivity is located in the middle between teachers who may feel that they are neutral on the subject, and those who feel like activists driven by a transformative mandate (Christodoulou, 2005; Zobnina, 2009).Gender sensitivity is a worldview that a teacher may hold to transform the way people view gender in the teaching of mathematics, and it involves a reflective and dialogical thinking that probes the premises of contemporary gender norms, roles and access to learning and resources (Valez & Ochoas, 2020). Gender sensitivity in mathematics encompass the activism that seeks to realise equal participation in the subject and fair distribution of resources, as a consequence of gender analysis and efforts to materialise gender inclusion. Gender sensitivity helps teachers of mathematics globally to appreciate how gender ascriptions have benefitted men more than women. Patriacrchy has tended to uphold gender inequality by making it difficult to unmask its operation from the household to the institution level (Valez and Ochoas, 2020).

Oakley (1972) defined sex as biological, while gender was perceived as a social construct. Although the conception of gender has remained largely binary, multiple studies today have emerged that are non -binary and based on poststructural theory which acknowledge multiplicity of sexes (Hall and Noren, 2021). Calls to trascend the binary divide and to include the transgender and queer has gone largely unheeded (Hall and Noren, 2021). Ovie-Ekpewu, (2011) conducted a study to determine the sensitivity of science and mathematics teachers in the learning of mathematics with a sample constituting 51 participants, and concluded that the sensitivity of both male and female teachers was low, and girl learners were not securing the encouragement to study science. Nevertheless the sensitivity of female teachers was observed to transcend that of male teachers.

Henningsen (2008) noted that mathematics education, learning environment and didactics, lacks sensitivity, and inadvertently militate against efforts to increase female participation in the field. In order to be sensitive mathematics has to take into account the different ways in which males and females approach learning in the subject, for instance females in general prefer to function in groups, as opposed to class teaching.Lack of sensitivity to gender is also perceived in the area of assessment where female students in general, tend to prefer project based assessments, in contrast to the traditional form of assessments. Even in the context of using technology, it is noted that male students prefer an experiential approach, while female students prefer a systematic approach.

Global gender differences in the learning of mathematics have been attributed to personal, family and biological factors in the teaching of mathematics in Nepal. Mathematics as a subject is viewed as a domain were boys exercise hegemony. Even at the level of secondary school in Nepal, the average score for boys was higher than that of girls, reflecting the gendered nature of the subject (Hall et al., 2019). Boys tend to receive affirmation to study and excel in mathematics, while girls generally lack such form of support and validation. There is empirical evidence that gender bias continues to proscribe the participation of females in the mathematics (Hall et al., 2019).

Lubienski and Pinheiro (2020) averred that gender gap tended to be wider in schools that are designated as high socio-economic schools than those than are termed as low socio- economic status. Parental influences are viewed as more important for the subject a child can undertake in primary school. It was noted that parents of learners in primary school, who attended high socio-economic status school, encouraged them to engage in parent orchestrated activities that are traditional and gendered, such as dance lessons for girls and sporting events for boys. Based on this observation, it is therefore alluded that parents may engage their children in more gendered activities that impacts on their choice of subjects.

A study on PhD women who enrolled in mathematics displayed that woman in mathematics, held sentiments that teaching mathematics was much more intriguing for them than engaging in research, and their withdrawal from the field of mathematics research, was chiefly due to family chores and duties, and they had a proclivity to be less confident in mathematics than men (Lubienski & Pinheiro, 2020). Most women would not hesitate to sacrifice their career in mathematics in order to focalize on their family.

Herzig (2010) postulated that it is important for women to experience a sense of belonging in the field of mathematics at university level. Lack of the nurturing of the sentiments that female students belong to a mathematical department and are a component of that research and learning community militates against their resilience even at doctoral level. The females students face impediments such as severe competition in mathematical department that are totaly devoid of a supporting culture and also insensive didactics, that do not meet the needs of students or inspire a passion for mathematics. Additional obstacles that female students have to confront included limited and unfriendly attitudes of supervisors and lack of support from faculty.

Sexist practices in mathematics have also been reported at university level, with some academic staff, blatantly denigrating the intelligence of women in mathematics and displaying unwanted sexual advances. Females who participant in mathematics at university level have articulated the importance of attending a class with a critical number of students of their gender in class in order to counter the sense of alienation. The number of women in mathematics declines as one moves up the academic hierarchy, with very few women attaining the doctoral level depriving female undergraduate the role models that they desperately need to inspire them. Some females are also socialised to value validation and reinforcement, traits that can work against them in an extremely competitive spaces such as mathematics. This desire for belonging can even reinforce the stereotype that they lack mathematical aptitude. All the forces that are explored above tend to intersect and buttress the sentiments that women are not welcome in mathematical spaces at university level, which are male -dominated.

Nollenberger et al.,(2014) expounded that the math gender gap has been postulated to be an outcome of either environmental or biological factors. James (2007) enumerated the biological differences as:

- Women tends to display sharp hearing ability when compared with males;
- Men display sharp vision when compared to girls;
- Women show an uncanny ability to read non-verbal signals;
- Most men would like to learn by doing, while women learn from observing;
- Women are less likely to experience a dysfunction termed dyscalculia;
- Men are likely to recall what they see, while women are likely to recall words and places where they placed objects;
- Women may display perceptual skills;

Dickerson et al., (2015) expounded that there is insufficient empirical data to question the genetic explication of gender differences. The same author submits that gender gaps tend to differ broadly among individuals who have different characteristics both within communities and within nations and the biological difference becomes insignificant in some section of the population. As a corollary, biological theories must be able to explain these differences by providing a resolution of the nature versus nurture discourse as opposed to adopting a monocausal explication. It is for that reason that the societal explanation tends to be more convincing, because it is apparent performance in maths tends to vary according to gender and that boys display higher test scores. It is also attested that parents may send a boy child to a school that offers high standards of education, while sending a girl child to an average school, and teachers may practice discriminatory practices in the allocation of school resources.

Mejia-Rodriguez et al.(2021) underscores the role of the parent in influencing subject choice through socialisation. Parents have a major effect on the child's self-concept and through socialisation impacts a child values, beliefs and behaviour. The same author submit that focalising on gender gap at university level is a bit too late, as the differences begin to emerge in primary school. Self- concept explores an individual's perception of their performance in mathematics. Learner's academic self- concept is instituted early in childhood (Herbert & Stipek, 2005;). Higher or lower levels of self- concept are a function of interactions of a child with parents and teachers. Gender differences in parental self –concept may be accounted for by stereotypical gender views that perceive mathematics as a male domain. Gender streotypes can be conveyed by individuals that are more important in the life of the child such as a parent or an educator.Consequently, the school and the home exercises primacy in transmitting gender stereotypes (Lee and Anderson, 2015).

Children who are actively engaged in mathematic affiliated games would display a propensity to like mathematics. The nature of the game and the frequency of exposure is also paramount as research also demonstrates that some parents tend to encourage boys to participant in numeracy related games, while girls are urged to undertake literacy activities. Stereotypical activities may have the effect of underming the self-concept of girls in mathematics (Maloney et al., 2015). United Nations Scientific and Cultural Organisation (UNESCO,2016) encapsulates the constrictions that female students confront in learning mathematics and STEM as:

- Sociocultural variables and expectations that young people have of securing jobs particularly in areas that are considered appropriate for women;
- Training of teachers in STEM that is not gender and culturally responsive;
- Teaching and learning resources that are upholds gender inequality;
- Women tend to receive less support from parents, guardians and teachers to study mathematics;
- The reduced number of women in STEM translate to a limited number of female role models in STEM;
- .Gender stereoptypes culminates in women being predominant in some areas as opposed to others;

Some studies have espoused the following reasons to account for the gender gap in mathematics:

- Men display a tendency to learn from exploring concepts for example testing gravity playfully in an elevator by jumping up and down. Women would perceive such activities as disruptive;

- It is also submitted men during childhood tend to be more robustious than women and physical;
- Men also displayed a tendency to be prone to engage in jokes in the learning space and a fascination with science fiction when compared to women;

Leder (2015) echoes that in Africa the impediments to the progress of women in mathematics are early marriages practices, paucity of financial resources at family level, obstacles that girls face at institution level, and dismal performance performance. Students in Africa face households chores, curriculum that lack gender sensitivity, inadequate school facilities, dearth of skills among the mathematics teachers, lack of encouragement from parents or gurdians and lack of confidence among girls.

It is important to note that there are no fixed identities of women and males and that identities change over time (John-Steiner, 2014). Human beings are constituted and also constitute themselves through language. Carlana (2019) averred that the gender gap between males and females in standardised tests, expand when students are allocated to teachers who evince strong gender streotypes. The same study observed that students tend to underperform when they take standardised tests administered by gender biased teachers. It is concluded that gender streotypes tend to undermine the performance of girls. The next section explores gender and identity in order to illuminate the discussion on gender sensitivity and teaching of mathematics.

GENDER AND IDENTITY

Gender is a vital component of gender identity (Chakabwata, 2023;Cosker-Rowland, 2023). Baams & Kaufman (2023) observed that from the 1980s, to the present there is a growing consciousness among researchers of the polyvalent and multiplicity of sexual orientations. A study that was conducted in England reflected that between 2009 and 2017, there was a decline in the number of adults who identified as heterosexual and a growth in the number of people who declared to be gay/lesbian or bisexual (MacCarthy et al, 2020; Jenkins, 2018).

An individual gender identity may vary from their gender expression. Gender in academic circles is being recognised as a continuum. Some studies indicate that gender is multidimensional and fluid (Bishop et al,2020). Luis (2017) noted that contemporary views of mathematics as a gendered space can be expanded by the intersectional lens that examines ways in which gender interplays with other forms of identities to create interlocking disadvantages in the learning of mathematics (Chakabwata, 2023; Lykke, 2010; Salem, 2018). Intersectionality is a lens that rejects a monocausal explanation of phenomenon (Crenshaw, 1991). Women of colour have to contend with additional adversarial forces which included constricted access to university, lack of access to informal social networks that can support learning, being treated as invisible in learning spaces such as laboratories and science department and also challenges of securing mentors and seeing their contribution in learning spaces trivialised.

GLOBALISATION OF MATHEMATICAL LEARNING

Skovsmose (2008) explicated globalisation as the spread of neoliberalism globally and the growth in the interconnected world through technology. In the context of mathematics, it is indicated that globalisation, has created a curriculum for mathematics that is almost uniform for all countries. The similarities were identified globally, in areas which include curriculum design, nature of content, organisation of content, and the theories shaping didactic practices.

Mathematics praxis has tended to be stable over multiple years with a few changes in one nation leading to changes throughout the world. Reforms advocated in Britain and the United States in the 1960s in the area of assessments and standards, were soon spread across the world. Globally mathematics education is given pre-eminence which may only be equal to language education. The global importance of mathematics was evidenced by the declaration of the year 2000 as the international year of mathematics.

Downey et al., (2018) asserted that universities have always manifested a global nature as they permit the movements of ideas and knowledge across borders. Movement of people from one state to a university in another state is common due to globalisation. The exponential movement of people from one state to another and the swift changes in communication and technology facilitated by the internet means that the higher education has become globalised. Robitaille and Tavers (1992) observed that mathematics is one of the mostly globalised subjects in the world determined by the research and conferences held annually. Skovsmose (2007) expound that globalisation is subsumed by the following points:

- Functions of globalisation are made possibly by technology;
- Globalisation is intimately connected to capitalism and the opening up of new markets;
- The processes of globalisation do not developed in a linear way;
- Globalisation is related to the distribution of goods and services;
- The expansion of capitalism is related to the growth of poverty;
- It is also linked to the development of armed societies such as the expansion of the military.

Globalisation is linked to neoliberalism. It is in that context that a person is capable of thinking globally in terms of social justice such as marginalisation of women in the teaching of mathematics. Mathematics and sciences are the two of the highly globalised subjects, due to the opportunity they present in terms of employment. Mathematics forms a nexus with power, which has a political, technological and economic ramifications (Skovsmose, 2007). Globalisation has led to the internationalisation and universalisation of mathematic education. Atweh, et al., (2077) explains the preceding 100 years has demonstrated a tremendous interactions and collaboration among academics globally at conferences and also through research projects. These engagements also included sharing of curricular resources, taking part in common professional development programmes, and undertaking of numerous cross- comparative studies. All these interactions are transpiring in the ambit of globalisation, which has tended to exercise hegemony over personal, intellectual and social contexts. These two subjects tends to be premised on the contested basis of objectivity and neutrality. These assumptions are extended to teacher professional development, curriculum development and innovations, on-going professional development and research work.

CONTEMPORARY SITUATION AND BEST PRACTICES

Mejias et al., (2021) noted that recently the gap between male and female students in the field of mathematics has started to close or even shift in some cases in favour of females. However in numerous countries, performance of men, continue to surpass that of women. Research indicated that gender gaps continue after high school for instance in the United States, men typically perform better than women on the Scholarstic Aptitude Test (SAT), a standardised test employed to determine college registration over the last 50 years (Chubbuck et al, 2016). The same trend was observed in other countries, for instance in Sweden, Scholastic Aptitude Test, men consistently outperform women by a third of the standard deviation. In Turkey, men surpass women in all subjects assessed. In addition among the 540,000 graduates from more than 200 countries who wrote the Graduate Record Examination (GRE) between 2013 and 2018, men surpassed women graduates on the quantitative section of the test. Even in OECD countries gender gaps which favour men persist (Organisation for Economic Co-operation and Development, 2016; Educational Testing Services, 2018; Saygin, 2020; Graetz and Karimi, 2019).

POSTSTRUCTURAL FEMINIST THEORY

In this chapter a poststructural theory is employed in order to explicate the limited enrolment of women in Mathematics globally. Damla (2022) postulated that post structural theory eschews efforts to discuss gender from a lens that is reductive and generalised theory as encapsulated in earler theories such as radical and liberal feminism. Angermuller (2020) explicates that poststructural feminism is located at intersection of disciplines such as aesthetics, politics and social sciences. It is a frame that is linked to indisciplinary studies that is opposed to binary lens,and is dialogical and post –humanistic. With its emphasis on the chiasm between language and power, poststructuralism has been employed in many disciplines including in discourse analysis and discourse theory.

Poststructural feminism focuses on discourse that is located in a particular historical context. In addition, the theory inquires into the way gender is constituted and how people construct their own sexual identities. Angermuller (2020) averred that poststructural feminist theory uses language and is averse to binary conceptualisation of issues on agency and identity. In particular poststructuralism seeks to probe how discourse is made or constituted or unestablished in discursive situations. Poststructural theory perceives discourse as a practice that exemplify the social, but also helps to manifest it.

Poststructural theory averred that there are no static and rigid forms of identities (John-Steiner, 2014). Individuals evolve in the context of multiple relationships, customs, chores, transitioning within and across time and different path our lives assume. People find themselves situated in a social order that is unjust and where their identities are constantly changing. People's identities or positionalities embrace variables which include race, sexual orientation, age and work. Hall and Noren, (2021) postulated that multiple researchers still conceptualise gender in a binary way as being constituted of men and women only. Contemporary research is shifting the focus from being binary to embrace other diverse forms of sexuality.

SOCIALCULTURAL THEORY

Socialcultural theory is employed in this study to illustrate how young people develop cognitively by engaging with significant people in their lives. Sociocultural theory postulated by Lev Vygotsky asserted that cognitive thinking takes place in a social cultural context. Vygotsky distinguished between development and learning and he explicated development as constituting internal individual processes and learning as external process. Vygotsky (1978) opined that learning and development constitute two mutual factors acting simultaneously. Interaction with the minds of others is paramount for cognitive development. Parents and educators interact with children and shape their perceptions of gender stereotypes in a way which is either negative or positive. Wass (2012) opinned that the socialcultural theory foregrounds the development of higher level cognitive thinking in terms of activity that occurs in a social and cultural context. The Zone of Proximal Development is viewed as one of the landmark concepts postulated by Vygotsky.

ZPD which represents an area between what students are able to do on their own and the potential that they have to acquire knew knowledge with the assistance of a competent adult (Woods et aal., 2020). Vygotsky (1978) postulated that all learning is social in nature and has an effect on the way people think. In Vygotsky perspective the development of individuals occurs due to the nexus of individual,cultural-historical and social variables. John-Steiner (2014) opined that the asymmetry that is noted in men and women students may be traced to socialisation during early stages of development, the way in which socialisation displays unequal power relationships that are regular and based on customs, use of language to negotiate meanings and to create discourses and how the self is mutually constituted at the level of self and also community.

SITUATED LEARNING

The term situated learning has its origin in the work of Lave and Wenger (Lave,2008). Lave and Wenger are also associated with the term community of practice, whose meaning is highly contested, and has been explicated to refer to a conceptualization of learning as a dynamic and social activity, which transpires as individual human beings are always taking part in a variety of enterprises which entails interacting with other people (Lave & Wenger,1991). However, students gingerly move from the community of practice, to participate in more complex and vigorous activities. The process can transpire in an imperceptible way (Besar, 2018). The situated learning theory is based on the premise that learning that occurs and is implanted in a real context is different from learning that is non-situated. Lave & Wenger (1991)

Lave & Wenger (1991) argued that learning transpires in a community which ensure that members are also acculturated and simultaneously participant in the spreading and proliferation of the ideas or artefacts. Being knowledgeable entails being able to take part in the interlocking networks of relationships in a community. In the case of this chapter, situated learning accounts for the way gender stereotypes may be propagated in communities, at family or school level. Students may be impacted by stereotypical views regarding mathematics as a masculine subject and begin to disseminate them without probing the implicit bias. This is even more applicable to this study, particularly the view that learning is a social activity. Gender values are transmitted in a cultural context where values are framed in a community setting (Abdel-Hafez Mahoud, 2021). Students fashion their own understanding through interactions with other learners.

The theories of situated learning have been postulated to account for female performance in mathematics by virtue of the way they emphasize learning as occurring in a social context (Herzig, 2010). Situated learning is also premised on the understanding that gender doesn't exist in the learners' mind but is an outcome of the function between the mind and world. An individual develops conceptual knowledge when they engages with the world and then construct meaning (Abdel-Hafez Mahoud, 2021). Knowledge is impacted by what a person does, the context and culture where it is employed. This means that construction of meaning occurs as a person engages actively in a situation and then mental models are developed leading to generalization.

Many scholars have probed the issue of student sense of belonging in mathematical spaces even at K-12. It is further added that female students experience a sense of exclusion, which if it is superseded, then the goal of promoting access to mathematics for all gender would have been attained (Herzig, 2010). Most universities have focalized on ensuring that all students experience a sense of belonging in mathematics. The same argument has been extruded for doctoral students in mathematics that their resilience depends on the degree to which they are integrated into the community of learning mathematics at university level.

GLOBAL TRENDS ON GENDER EQUALITY IN MATHEMATICS

It is critical that women be empowered in order to actively participant in mathematics across the school curriculum from the primary to university level. Empowerment leads to more effective decision -making and autonomy for women in terms of academic and career choices, thus promoting equality of opportunity (Poudel, 2017). In order to address the obstacles that women face in mathematics a number of interventions were proposed such as the setting up of a ministerial special committee towards Education for All (EFA). African Union (AU) has adopted gender mainstreaming as an appropriate response to gender inequality. AU embarked on a gamut of responses to enhance the participation of girls in Science, Mathematics, Engineering and Technology (STEM). Some of these initiatives include:

- Special Project on Scientific, Technical and Vocational Education of Girls in Africa UNESCO medium-term strategy (1991-2001)

- Science, Technology and Engineering Education in the UNESCO medium -term 2008-2013;
- International Conference on Gender Science and Technology held in Beijing;
- Girls STEM initiative held in Abuja;
- New Partnership for Africa's development sponsored by the World Bank.

In spite of all these efforts, mathematics as well as other science fields are still very proscribed areas for women. The constricted nature of female participation in mathematics globally, is perceived to be a function of recalcitrant social-cultural obstacles, dearth of well articulated policy frameworks to improve the enrolment of female students, absence of program evaluations and meagre capacity to conduct gender analysis (Leder, 2015).

Kavatsyuk et al.,(2021) postulated that a study was conducted at University College Groningen in order to determine reasons for underrepresentation of females in STEM. The study led to the implementation of evidence- based practices to build the identity of female students in STEM. In the same study students also indicated the significance of creating a safe learning environment in which students can engage in learning with no trepidation of being harassed. Fernandez et al., (2023) postulated that although women may outnumber men in many other programmes they tend to be outnumbered in STEM. Global trends in mathematics reflects the existence of a persistent gender disparity in the enrolment of mathematics in different countries (Arroyo-Barrignete et al.,2023). Arroyo-Barrignete et al., (2023) postulated that in Spain in primary and secondary schools the gap is still defying efforts to close it in mathematics at all levels. The gender gap has implications in terms of opportunities for females in the labour market in the area of STEM.

DIGITAL TECHNOLOGY IN THE TEACHING OF MATHEMATICS

Roth & Ames (2014) indicated that there is a general perception that the use of digital technology had a positive effect in the learning of Mathematics.However the research on the teaching of mathematics, has not been synthesised in a meaningful way. A number of studies have declared that the use of technology in the teaching of mathematics had a positive effect on the subject (Cheung & Slavin, 2011 (Li & Ma, 2010).

United Nations Development programme (2021) noted that digitalisation is not gender neutral. Gender stereotypes also among other factors continue to impeded women's progress in mathematics education. The same source submittted that societal norms, lack of role models and lack of support from families impede women's access to technology.

DISCUSSIONS

Sevilla & Cue-Ruiz, (2024) noted that underperformance by females in comparison fo their male counterpart has an effect on the labour market outcomes. In order to address the gender in mathematics at a global level, it is imperative to probe stereotypes and how they impact performances in mathematics and on labour choices. It is noticeble that in spite of the progress women have made over more than seventy years, there still remains marginalised in mathematics as well as in STEM (Cavaglia et aal.,2020). This gap in mathematics translate to labour market outcomes, where women are more prone to select courses that will make them earn less than 6% of what male graduates earn and also 10% of the highest earners (Bertrand, 2018).

Empirical data collected from several nations shows that the gender gap in mathematics in primary school seem almost inconsequential in primary school, but is prone to triple between the ages of 9-10 and 15 and 16. This gap has been characterised as equivalent to girls being absent from school for four month of learning (Bharadwaj et aal.,2016). Bertrand (2018) submitted that there is no biologiocal difference between men and women when it comes to performance in mathematics. Even when sex differences are taken into account such as physical aggression and sensation seeking, and spatial capabilities it is becoming very implausible to justify differences as being attributed to biology only. Evidence emerging from the field of neuroscience indicates that the brain is able to change and modify according to the social environment (Rippon, 2019).

Gender gaps in Mathematics have been sustained more by stereotypes rather than science. These stereotypes have been extruded as girls are inherently incapable of engaging in mathematics than boys, and girls are less likely to engage in risk behaviour than boys. These stereotypes have been employed to offer a plausible explanation of the gender gap as opposed to biology driven explanations.

Carlana (2019) opined that stereotypes have an effect of constricting the choices of female students, even if they do not ascribe to them. The same author noted that in one study which measured Gender- Science Implicit Test Association among educators noted that teachers who have prejudice towards women engaging in mathematics tended to advice girls not to take up mathematics. The same study noted that girls who went through instruction under such maths teachers tended to display less confidence in their efficacy in mathematics. This process transpires under the conditions where parents or teachers assume that they have adequate knowledge about their children or students (Dossie et aal,. 2021).

Grils may receive less input in mathematics from teachers or parents when they insist in engaging in subjects that are perceived as male dominated such as mathematics or girls imbibe the gender stereotype, which culminates with them electing subjects that are perceived as appropriate for women. Some researchers have also alluded to the fact that women detests extremely competitive environments and that if the assessments context is made less competitive then they tend to excel better than men.

Women have not only been absent in mathematics across the learning spectrum, but also in STEM in general (Fernandez et al., 2023). Parmaxi, et al., (2023) opined that even at the workplace women are discriminated against for being mothers in favour of men or women who do not have a child.

RECOMMENDATIONS

Given the pervasive nature of gender stereotyping that exist in education, it is important to scrutinize policies that shape the learning landscape for females in higher education. One way is to engage in interventions that assist in the identification of stereotypes in order to inhibit them from being a component of decision -making. One vital way of realizing that goal is to employ the use of empirical tool which include the implicit association test (IAT), a computer software that is designed to detect implicit bias. The utilization of IAT to detect educators' implicit bias, can help to a great extent to expose implicit biases. Higher education institutions can also engage in on-going professional development activities that are designed to help their staff identify implicit biases (Alesina et al., 2018).

Some research explains the gender difference by appealing to the nature of the assessment environment which appears to be too competitive (Niederle & Vesterlund, 2010). In that case, it is vital to ensure that test taking environments take into account the tendency of female students to withdraw from situations where they have to engage in extreme competition. The manner in which tests are orchestrated, and marks are given in academic context makes females to fail. Policy mediation to tackle gender gap in mathematics at university level has to start early in primary schools, where ensuring that boys and girls are made immune to gender stereotypes and that there are multiple role models for women students in the field of mathematics.

Some research has also supported the instituting of women only mathematic classes. Empirical evidence supports the view that when women study in an all women class at university they are unlikely to suffer attrition (Booth et al., 2018). Single sex classes have been employed in some states to help improve female learning outcomes. Some institutions have recommended the introduction of women mentoring programs that are designed to enhance their performance in given courses such as mathematics. Other efforts that have been mooted in schools and colleges are programs that are designed to develop resilience and mettle in women students. This can be done early in school where children are exposed to the part effort plays in attaining goals and also the importance of perseverance, the gender gap will decline.

It is also becoming apparent that stereotypes are vulnerable to changes and it may be even fruitful to investigate the conditions and factors that help facilitate that change.

John-Steiner (2014) adduces that some of the ways that have been advanced to ameliorate the gender gap in mathematics include crafting a new paradigm for women socialisation, distribution of domestic chores, and the eradication of sexist techniques in the teaching of mathematics. Kavatsyuk et al.,(2021) observed that some of the interventions that may be invaluable in enhancing gender equality in mathematics, included creating safe learning environment where students are free to explore and ask questions, build confidence of women students that impede them from taking mathematics due to fear of failing, tranforming students' and educators negative beliefs regarding the learning of mathematics and ensuring that students' learning in mathematics is directly related to the authentic experience since women students are fascinated with the social dimension of learning. Other measures that can enhance the learning experience of students in mathematics incorporate the use of role models, use of peer mentoring, redesigning the curriculum and a deliberate plan to eliminate biases from the learning resources.

CONCLUSION

In this chapter it was noted that there are global gender differences in mathematics at university at all levels. The study noted that masculine perspective has a dominant effect on the number of women who earn doctoral programs. In some countries the enrollment of women tends to be higher in the first-degree level and then they decline by the time they reach doctoral level. Other explications have attributed the gender gap to biological factors. The sociocultural, the poststructural feminist and situated learning theories were employed to illuminate the chapter. It was also noted that a host of interventions have to be employed in order to address the gender gap in mathematics which include probing the way boys and girls are socialized and also addressing college related factors that shape the experiences of females in mathematics.

REFERENCES

Abdel-Hafez Mahoud, H. (2021). Using a situated learning-based strategy to develop some critical thinking and enhance English Language Learning Interests at Preparatory Stage Pupils. *Journal of Education*, 87(87), 1–48. 10.21608/edusohag.2021.174506

Abraham, A., & Dingel, M. (2021). Exploring the relationship among mathematics attitude, gender, and achievement of undergraduate Health Science Students. *Mathematics teaching Research Journal, 13.*

Alesina, A., Carlana, M., La Ferrara, E., & Pinotti, P. (2018). *Revealing streotypes: Evidence from immigrants in schools.* National Bureau of Economics Research. 10.3386/w25333

Angermuller, J. (2020). *Poststructural discourse studies: From Structure to practice. The Cambridge Handbook of Discourse Studies.* Routledge.

Arroyo-Barrignete, J., Carabias-Lopez, S., Borras-Pala, F., & Martin-Anton, G. (2023). *Gender differences in Mathematics Achievement: The case of a business school in Spain. Sage Open, 1-14.*

Baams, L., & Kaufman, T. (2023). Sexual orientation and gender identity/expression in adolescent research: Two decades in review. *Journal of Sex Research*, 60(7), 1004–1019. 10.1080/00224499.2023.221924537307300

Bertrand, M. (2018). Course lecture-The glass ceiling. *Economica*, 85(338), 205–338. 10.1111/ecca.12264

Besar, P. (2018). The situated learning theory: The key to effective classroom teaching? *International Journal for Educational, Social, Political &. Cultural Studies*, 1(1), 49–60.

Bhandary, S. (2017). *Understanding some cultural barriers to women's access to education: A study in rural Nepal* [Madter's thesis, Norway: MF Norwegian School of Theology].

Bharadwaj, P., De Giorgi, G., Hansen, D., & Neilson, C. (2016). The gender gap in mathematics: Evidence from Chile. *Economic Development and Cultural Change*, 65(1), 1–166. 10.1086/687983

Bishop, M., Fish, J. N., Hammack, P., & Russell, S. (2020). Sexual identity development milestones in three generations of sexual minority people: A national probability sample. *Developmental Psychology*, 56(11), 2177–2193. 10.1037/dev000110532833471

Booth, A., Cardona-Sasa, L., & Nolen, P. (2018). Do single sex classes affect academic achievement? An experiment in a coeducational university. *Journal of Public Economics*, 168, 109–126. 10.1016/j.jpubeco.2018.08.016

Carlana, M. (2019). Implicit streotypes: Evidence from teachers' gender bias. *The Quarterly Journal of Economics*, 134(3), 1163–1224. 10.1093/qje/qjz008

Cavaglia, C., Mavhin, S., McNally, S., & Ruiz-Valenzuela, J. (2020). Gender, achievement, and achievement, and subjet choice in English education. *Oxford Review of Economic Policy*, 36(4), 816–835. 10.1093/oxrep/graa050

Chakabwata, W. (2023). *Intersections of gender, identity and subject choice in shaping student learning experience at a technical teacher training College in Zimbabwe.* [Thesis, University of South Africa].

Cheung, A., & Slavin, R. (2011). *The effectiveness of education technology for enhancing reading achievement: A meta-analysis.* Best Evidence. https://www.bestevidence.org/reading/tech/tech.html

Chipman, S. F. (2005). Research on women and mathematical issue. In Gallagher, A., & Kaufman, J. (Eds.), *Gender differences in mathematics: An intergrative psychological approach* (pp. 1–24). Cambridge University Press.

Christodoulou, J. (2005). *Glossary of gender-related terms.* Mediterranean Institute of Gender Studies.: European Institute for Gender Equality. https://medinstgenderstudies.org/wp-content/uploads/Gender-Glossaryupdated_final.pdf

Chubbuck, K., Curley, W., & King, T. (2016). Who's on first? Gender differences in performance on the SAT test on critical reading intems with sports and science content. *ETS Research Report Series*, 2(2), 1–116. 10.1002/ets2.12109

Cimpian, J. R., Luienski, S., Timmer, J., Makowski, M., & Miller, E. (2016). Have gender gaps in math closed? Achievement, teacher perceptions, and learnining behaviours across two ECLS-K chorts. *AERA Open*, 2(4), 1–19. 10.1177/2332858416673617

Colineauty, H., Soulier, A., Lepage, B., & Kelly-Irving, M. (2022). Considering sex and gender in Epidemiology: a challenge beyond terminology . From conceptual analysis to methodological strategies. *Biol Sex Differ. 13*(23), 1-10.

Cosker-Rowland, R. (2023). Recent Work on gender identity and gender. *Analysis*, 83(4), 801–820. 10.1093/analys/anad027

Crenshaw, K. (1991). Mapping the margins: Intersectionality politics, and violence against women of colour. *Stanford Law Review*, 43(6), 1241–1299. 10.2307/1229039

Damla, I. (2022). A poststructural feminist approach to the acqusition of gender roles in early childhood. *İstanbul Üniversitesi Kadın Araştırmaları Dergisi, 25,* 141-154.

Dickerson, A., McIntosh, S., & Valente, C. (2015). Do the mathematics: An analysis of the gender gap in mathematics in Africa. *Economics of Education Review*, 46, 1–22. 10.1016/j.econedurev.2015.02.005

Dossie, G. (2021). fIGLIO, D., Giuliano, P., & Sapienza, P. (2021). The family origin of the math gender gap is a white affluent phenomenon. *AEA Papers and Proceedings. American Economic Association*, 111(May), 179–183. 10.1257/pandp.20211124

Downey, G., Gray, T., Hall, T., & Singh, M. (2018). Developing global perspectives: Responding to the state of international education in Australian Universities. In T. Hall, T. Gray, Downey, G, & M. Singh's, *The Globalisation of Higher Education Developing Internationalised Education Research and Practice* (pp. 3-20). Western Sydney University: Palgrave/ MacMillan.

Educational Testing Services. (2018). *A snap of the individuals who took the GRE general test 2013-2018.*

Esen, Y. (2013). Making room for gender sensitivity in pre-service teacher education. *Evropejskij Issledovatel*, 61(10), 2544–2554. 10.13187/er.2013.61.2544

Fernandez, D., White, S., Smith, H., Conner, P., & Ryan, M. (2023)). Gender inequality in science, technology, and mathematics, gendered time disparities in perceived and actual time spent in perceived and actual in practical laboratory based activities . *Front. Educ. 8*, 1-10.

Ganley, C., & Lubienski, S. T. (2016). Mathematics confidence, interest and performance: Examining gender patterns and reciprocal relations. *Learning and Individual Differences*, 47, 182–193. 10.1016/j.lindif.2016.01.002

Graetz, G., & Karimi, A. (2019). *Examining gender gap variation across Forms* (Working Paper 8). Institute for Evaluation of Labour Market and Education Policy (IFAU).

Hall, J., & Noren, E. (2021). Innovations in "gender issues" research in mathematics education. *Mathematics Education Research Journal*, 33(4), 787–791. 10.1007/s13394-021-00404-8

Hall, J., & Noren, E. (2021). Innovations in "gender issues" research in mathematics education. *Mathematics Education Research Journal*, 33(4), 787–791. 10.1007/s13394-021-00404-8

Hall, J., Robinson, T., Flegg, J., & Wilkinson, J. (2019). Students' gendered experiences in undergraduate programs in University Mathematics Departments. In M. Graven, H. Venkat, A. Essien, & P. Vale, *Proceedings of the 43rd Conference of the International Group for the Psychology of Mathematics Education Vol* (pp. 320-327). Pretoria: PME.

Harding, S. (1996). Gendered ways of knowing and the "epistemology crisis" of the west. In Goldberger, J., & Tarule, B. (Eds.), *C, & M. Belensky, Knowledge difference and power: Essays inspired by Women's ways of knowing* (pp. 431–454). Basic Books.

Henningsen, I. (2008). Gender mainstreaming of adult mathematics education: Opportunities and challenges. *ALM-IJ*, 3(1), 32–40.

Herbert, J., & Stipek, D. (2005). The emergence of gender differences in children's perceptions of their academic competence. *Journal of Applied Developmental Psychology*, 26(3), 276–295. 10.1016/j.appdev.2005.02.007

Herzig, A. (2010). Women belonging in the social worlds of graduate mathematics. *The Mathematics Enthusiast, 7*(2).

James, A. (2007). *Teaching the male brain: How boys think, feel, and learn in school*. Corwin Press.

Jenkins, K. (2018). Towards an account of gender identity. *Ergo (Ann Arbor, Mich.)*, 5(27), 713–744.

John-Steiner, V. (2014). *Sociocultural and feminist theory: Mutuality and relevance*. Research Gate. https://www.researchgate.net/publication/241344833

Kavatsyuk, O., Ioannou, M., & Avraamidou, L. (2021). Gender-Inclusive instructional practices in university mathematics classes. *International Journal of Gender, Science and Technology*, 14(2), 1–16.

Kibrislioglu, N. (2015). An investigation about 6th students' attitude towards Mathematics. *Procedia: Social and Behavioral Sciences*, 186, 64–69. 10.1016/j.sbspro.2015.04.024

Lave, J. (2008). Situated learning and changing practice. In Amin, A., & Roberts, J. (Eds.), *Community, Economic and Organization* (pp. 283–296). Oxford University Press.

Lave, J., & Wenger, E. (1991). *Situated learning: Legimate peripheral participation*. Cambridge University Press. 10.1017/CBO9780511815355

Le, A., Bui, T., Tran, B., & Phung, T. (2023). Gender gap mathematics achievement: Vietnamese students in America mathematics competitions. *Issues in Educational Research*, 33(1), 137–138.

Leder, G. (2015). Gender and mathematics review revisted. *The Proceedings of the 12th International Congress on Mathematical Education* (pp. 145-170). Melbourne: Monash University, Melbourne.

Lee, K., & Anderson, J. (2015). Gender differences in mathematics attitudes in coeducational and single sex secondary education. In M. V. Geiger, & A. Bennsion, *Mathematics education in the margins (Proceedings of the 38th Conference of the mathematics education research group of Australasia* (pp. 357-364). Sunshine Coast: MERGA.

Li, Q., & Ma, X. (2010). A meta-analysis of the effects of computer technology on school students' mathematics learning. *Educational Psychology Review*, 22(3), 215–243. 10.1007/s10648-010-9125-8

Lubienski, S., & Pinheiro, W. (2020). Gender and mathematics: What can other disciplines tell us. *Journal of Urban Mathematics Education*, 13(1), 1–14. 10.21423/jume-v13i1a377

Luis, A. (2017). Unpacking the male superiority myth and masculinisation of mathematics at the intersections: A review of Research on Gender in mathematics education. *Journal for Research in Mathematics Education*, 48(4), 397–433. 10.5951/jresematheduc.48.4.0397

Lykke, N. (2010). *Feminist studies*. Routledge. 10.4324/9780203852774

MacCarthy, S., Saunders, C., & Elliot, M. (2020). Increased reporting of sexual minority orientation from 2009 to 2017 in England and implications for measuring sexual minority health disparities. *LGBT Health*, 7(7), 393–400. 10.1089/lgbt.2019.018132758045

Maloney, E., Rmirez, G., Gunderson, E., Levine, S., & Beilock, S. (2015). Intergenerational effects of parents math anxiety on children's math achievement and anxiety. *Psychological Science*, 26(9), 1480–1488. 10.1177/0956797615592630 26253552

Mandina, S., Mashingaidze, S., & Mafuta, J. (2013). Increasing female participation in advanced level mathematics: A perspwective from teachers and students in Zimbabwe. *African Educational Research Journals*, 1(3), 183–190.

Marcus, S., & Joakim, S. (2016). Gender differences in boys and girls and girls' perception of teaching and learning mathematics. *Open Review of Educational Research*, 3(1), 18–34. 10.1080/23265507.2015.1127770

Mejia-Rodriguez, A., Luyten, H., & Meelisen, M. (2021). Across the world: An exploration of student and parental data of TIMSS. *International Journal of Science and Mathematics Education*, 19(1), 1229–1250.

Mejias, P., McAllister, D., Diaz, K. G., & Ravest, J. (2021). A longtudinal study of the gender gap in mathematics achievement: Evidence from Chile. *Educational Studies in Mathematics*, 107(3), 583–605. 10.1007/s10649-021-10052-1

Mkhize, Z. (2022). 'They are just women, What do they know?'The lived experiences of African women doctoral students in the mathematics discipline in South African Universities. *Transformation in Higher Education*, 7(0), 1–11. 10.4102/the.v7i0.21836124010

National Centre for Science and Engineering Statistics. (2019). *Women minorities, and persons with disabilities in science and engneeering (Special Report NSF 19-304).* National Science Foundation, Directorate for Social, Behavioural and Economic Science.

Niederle, M., & Vesterlund, L. (2010). Explaining the gender gap in math test scores: The role of competition. *The Journal of Economic Perspectives*, 24(2), 129–144. 10.1257/jep.24.2.129

Nollenberger, N., Rodriguez-Planas, N., & Sevila, A. (2014). The maths gender gap: The role of Culture. *Discussion Paper No. 8379*. Bonn, Germany.

Oakley, A. (1972). *Sex,gender and society*. Temple Smith.

OECD. (2014). *PISA 2012 results: What students know and can do: Students performance in mathematics, reading and science*. OECD.

OECDiLibrary. (2024). *OECDiLibrary*. Indicators B7. https://www.oecd-ilibrary.org/sites/8389c70e-en/index.html?itemId=/content/component/8389c70e-en

Oksala, J. (2017). Feminism and Power. In A. Garry, S. Khader, & A. Stone, *The Routledge Companion tofeminist Philosophy* (pp. 678-688). Routeledge: New York. 10.4324/9781315758152-55

Organisation for Economic Co-operation and Development. (2016). *Excellence and equity in education*. OECD: OECD Publishing.

Organization for Economic Cooperation and Development (OECD). (2014). PISA 2012 results: What students know and can do: Student performance in mathematics,reading and science. *Rev ed Vol 1*.

Ovie-Ekpewu, R. (2011). The sensitivity of science and mathematics teachers to gender issues in the classroom. *Journal of Research in National Development*, 8(2). 10.4314/jorind.v8i2.66848

Ozumba, G. (2005). Gender-sensitivities in Igbo culture: A philosophical reappraisal. *Quidlibet Journal* 7(2).

Parmaxi, A., Christou, E., Valdes, J. F., Hevia, D., Perifanou, M., Economides, A., & Manchenko, M. (2023). Gender equality in science, technology, engineering and mathematics: Industrial vis-a-vis academic perspective. *Discover Education*, 3(3), 1–13.

Poudel, K. (2017). *Women empowerment through English language learning* [Dissertation, Kathmandu University: Kathmadu University and School of Education].

Rippon, G. (2019). *The gendered Brain: The New Neuroscience that shatters the myth of the female brain*. The Bodley Head.

Robitaille, D., & Tavers, K. (1992). International studies of achievement in mathematics. In Grouws, D. (Ed.), *Handbook of research on mathematics education* (pp. 687–709). Macmillan.

Roth, J., & Ames, J. (2014). *Beiträge zum Mathematikunterricht*. WTM-Verlag.

Ruhle, R. (2022). Mind the gap-an analysis of gender differences in mathematics and science achievement in South Africa. *Stellenbosch Economic Working Papers: WP04/2022*.

Salem, S. (2018). Intersectionality and its discontents: Intersectionality as traveling theory. *European Journal of Women's Studies*, 25(4), 403–418. 10.1177/1350506816643999

Sarmah, A., & Puri, P. (2014). Attitude towards Mathematics of the students studying in diploma in Engineering Institute (Polytechnic) of Sikkim. *Journal of Research & Methods in Education*, 4(6), 12–18. 10.9790/7388-04630610

Saygin, P. (2020). Gender bias in standard tests: Evidence from a centralised college admission system. *Empirical Economics*, 59(2), 1037–1065. 10.1007/s00181-019-01662-z

Sevilla, A., & Cue-Ruiz, P. (2024). How can we reduce gender gaps in mathematics? In *B. Atweh, A. Barton, M. Borba, N. Gough, C. Keitel, V. Vistro-Yu, & R. Vithal, Mathematics Literacy and Globalisation* (pp. 2–17). Springer.

Skovsmose, O. (2008). Mathematical literacy and globalisation. In *B. Atweh, A. Barton, P. Ernest, N. Gough, C. Keitel, C. Vistro-Yu, & R. Vithal, Internationalisation and Globalisation in Mathematics and Science Education* (pp. 3–18). Springer.

Syyeda, F. (2016). Understanding attitudes towards mathematics (ATM) using modal Model: An exploratory case study with secondary school children in England. *Cambridge Open-Review Educational Research e-Journal 3*, 32-62.

UNESCO. (2004). *A training manual for sensitising education managers, curriculum and material developers and media professionals to gender concerns*. UNESCO.

United Nations Scientific and Cultural Organisation (UNESCO). (2016). *Closing the gap in STEM: Drawing more girls and women into Science, Technology, Engineering and Mathematics*. UNESCO.

Valez, J., & Ochoas, S. (2020). Gender sensitivity and Its relation to gender equality. In W. Leal F, A. Azul, L. Brandli, A. Lange S, & T. Wall, *Gender Equality, Encyclopedia of the UN Sustainable Development Goals*. Switzerland: Springer Nature Switzerland. 10.1007/978-3-319-70060-1_46-1

Vasanthan, R., & Ancymol, J. (2021). Literature for gender sensitivity: An Overview. *Journal for Educators. Teachers and Trainers*, 13(1), 61–69.

Vygotsky, L. (1978). *Mind in Society: The development of higher pyschological processes*. Harvard University Press.

Wass, R. (2012). *Developing critical thinkers in Higher: A Vygotskian Perspective.* [Thesis, New Zealand: University of Otago].

Woods, J., Murzi, H., Greene-Havas, M., Diaz-Strandberg, A., & Vincente, S. (2020). *Role of social interaction in barriers facing the first year -international students in the United States.* American Society for Engineering Education.

Zobnina, A. (2009). *Glossary of gender-related terms.* Mediterranean Institute of Gender Studies, European Institute for Gender Equality. https://medinstgenderstudies.org/wp-content/uploads/Gender-Glossary-updated_final.pdf

ADDITIONAL READING

Alsina, A., & Mula, I. (2019). Advancing towards a transformational professional competence model through reflective learning and sustainability: The case of mathematics education. *Sustainability (Basel)*, 11(15), 1–17. 10.3390/su11154039

Ayalon, H., & Livneh, I. (2013). Educational standardization and gender differences in mathematics achievement: A comparative study. *Social Science Research*, 42(2), 432–445. 10.1016/j.ssresearch.2012.10.00123347486

Bernaldo, M., & Fernandez, S. G. (2021). *Globalisation and Education: Trends towards sustainability*. Intech Open. https://www.intechopen.com/chapters/78595

Bertoletti, A., Cannistrà, M., Lema, M., Masci, C., Mergoni, A., Rossi, L., & Soncin, M. (2023). The determinants of mathematics achievement: A gender perspective using multilevel random forest. *Economies*, 11(2), 1–30. 10.3390/economies11020032

Furuta, J. (2019). Liberal individualism and the globalization of education as a human right: The worldwide decline of early tracking, 1960–2010. *Sociology of Education*, 3(1), 1–9. 10.1177/0038040719873848

Grossen, M., & Mirza, N. (2020). Talking about cultural diversity at school dialogical tensions and obstacles to secondarisation. *European Journal of Psychology of Education*, 35(2), 243–264. 10.1007/s10212-019-00442-8

Lewis, S. (2021). *Globalization and education*. 10.1016/B978-0-12-818630-5.01001-0

Pashby, K., Costa, M., Stein, S., & Anreotti, V. (2020). A meta-review of typologies of global citizenship education. *Comparative Education*, 56(2), 144–164. 10.1080/03050068.2020.1723352

Santamaría-Cárdaba, N., Martins, C., & Sousa, J. (2021). Thematics teachers facing the challenges of global society: A study in primary and secondary education in Spain. *Eurasia Journal of Mathematics, Science and Technology Education*, 17(4), 1–9. 10.29333/ejmste/10806

Stavre, I. (2020). Higher education and globalization in the context of the COVID-19 crisis. *European Journal of Education*, 3(2), 34–48. 10.26417/812dro50g

KEY TERMS AND DEFINITIONS

Bias: A tendency to show favour towards something.

Feminist: Men and women who are commited to equality between sexes.

Gender: Refers to way boys or girls are socialized in the society.

Poststructural Theory: Explanations that go beyond modernism and scrutinize ways in which language is used to display power and inequality.

Sexist: Attitudes that show discrimination against one sex.

Situated Learning: Instruction which takes into account the context.

Stereotype: An attitude that may show negative or positive bias.

Zone of Proximal Development: An imaginary gap between what learner is able to do on their own and what they can potentially do with the help of an adult.

Chapter 8
Gender and Innovations in Mathematics Education

Wilfred Monyoro
Kisii University, Kenya

ABSTRACT

From 2012 to 2018, pupils taking the Kenya Certificate of Secondary Education (KCSE) routinely performed poorly in Mathematics on a national level in Kenya. In KCSE Test data from 2012 to 2018, it was noted that probability was one of the mathematical concepts that the majority of pupils found challenging. A novel approach that advocates for the incorporation of computer assisted learning methods like simulation has been put out to address the issue. In this study, learner achievement in probability in Mathematics will be compared to computer-based simulation (CBS) effect in public secondary schools in Kisii County. The objective of the study will be to determine the effect of computer simulation (innovation) on students' achievement in probability by sex (mathematics education).

BACKGROUND TO THE STUDY

Teachers have different attitudes toward boys and girls, according to Bachar (2012), with boys being given more time than girls and girls being handled with compassion and forgiving emotions. Boys credit their success to personal abilities and capabilities, but girls credit it to investment, an easy test, a competent instructor who clearly explains the content, luck, group work, and social learning (Bachar, 2012).

This chapter aims to add to the ongoing discussion by examining the impact of computer-based simulation on student learning and accomplishment in probability based on gender. Girls underachieve in science and Mathematics, Eshiwani (1982), and this under-performance is largely due to teachers' bias in the teaching and learn-

DOI: 10.4018/979-8-3693-2873-6.ch008

ing process. Teachers tend to focus on boys more than girls, placing them behind. FAWE (1999) found that girls' achievement in science and Mathematics is poorer, in part due to their negative attitude. According to Erwin (1993), boys are more interested in Mathematics than girls in secondary school. According to research, there is a gender imbalance in school sciences, with more boys taking science and Mathematics topics than girls (Rostvik, & Fyfe, 2018). However, while there are no gender differences in overall (IQ), there are significant variances in cognitive ability between boys and girls on tests. According to Trowbridge, Bybee, and Powel (2004), gender differences in physics are significant, while in Mathematics they are minor or non-existent. Kans, & Claesson, (2022) claims that gender inequalities in science interest are primarily related to boys' interest in physical sciences against girls' interest in biological sciences. EACEA (2010), the European Commission's executive agency for education, audio-visual, and culture, found that gender inequalities in science were significant within schools or programmes in European countries.

Girls have six points higher Mathematical achievement than boys in grade 8 science students throughout the trends in international Mathematics and science the survey, TIMSS 2007 countries, according to Martin, Mullis, and Foy (2008). Professor Hazan (2010) claims that when girls learn in classes with other girls rather than boys, their performance improves. Girls are aware of the notion that they have poor Mathematical aptitude. According to Banks (2013), girls who attend classes without boys obtain more success. Separate classes for girls and boys in schools with various teaching approaches for each gender, according to Miller (2009), a trained psychologist, is one of the options to ensure that girls compete favourably with boys in Mathematics and other subjects. Competitiveness and achievement-oriented learning are less effective for girls than cooperation and personal communication (Kiran, 2009).

Gender was supposed to be a social regulating system that dominated feminine and masculine conceptions and controlled a specific cognitive organization based on biological distinctions (Ojeda & Flores, 2008). Gender, according to Wangid & Widyastuti, (2021), refers to the traits and behaviours that society considers proper for men and women, whereas sex refers to biological distinctions between females and males. Through interactions with peers, instructors, and the environment, children begin to form sex schemas or ordered networks of knowledge about what it means to be a boy or a girl. This socio-cultural construct categorized men's and women's social relationships, social practices, and how individuals interact with physical and cognitive objects. This concept had been one of the foundations for the study of human behaviour and predispositions for a long time (Ogembo, 2017). Assuming that the goal of adding technology into Mathematics classrooms is to promote all students' learning, it was critical to determine whether it benefited all students or the vast majority, or if it benefited only select groups, particularly those based on sex.

It was particularly vital to track students' participation in Mathematics instruction and learning, as well as their attitudes regarding computer-based Mathematics and, most significantly, the outcomes of both situations. The purpose of this study is to see how technology affected these dimensions: did it promote gender equality or did it worsen the gap between boys and girls?

n a quasi-experimental study aimed at evaluating the effect of Mathematical innovation and technology on students' academic performance. Oginni and Popoola (2013) discovered a significant difference in performance depending on gender in favor of boys. Bandura, Barbaranelli, Caprana, and Pastorelli (2001), as well as Eccles and Wigfield (2001), came to similar results. On the other hand, other studies have identified no significant differences in learner achievement between boys and girls who have been exposed to CBS. For example, Ursini, Sanchez, Orendai, and Butto (2004) observed that incorporating technology into Mathematics classrooms appeared to increase gender parity in student conduct. Yusuf and Afolabi (2010) showed that students exposed to Computer-Based Simulation (CBS) performed better individually or jointly than their peers exposed to a traditional classroom education in a study of the effects of CBS on secondary school student's performance in biology. In either individual or cooperative tasks, there was no significant difference in performance between male and female students exposed to CBS.

Early research revealed that teachers regarded girls and boys differently, with males' arithmetic ability often being rated higher than girls' (Upadyaya & Eccles, 2014), and boys receiving higher demands and more specific feedback (Sadker & Sadker, 1986). There is a conventional gender imbalance in favour of boys, according to research on arithmetic achievement (Aunola, Leskinen, Lerkkanen & Nurmi, 2008, Martin & Cheng, 2008).

Other researchers, on the other hand, have concluded that gender differences in Mathematics achievement are insignificant (Lindberg, Hyde, Peterson, & Linn, 2010), corroborating Robinson and Lubienski (2011)'s findings that girls have received slightly better grades in Mathematics over the last four decades than boys, findings that Brown and Kanyongo (2011) also supported.

Teacher and peer support are favourably associated with academic attitudes, achievement, emotions, learning, motivation, and self-efficacy, according to a study (Danielsen, Wiium, Wilhelmsen & Wold, 2010). In Israel's elementary school mathematics classes, boys and girls study together in one classroom, even though their achievement levels differ (Bachar, 2012).

In the Meyzav test, a national examination of students' achievement in elementary school that revealed an apparent difference between boys and girls, with girls' scores being 12 points lower than boys' scores and the gap existing throughout Israel, Ayalon and Livnah (2013) addressed the gender gap in Mathematics and science achievement

CBS will help girls learn better by empowering them and allowing them to compete. However, Bachar (2012) attributes boys' success in Mathematics to their abilities and talents. Bacher (2012) goes on to say that girls prefer group learning and social learning, whereas Mathematics is seen as a competitive and achievement-oriented subject. Any change in strategy will have a significant impact on the attitudes of female students and teachers.

Fear of Mathematics and anxiety about completing arithmetic are low among students who are terrified of Mathematics, according to Beilock, Gunderson, Ramire, and Levine (2010). This is because the majority of teachers in lower early childhood education are women who have a negative attitude toward Mathematics and hence discourage young children from learning Mathematics at an early age (Beilock, et al., 2010).

According to a meta-analysis, boys outperform girls in problem-solving Mathematics assessments (Li, Zhang, Liu, & Hao, 2018) while comparatively in another research project girls are stronger in computation, but there is no substantial difference in understanding Mathematics topics between the sexes. According to another study, girls in Mathematics tend to do better than boys (Downey & Vogt Yuan, 2005).

Mazana, Suero Montero, & Olifage, (2019), highlighted attitudes about the usefulness and confidence in learning Mathematics as crucial, with boys demonstrating that they were more confident and believed that Mathematics would be more beneficial to them than females. Another study, which looked into the factors that influence arithmetic achievement in Mathematics classes with a gender disparity, found that boys outperformed girls (Van Mier, Schleepen, & Van den Berg, 2019). Hall et al (1999) looked at gender disparities in Mathematics achievement among American children in grades 5 through 8 and found no significant differences.

LITERATURE REVIEW

Teacher's Gender and CBS Integration

According to Russell and Bradley (1997), there is a link between gender and computer anxiety, with female teachers reporting higher levels of anxiety than male teachers. This is because male teachers have a more positive attitude toward CBS than female teachers. Teachers who are female have a negative attitude toward the usage of CBS (Ogembo, 2017). This is due to female Mathematics teachers' ineffective use of CBS, which is caused by an excessive workload and a lack of self-confidence in using CBS (Salau, 2000). Male secondary school instructors have more confidence in computers than female teachers, according to Bovée, Voogt, & Meelissen, 2007).

Male instructors are more efficient and have better CBS skills than female teachers, according to Vekiri, & Chronaki, (2008). Female instructors' usage of technology in the classroom has been reported to be low, owing to a variety of causes including limited access to tools, inadequate abilities, and a lack of enthusiasm, according to reports (Christensen, & Knezek, 2009). Early research on female teachers' usage of CBS found that they do not appreciate technology and would support the status quo for fear of being made obsolete, according to Ogembo (2017). Female teachers' use of CBS in the classroom was less than male teachers, according to Sanders, 2005). However, Steele (2006, March) found that female instructors' self-perceptions of technology competence improved while their male counterparts stayed unchanged. Females were more likely than males to use CBS tools in the classroom, according to Sanders, J. (2005). Furthermore, Gordon (2003) found that gender is not a predictor of CBS integration in the classroom.

Computer Anxiety and Use of CBS

Ahiatrogah, et al., (2013) linked anxiety with technology integration in classroom teaching and learning. He says that anxiety is a chronic physiological condition. Teachers' levels of computer fear were linked to their judgments of the tool's utility, according to Russell and Bradley (1997). Tsai and Tsai (2003) discovered a strong link between metacognitive skills, computer achievement, and computer anxiety in students. Other studies have discovered a strong link between computer anxiety and age (Namlu & Ceyhan, 2002), computer experience (Yaghi & Ghaith, 2002), frequency of computer use (Necessary & Parish, 1996), neuroticism (Anthony, Clarke & Anderson, 2000), and a person's assessment of a computing scenario (Anthony, Clarke & Anderson, 2000).

According to Ahiatrogah, et al., (2013), computer anxiety is a significant predictor of teachers' attitudes toward interactive computer technology, and one might be prompted to conclude from the study's findings that, because the majority of secondary school teachers appear to be technophobic, their uptake of computer technology skills for onward transfer to the education system is slow, low, and insignificant. Russell and Bradley (1997) cited Pelgrum and Plomp (1991), while Papanastasiou, & Angeli (2008) made similar observations. The ability of CBC to teach difficult or complex concepts has been acknowledged (Namlu & Ceyhan, 2002). Computer-based simulations (CBS) can help students understand dynamic and complex topics that are difficult to explain using words, equations, or class activities.

The goal of the study was to see how computer anxiety affected Mathematics teachers' incorporation of CBC in probability teaching and learning. In today's Mathematics education and learning, CBC is extremely significant (Cheng, 2009). Computers can be used as a complement to help students attain their educational goals. It is said that using computers boosts student confidence and aids in the discovery of interconnections between the components of a complicated system (Namlu & Ceyhan, 2002).

METHODOLOGY

Research methodology according to Kohari (2004), is a way to systematically solve the research problem. In this chapter, it outlined the research design, the research variables, study location, target population, sampling procedure and sample size. Also included in the chapter were instruments for data collection, the pilot phase for pretesting the instruments, reliability and validity, procedure used for data collection, methods of data analysis and issues of ethical consideration.

Methodology is a formal plan of action for a research project. A mixed method design comprising both quantitative and qualitative research designs was used for this study. According to Creswell (2009), the problems addressed by social and health science researchers are complex and use of either quantitative or qualitative approaches by themselves is inadequate to address this complexity. (Cohen, Marrison, 2004; Greene, Caracelli & Graham, 1989; Strauss & Corbin, 1990) argue that the use of both quantitative and qualitative data and data analysis allow researchers to simultaneously make generalization about a population from the results of a sample and to gain a deeper understanding of the phenomena of interest.

The study's main objective was to assess the effects of computer-based Simulation (CBS) on secondary school students' achievement in probability and mathematics in general. The researcher employed a Solomon's four group type quasi-experimental design to accomplish this. In this study, the students were placed into four groups, two of which were experimental and two of which were control. The first group consisted of students who were open to new experiences. It was assessed both before and after treatment, whereas the control group was assessed both before and after no treatment.

The third group was also a test group. This cohort was only examined once after getting therapy. Finally, the fourth group (a different control group) was examined only once, with no treatment or pre-testing. The Solomon's four-group model can be expressed visually as follows:

Group Pre-Test Treatment Post-Test

$\mathbf{E_1}\ O_1\ X\ O_2$
$\mathbf{C_1}\ O_3\ O_4$
$\mathbf{E_2}\ X\ O_5$

$C_2 O_6$

KEY: E_1: Experimental group 1, C_1: Control group 1, X: Treatment, O: Testing

This approach provided considerable evidence for intervention while also allowing for the evaluation of both testing effects and confounding variables (Johnson and Onwuegbuzie 2004).

Study Variables

The independent variable, the variable that causes or affects outcomes, in this study was a CBS computer simulation approach. The dependent variable or outcome variable was the variable that depends on the independent variable. In this study, the dependent variable was secondary school students' achievement in probability. The integration of CBS in Probability teaching and learning was the method to be examined here. As a result, the experimental groups received the CBS for Probability instruction and learning, while the control groups received no treatment. Instead, the control group was allowed to continue with their current method of teaching and studying probability. Conventional strategies were used in this case. Thus, teaching methodologies (CBS or conventional) were the study's independent variables, while student accomplishment in Probability was the study's dependent variable.

Apart from the established independent variable, instructional technique, numerous additional factors influenced student achievement in Probability (the dependent variable). This was a set of control variables (intervening variables). Students' attitude toward Mathematics, student gender, accomplishment level, and students' aptitude in Mathematics were among the control variables studied by the researcher.

Location of the Study

The study was carried out in Kisii County, one of Kenya's forty-seven counties. This was necessitated by the fact that no other similar studies have been carried out in the county, poor performance in mathematics in the county and that most schools are in close proximity to the road network and the researcher.

Target Population

Probability was taught in the form three Mathematics Syllabus, the study's target group was form three students. According to data acquired from Kisii County's Director of Education office, the target population consisted of 6,038 form three students and 136 Mathematics teachers in 67 public secondary schools having ICT infrastructure for CBS integration.

Table 1. Sampling grid for students

	School type	Number of schools	Streams sampled	Students sampled
Schools	County	17	2	90
	Sub-county	38	2	90
Total		55	4	180

Sampling was done for both county and sub-county public schools. Two county schools purposely sampled with 90 students and two subcounty schools with 90 students.

Research Instruments

The study employed four instruments in order to achieve the objectives.

- Researcher made Pre-test Students' Mathematics Test (PRESMT),
- Post-test Students' Mathematics Test (POSMT),
- Students' Questionnaire (SQ).

The instruments utilized in this study were created with the objectives in mind. The study's major objective was to determine the impact of the CBS teaching style on student probability performance. The study also attempted to determine the effects of CBS on the performance of children with varying levels of mathematics achievement as well as their attitudes toward probability.

Validity

Threats to internal validity, according to Trochim (2001) prevent researchers form establishing the real causal relationship of a study. Gay & Airasian (2000) propose that random selection of participants and random assignments to treatment and control groups are powerful approaches to overcoming internal threats to internal validity.

Reliability

Reliability refers to how consistent the results from the test are, or how well the test is actually measuring what you want it to measure (Bonnet,2003). Internal consistency techniques were used to assess the instruments' reliability in this study. Because the items were of varying degrees of difficulty, the Kuder-Richardson formula 20 or (KR-20) was used to estimate the instrument's reliability:

KR-20 = n 1- $\sum$p. q

n-1 s^2x

Where KR20 = Kuder Richardson reliability coefficient
n = Number of items used to measure the concept
p = proportion of people passing the item
q = proportion of people failing the item.
S^2 = Variance of all scores
s^2x = Variance of individual scores.
Source: Mugenda and Mugenda (1999: 99)

A minimum coefficient of 0.5 was allowed for this instrument. A Kuder Richardson coefficient of 0.5 to 0.7 was rated acceptable by Bonnet (2003), while one greater than 0.7 was considered excellent.

RESULTS AND FINDINGS

General and Demographic Information

Solomon's four quasi-experimental research designs were used to attain the study's goal. 198 from three students and four public secondary schools (sub-county coeducational schools) participated in the study. In comparison to girls, there were slightly more boys (102 boys and 96 girls). The students were divided into four groups, with a study unit consisting of a group of students from each school who were randomly allocated to either the experimental or control groups (two experimental and two control groups). Treatment was delivered (application of CBS in teaching and learning probabilities to the two experimental groups and use of conventional teaching and learning methods to the two control groups) and all four groups were then post-tested.

Presentation of Findings and Interpretations

This part covers the findings of the study in the order in which they were starting with descriptive statistics, hypotheses testing, qualitative data, and finally, discussions.

CBS and Students' Achievement in Probability

The Solomon's 4 matrix was used for this study, and it was sub-divided into two main groups: experimental group and control group. The experimental group was further subdivided into two; experimental group one (E1) and experimental group two (E2). Similarly, the control group was sub-divided into two; control group one (C1) and control group two (C2). E1 group was subjected to a pre-test then they were subjected to the treatment with computer simulation on teaching and learning probability, then finally subjected to a post-test. C1 group was given a pre-test and they continued learning probability without using Computer simulation, then finally subjected to a post-test. E2 group was subjected to the teaching and learning of probability rising Computer simulation and then later subjected to a post-test while C2 group was not subjected to the use of Computer in teaching and learning probability but were tested with a post-test.

The initial goal was to see if there was a difference in probability achievement between students who were taught using Computer simulation and those who were not. To do this, a control (C1) and experimental (E1) set of students were pre-tested with designed pre-test, the intervention was provided, and the students were then post-tested to measure the effect of the intervention.

The pre-test test administered to two test groups C_1 and E_1 yielded results as presented in Table 2.

Table 2. Students' achievement in mathematics – pre-test

	Test groups composition				Total	
	E1		C1			
	f	%	f	%	f	%
Less than 20	35	35.4	29	29.3	64	64.6
20 – 40	11	11.1	17	17.2	28	28.3
Above 40	3	3.0	4	4.0	7	7.1
Total	49	49.5	50	50.5	99	100
Mean	14.63		17.58		16.12	
SD	11.39		11.07		11.27	

The results suggest that students in the control group C1 (M=17.58, SD=11.07) had higher achievement than those in the experimental group E1 (M=14.63, SD=11.39). More than half of the students (64.6%) scored below 20% on the test, with E1 students (35.4%) outnumbering C1 students (29.3%), showing that the experimental group performed below average on the pre-test than the control group. Students from C1 scored somewhat higher (17.2%) than those from E1 in the 20–40% range (11.1%). A similar pattern was observed in individuals with a score of 40 or above, with 4.0% coming from C1 and 3.0% from E1. In general, the results show that students in the control group C1 fared better in the pre-test examination than their experimental group E1 peers. This could be attributed to the instructional methodology of conventional approach (not using CBS simulation) in teaching and learning probability for both experimental group and the control group before treatment.

CBS and Students' Achievement in Probability Based on Their Gender

The study sought to determine the of CBS on student achievement based on their gender. The study's hypothesis, Ho_3, was to observe if there was a statistically significant difference in achievement between boys and girls CBS subjects. Data on students' sex were presented in Figure 1.

Figure 1. Student's gender

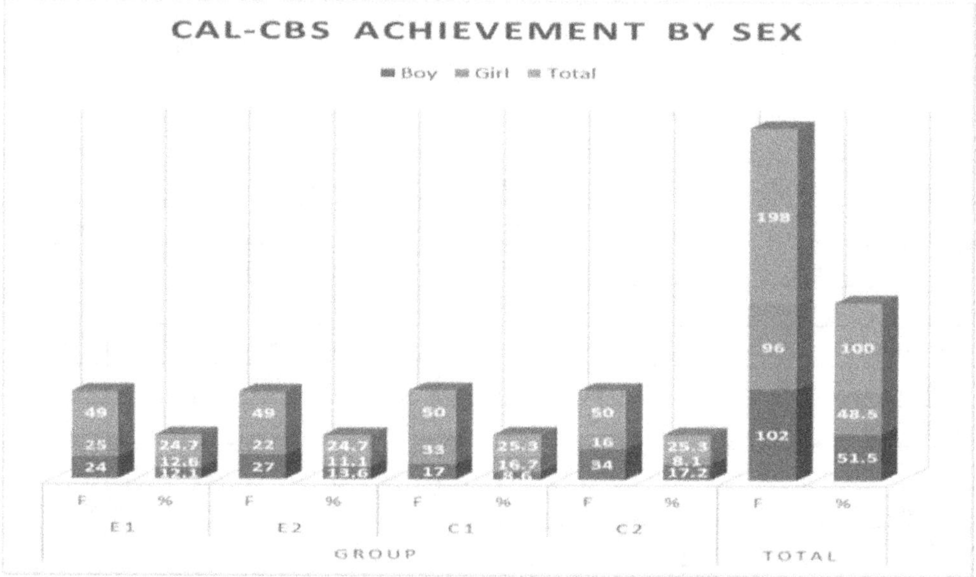

According to the data in Figure 1, each of the four test groups contained 102 boys and 96 girls. For the test of homogeneity of the composition of the test groups in terms of gender, ANOVA statistics, $F(3,194) = 1.67$, $p = 0.235$, indicated insignificant mean differences at the 0.05 level of significance, allowing continued analysis. The results of the pre-test are provided in Figure 2.

Figure 2. Students' achievement in pre-test by sex

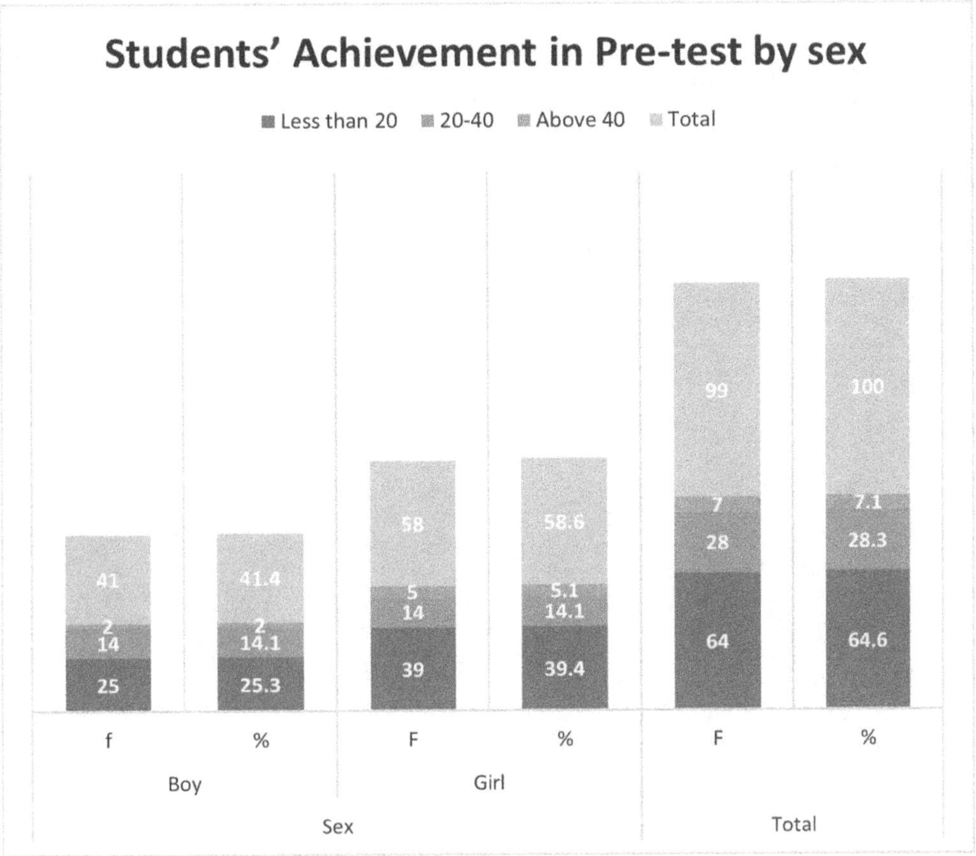

The data revealed that the majority of the students scored less than 20% on the pre-test. Females made up 39.4% of the population, while males made up 25.3%. In the pre-test, an equal proportion of male and female students scored between 20 and 40%, but twice as many female students 5.1% than male students 2.0% scored above 40%. In the pre-test, an independent sample t-test was utilized to evaluate male and female students' success levels. Female students' mean achievement (M=16.38) was somewhat higher than male students' (M=15.76). This suggests that female students outperformed male students by a small margin. The results of these students' exposure to CBS and subsequent post-testing are shown in Figure 3.

Figure 3. Students' achievement in post-test by sex

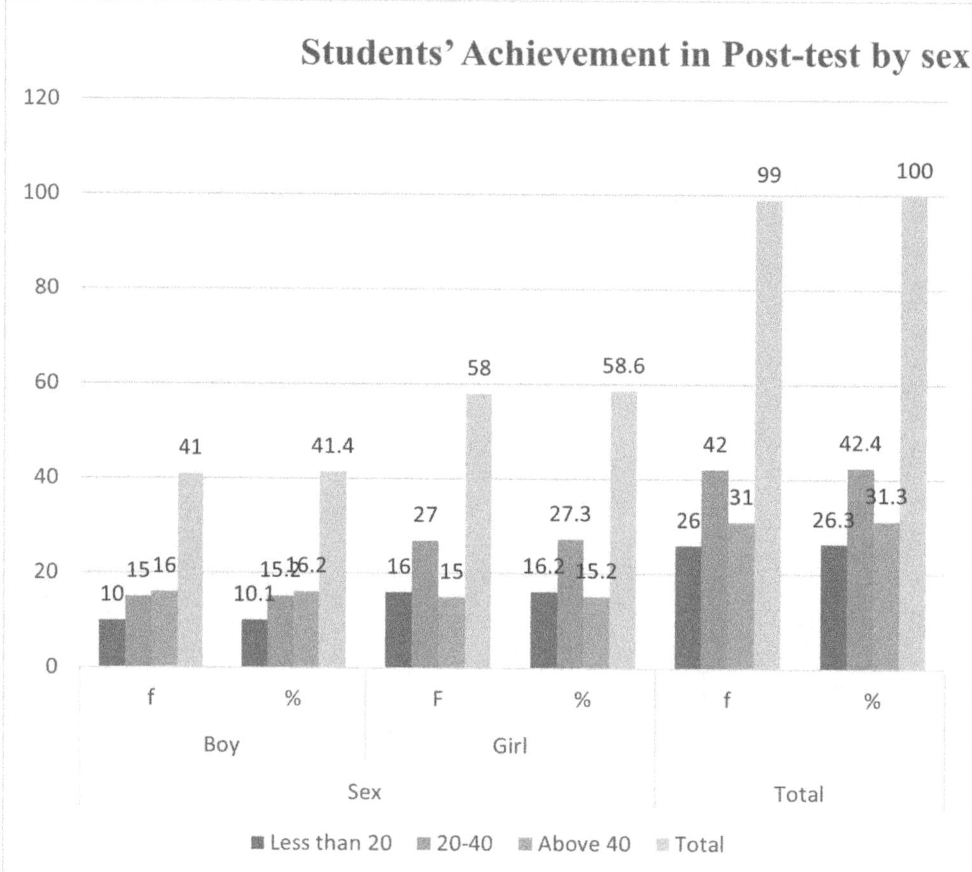

Figure 3 shows that a quarter of students 26.3% scored less than 20% on the post-test, with somewhat more females scoring less than 20% than males. When it came to individuals with scores between 20% and 40%, a similar trend was seen, with more females than males scoring between 20% and 40%. However, there were more male students (16.2%) than female students (15.2%) among those who scored above 40%.

The study also aimed to look into CBS on student achievement based on gender. The study's third hypothesis, Ho_3, intended to determine whether there is a statistically significant difference in achievement in Probability between male and female students when they are exposed to CBS. To examine the impacts of CBS on students' achievement based on sex, a two-way analysis of variance (ANOVA) test was used, and the results are shown in Table 3.

Table 3. Effect of CBS on students' achievement based on gender

Source	Type III Sum of Squares	df	Mean Square	F	Sig.	Partial Eta Squared
Corrected Model	10852.647[a]	3	3617.549	15.980	.000	.335
Intercept	105703.031	1	105703.031	466.928	.000	.831
Gender	190.619	1	190.619	.842	. 361	.009
Gp	9570.802	1	9570.802	42.278	.000	.308
Gender * Gp	1346.969	1	1346.969	5.950	.017	.059
Error	21506.080	95	226.380			
Total	145582.000	99				
Corrected Total	32358.727	98				

a. R Squared = .335 (Adjusted R Squared = .314)

Significant disparities in achievement levels were detected between students in the experimental groups E1 (M=43.70, SE=2.11) and C1 (M=23.48, SE=2.20), which were similar to those observed between E2 (M=51.45, SE=2.12) and C2 (M=31.75, SE=2.23). Students in E1 had substantially different exam scores from those in C1 and C2, while students in E2 had significantly different test scores from those in C1 and C2. F (3,198) =33.05, p.001, p2 =0.343, F (3,198) =33.05, p.001, p2 =0.343, F (3,198) =33.05, p.001, p2 =0.343, F (3,198) =33.05, p.001,

A post hoc (Tukey HSD) test revealed that there is a substantial mean difference between E1 and C1 and C2 students' achievement levels. The achievement levels of E2 students, as well as those in C1 and C2, followed a similar pattern.

Male students had a comparatively greater mean achievement in Probability (M=37.39) than female students (M=31.29) based on gender. F (1,198) =.273, p =.602, p2 =0.001, although the differences were not significant, male students in E1 had a higher mean (M=48.91) than female students in the same group (M=38.48) within the groups. In C1, the mean achievement of female students (M=25.85) was higher than that of male students in the same group (M=21.12).

Though the use of CBS in instruction comparably influenced students of different gender, resulting in a negligible effect across gender, the difference in achievement levels between the experimental and control groups demonstrated its superiority as a method of instruction for Probability and Mathematics in general. As a result, the data gained failed to refute the third hypothesis, $H0_3$, which suggested that when Probability is taught by CBS, computer simulation, there is no significant difference in achievement by gender and either sex would use CBS, Computer simulation in learning probability which is in line with the globalization on gender and innovation in mathematics education.

DISCUSSIONS AND RECOMMENDATIONS

The findings of this study showed the insignificant effect of the use of CBS on the achievement of students based on gender. This implies that the use of CBS affects the achievement of boys and girls equally. It is consistent with that of Ursini, Sanchez, Orendai, and Butto (2004) who observed that incorporating technology into probability classrooms appeared to increase gender parity in student conduct. However, it contradicts that of Oginni and Popoola (2013) who discovered that there is a significant difference in performance depending on gender in favor of boys than girls. Similar results were arrived at by Bandura, Barbaranelli, Caprana, and Pastorelli (2001) as well as Eccles and Wigfield (2001).

The objective set out to examine the effect of CBS on achievement of students based on their gender. Results from the post-test showed that male students in experimental group posted a higher mean achievement compared to their female counterparts, the mean difference though being insignificant implying that the use of CBS in instruction in probability changes students of different gender similarly. It can therefore be concluded that CBS teaching strategy is an effective teaching strategy for both male and female learners.

From descriptive findings, Male students had a comparatively greater mean achievement in Probability (M=37.39) than female students (M=31.29) based on gender. However, it contradicts the findings of Robinson and Lubienski (2011), who found out that girls received slightly better grades in Mathematics over the last four decades than boys, findings that Brown and Kanyongo (2010) also supported. In a mixed research, Ayalon and Livnah (2013) addressed the gender gap in Mathematics achievement in the Meyzav test, a national examination of students' achievement in elementary school that revealed an apparent difference between boys and girls, with girls' scores being 12 points lower than boys' scores and the gap existing throughout Israel.

Hudig, Scheepers, Schippers, & Smeets, (2021) affirms that girls prefer group learning and social learning using CBS and do well than boys. However, their investment in Mathematics and an easy test, a good teacher who explains Mathematics well, and good luck are all attributes of good performance in Mathematics. It is inconsistent with that of Hall *et al* (1999) who looked at gender disparities in Mathematics achievement among American children in grades 5 through 8 and found no significant differences.

CONCLUSION

The objective set out to examine the effect of CBS on achievement of students based on their gender. Hypothesis of the study sought to determine whether there exists a statistically significant difference between achievements of male and female exposed to CBS. Descriptive findings illustrated a fair distribution of students by gender in control and experimental groups. Data obtained showed that achievement level of female students (M=16.38) was slightly higher than that of male students (M=15.76). This means that female students performed better than their male in the pre-test. Results from the post-test showed that male students in experimental group posted a higher mean achievement compared to their female counterparts, the mean difference though being insignificant. This imply that the use of CBS in instruction in probability impacts students of different gender similarly thus the insignificant effect across gender.

Results from the post-test showed that male students in experimental group posted a higher mean achievement compared to their female counterparts, the mean difference though being insignificant implying that the use of CBS in instruction in probability impacts students of different gender similarly. It can therefore be concluded that CBS teaching strategy is an effective teaching strategy for both male and female learners.

Recommendations

It is suggested that further research be undertaken in the following areas:

i. A longitudinal study of the effectiveness of CBS on teaching and learning other topics in mathematics in secondary schools in Kisii County, Kenya.
ii. A comparison of primary school pupils' achievement in mathematics using CBS driven curriculum, a traditional driven curriculum and a blend of CBS and traditional approach curriculum in public and private schools in Kisii County, Kenya.
iii. Investigation of how CBS approach can be used to change the perceptions of mathematics teachers about teaching and learning probability (Mathematics).

REFERENCES

Afshari, M., Bakar, K. A., Luan, W. S., Samah, B. A., & Fooi, F. S. (2009). Factors affecting teachers' use of ICT. *Online Submission, 2*(1), 77-104. 201

Ahiatrogah, P. D., Madjoub, M. B., & Bervell, B. (2013). *Effect of Computer Assisted Instruction on the Achievement of Basic School Students in Pre- Technical Skills.* Academic Journal of Interdisciplinary. 10.5901/ajis.2013.v2n1p77

Ahiatrogah, P. D., Madjoub, M. B., & Bervell, B. (2013). Effects of computer assisted instruction on the achievement of basic school students in pre-technical skills. *Academic Journal of Interdisciplinary Studies*, 2(1), 77–86. 10.5901/ajis.2013.v2n1p77

Alessi, S. M., & Trollip, S. R. (2001). Simulations. *Multimedia for Learning (Methods and Development)*, 213-269.

Andriani, D., Widada, W., Herawaty, D., Ardy, H., Nugroho, K. U. Z., Ma'rifah, N., & Anggoro, A. F. D. (2020). Understanding the number concepts through learning Connected Mathematics (CM): A local cultural approach. *Universal Journal of Educational Research*, 8(3), 1055–1061. 10.13189/ujer.2020.080340

Banks, J. A. (2013). The nature of multicultural education. *Multicultural education: Issues and perspectives*, 3-24. Research Gate.

Brophy, J. (1999). Toward a model of the value aspects of motivation in education: Developing appreciation for. *Educational Psychologist*, 34(2), 75–85. 10.1207/s15326985ep3402_1

Brown, J. S., Collins, A., & Duguid, P. (1989). Situated cognition and the culture of learning. *Educational Researcher*, 18(1), 32–42. 10.2307/1176008

Brown, L. I., & Kanyongo, G. Y. (2010). Gender differences in performance in mathematics in Trinidad and Tobago: Examining affective factors. *International Electronic Journal of Mathematics Education*, 5, 113–130. 10.29333/iejme/253

Christensen, R. W., & Knezek, G. A. (2009). Construct validity for the teachers' attitudes toward computers questionnaire. *Journal of Computing in Teacher Education*, 25(4), 143–155.

Creswell, J. W. (2009). *Research design: Qualitative, quantitative and mixed methods approaches* (3rd ed.). Sage.

Denzin, N. K., & Lincoln, Y. S. (Eds.). (2000). *Handbook of qualitative research* (2nd ed.). Sage.

Divaharan, S., & Koh, J. H. L. (2010). Learning as students to become better teachers: Pre-service teachers' IWB learning experience. *Australasian Journal of Educational Technology*, 26(4). Advance online publication. 10.14742/ajet.1072

Downes, S. (2006). *An introduction to connective knowledge.* Downes. http://www.downes.ca/post/33034

EACEA. (2011). *Science Education in Europe: National Policies.* Practices and Research.

Erwin, J. (1993). *Friendship and peer relationship in children.* John Wiley & Sons Ltd.

Eshiwani, G. S. (1982). A study of Women's Access to higher Education in Kenya with a special reference to Mathematics and Science Education. Bureau of Education research, Nairobi.

FAWE. (1999). Improving Performance of Girls in Schools. A Report. Nairobi, Kenya.

Fischbein, E., & Schnarch, D. (1997). The evolution with age of probabilistic, intuitively based misconceptions. *Journal for Research in Mathematics Education*, 28(1), 96–105. 10.2307/749665

Garrison, J. (1998). *Toward a pragmatic social constructivism.* In M. Larochelle, N. Bednarz, & J. Garrison (Eds.), *Constructivism and education* (pp. 3-60). Cambridge, UK: Cambridge. 10.1017/CBO9780511752865.005

Gherasim, L. R., Butnaru, S., & Mairean, C. (2013). Classroom environment, achievement goals and maths performance: Gender differences. *Educational Studies*, 39(1), 1–12. 10.1080/03055698.2012.663480

Hansen, B. B., & Klopfer, S. O. (2006). Optimal full matching and related designs via network flows. *Journal of Computational and Graphical Statistics*, 15(3), 609–627. 10.1198/106186006X137047

Hykle, J. A. (2011). Interrelationships among Cognitive Controls, Gender Science Achievement, and Science Process Skills. University of Cincinnati. in Mathematics Achievement: A Comparative Study. „Social Science" 42, John, P. (2002). Quantitative methods. In Marsh, D., & Stoker, G. (Eds.), *Theory and methods in political science* (2nd ed., pp. 184–211). Palgrave Macmillan.

Johnson, D. W., & Johnson, R. T. (1996). Cooperation and the use of technology. In D. H.

Johnson, L., Smith, R., Willis, H., Levine, A., & Haywood, K. (2010). *The horizon report.* Horizon.

Johnson, R. B., & Onwuegbuzie, A. J. (2004). Mixed methods research: A research paradigm whose time has come. *Educational Researcher*, 3(7), 14–26. 10.3102/0013189X033007014

Jonassen, D. H., Carr, C., & Hsiu-Ping, Y. (1998). Computers As Mindtools For Engaging Learners In Critical Thinking. *TechTrends*, 43(2), 24–32. 10.1007/BF02818172

Jones, G. A., Langrall, C. W., & Mooney, E. S. (2007). Research in probability: Responding to classroom realities. *Second handbook of research on mathematics teaching and learning, 2*, 909-955.

Jules-Van-Belle & Soetaert. (2001). Breakdown into the virtual user- involved design and learning. *Journal of Technology and Teacher Education*, 8(2), 31–42.

Kulik, C., & Kulik, J. (1991). Effectiveness of computer-based instruction: An updated analysis. *Computers in Human Behavior*, 7(1-2), 75–95. 10.1016/0747-5632(91)90030-5

Li, M., Zhang, Y., Liu, H., & Hao, Y. (2018). Gender differences in mathematics achievement in Beijing: A meta-analysis. *The British Journal of Educational Psychology*, 88(4), 566–583. 10.1111/bjep.1220329265190

Linn, M. C., Chang, H. Y., Chiu, J., Zhang, H., & McElhaney, K. (2010). Can desirable difficulties overcome deceptive clarity in scientific visualizations? *Successful remembering and successful forgetting: A Festschrift in honor of Robert A. Bjork*, 239-262.

Martin, M. O., Mullis, S., & Foy, P. (2008). *TIMSS 2007 International Science Report: Findings from IEA's Trends in International Mathematics and Science Study at the Fourth and Eighth Grades*. TIMSS & PIRLS International Study Center, Boston College.

Mayer, R. E. (1996). Learners as information processors: Legacies and limitations of educational psychology's second metaphor. *Educational Psychologist*, 31(3/4).

Mazana, Y. M., Suero Montero, C., & Olifage, C. R. (2019). *Investigating students' attitude towards learning mathematics*.

McDowall, T., & Jackling, B. (2010). Attitudes towards the accounting profession: An Australian perspective. *Asian Review of Accounting*, 18(1), 30–49. 10.1108/13217341011045999

McLeod, D. B. (1992). Research on affect in mathematics education: A reconceptualization. In Grouws, D. A. (Ed.), *Handbook of research on mathematics teaching and learning* (pp. 575–596). Macmillan.

McNabb, D. E. (2004). *Research methods for political science: Quantitative and qualitative methods*. M.E. Sharpe, Inc.

Merriam, S. B., & Caffarella, R. S. (1999). *Learning in adulthood: A comprehensive guide (2nd.)*. Jossey-Bass Publishers.

Mezirow, J. (1991). *Transformative dimensions of adult learning*. Jossey-Bass.

Middleton, J. A., & Spanias, P. A. (1999). Motivation for achievement in mathematics: Findings, generalizations, and criticisms of the research. *Journal for Research in Mathematics Education*, 30(1), 65–88. 10.2307/749630

Midgley, C., & Urdan, T. C. (1995). Predictors of middle school students' use of self-handicapping strategies. *The Journal of Early Adolescence*, 15(4), 389–411. 10.1177/0272431695015004001

Mignani, S., Ottaviani, M. G., & Ricci, R. (2014). Teaching statistics at secondary education in Italy: Some issues on large scale standardized test results. *ICOTS9.*

Miheso, M. (2002). Factors Affecting Mathematics Performance Among Secondary Schools Learners, In Nairobi Province, Kenya. [Thesis, Kenyatta University, Nairobi].

Mihindo, W. J., Wachanga, S. W., & Anditi, Z. O. (2017). Effects of Computer-Based Simulations Teaching Approach on Students' Achievement in the Learning of Chemistry among Secondary School Students in Nakuru Sub County, Kenya. *Journal of Education and Practice*, 8(5), 65–75.

Miller, T. (2009). *Television studies: The basics*. Routledge. 10.4324/9780203854198

Moore, D. S. (1997). New pedagogy and new content: The case of statistics. *Internationa statistical review, 65*(2), 123-137.

Moore, P., & Fitz, C. (1993). Gestalt theory and instructional design. *Journal of Technical Writing and Communication*, 23(2), 137–157. 10.2190/G748-BY68-L83T-X02J

Morgil, I., Yavuz, S., Oskay, Ö. Ö., & Arda, S. (2005). Traditional and computer-assisted learning in teaching acids and bases. *Chemistry Education Research and Practice*, 6(1), 52–63. 10.1039/B3RP90001K

Morrison, G. R., Ross, S. J., Morrison, J. R., & Kalman, H. K. (2019). *Designing effective instruction*. John Wiley & Sons.

Moshman, D. (1982). Exogenous, endogenous, and dialectical constructivism. *Developmental Review*, 2(4), 371–384. 10.1016/0273-2297(82)90019-3

Mugenda, O., & Mugenda, A. (2003). *Research methods: Quantitative and Qualitative methods*. Acts Press.

Mugenda, O. M. (1999). *Research methods: Quantitative and qualitative approaches*. African Centre for Technology Studies.

Murphy, K. L., & Cifuentes, L. (2001). Using web tools, collaborating, and learning online. *Distance Education*, 22(2), 285–305. 10.1080/0158791010220207

Mwalongo, A. (2011). Teachers' perceptions about ICT for teaching, professional evelopment, administration and personal use. *International Journal of ducation and development using information and communication echnology, 7*(3), 36.

Namlu, A., & Ceyhan, E. (2002). *Computer anxiety: A study on university students*. Anadolu University Publishing.

National Council of Teachers of Mathematics. (2000). *Principles and standards for school mathematics*. Author.

Nicolaidou, M., & Philippou, G. (2003). *Attitudes towards mathematics, self-efficacy and achievement in problem solving. European Research in Mathematics Education III*. University of Pisa.

Ogembo, J. O. (2017). *Effects of Computer Assisted Learning on Instruction in Organic Chemistry in Public Secondary Schools in Kwale County, Kenya*. [Doctoral dissertation, Kenyatta University, Nairobi].

Oginni, O. I., & Popoola, A. A. (2013). *Effects of Mathematics Innovation and Technology on Students Academic Performance in Open and Distance Learning (ODL)*. ODL. oasis.col.org

Ojeda, L., & Flores, L. Y. (2008). The influence of gender, generation level, parents' education level, and perceived barriers on the educational aspirations of Mexican American high school students. *The Career Development Quarterly*, 57(1), 84–95. 10.1002/j.2161-0045.2008.tb00168.x

Olakanmi, E. E., Gambari, I. A., Gbodi, E. B., & Abalaka, E. A. (2016). Promoting intrinsic and extrinsic motivation among chemistry students using computer assisted instruction. *International Journal of Educational Sciences*, 12(2), 155–168. 10.1080/09751122.2016.11890422

Olga, P. (2008). *The effects of computer-assisted instruction on the Achievement, attitudes and retention of fourth grade Mathematics course*. [Thesis, Middle EastTechnical University].

Onatsu-Arvillomi, T. P., & Nurmi, J.-E. (2002). The development of achievement strategies and academic skills during the first year of primary school. *Learning and Instruction*, 12(5), 509–527. 10.1016/S0959-4752(01)00029-9

Ormrod, J. E. (1999). Human learning (3rd ed.). Upper Saddle River, NJ: rentice-Hall

Orora, W., Keraro, F. N., & Wachanga, S. W. (2014). Effects of cooperative e-learning teaching strategy on students' achievement in secondary school Biology in Nakuru County, Kenya. *Sky Journal of Educational Research. 2(1)*, 001-009.

Otach, O. (2008). *Abolishing school fees in Africa: Lessons from Ethiopia. Ghana, Kenya, Malawi and Mozambique.*

Otieno, K. O. (2010). Teaching/learning resources and academic performance in mathematics in secondary schools in Bondo District of Kenya. *Asian Social Science*, 6(12), 126.

Palloff, R. M., & Pratt, K. (1999). *Building learning communities in cyberspace.* San Francisco, CA.

Papert, S. (1993). *Mindstorms: Children, computers, and powerful ideas* (2nd ed.). Harper Collins Publishers.

Patel, D. B. (2013). Construction and effectiveness of computer aided instruction (CAI) programme for the units of science and technology of standard VIII. *International Journal of Research in Education*, 2(1), 112–115.

Pawlak, M. (2022). Research into Individual Differences in SLA and CALL: Looking for Intersections. *Language Teaching Research Quarterly*, 31, 200–233. 10.32038/ltrq.2022.31.14

Pedretti, E., Mayer-Smith, J., & Woodrow, J. (1999). Teaming technology enhanced instruction in the science classroom and teacher professional development. *Journal of Technology and Teacher Education*, 7(2), 131–143.

Pelgrum, W. J. (2001). Obstacles to the integration of ICT in education: Results from a worldwide educational assessment. *Computers & Education*, 37(2), 163–178. 10.1016/S0360-1315(01)00045-8

Pelgrum, W. J. (2005). ICT Integration, Data from International Cooperative Studies. *Computers & Education*, 37.

Pelgrum, W. J., & Plomp, Tj. (1991). *The use of computers m educauon worldwide Results from the lEA "Computers tn Educatton' survey m 19 educattonal systems.* Oxford Pergamon Press.

Pepper, S. C. (1942). *World hypotheses: A study in evidence*. University of California Press.

Pereira-Mendoza, L. (2002). Would you allow your accountant to perform surgery? Implications for the education of primary teachers. In Phillips, B. (Ed.), *Proceedings of the Sixth International Conference on the Teaching of Statistics*. Hawthorn, VIC: International Statistical Institute.

Pfannkuch, M., & Ziedins, I. (2014). A modelling perspective on probability. In Chernoff, E. J., & Sriraman, B. (Eds.), *Probabilistic thinking: resenting plural perspectives (pp. 01–116)*. Springer. 10.1007/978-94-007-7155-0_5

Piaget, J. (1972). *The principles of genetic epistemology* (Mays, W., Trans.).

Pratt, D., & Ainley, J. (2014). Chance re-encounters: Computers in probability education revisited. In Wassong, T. (Ed.), *Mit Werkzeugen mathematik and stochastik lernen–using tools for learning mathematics and statistics* (pp. 165–177). Springer Fachmedien. 10.1007/978-3-658-03104-6_13

Riegle-Crumb, C., Farkas, G., & Muller, C. (2006). The role of gender and friendship in advanced course taking. *Sociology of Education*, 79(3), 206–228. 10.1177/003804070607900302 20333274

Ringgold, T. L. (2001). *The evaluation of computer attitudes and the validation of computer self-efficacy scales of minority students at an urban community college in Baltimore City*. Morgan State University.

Robinson, J. P., & Lubienski, S. T. (2011). The development of gender achievement gaps in mathematics and reading during elementary and middle school: Examining direct cognitive assessments and teacher ratings. *American Educational Research Journal*, 48(2), 268–302. 10.3102/0002831210372249

Ross, C. E., & Broh, B. E. (2000). The roles of self-esteem and the sense of personal control in the academic achievement process. *Sociology of Education*, 73(4), 270–284. 10.2307/2673234

Rostvik, C. M., & Fyfe, A. (2018). *Ladies, gentlemen, and scientific publication at the Royal Society, 1945-1990*. Open Library of Humanities. 10.16995/olh.265

Rozell, E. J., & Gardner, W. L. (1999). Computer-related success and failure: a longitudinal field study of the factors influencing computer- related performance. *Computers in Human Behavior, 15*(1), 1-10.

Russell, G., & Bradley, G. (1997). Teachers' computer anxiety: Implications for professional development. *Education and Information Technologies, 2*(1), 17-30.

Sanders, J. (2005). *Gender and technology in education: A research review*. Seattle: Center for Gender Equity.

Scardamalia, M., & Bereiter, C. (1994). Computer support for knowledge-building communities. *Journal of the Learning Sciences*, 3(3), 265–283. 10.1207/s15327809jls0303_3

Shaughnessy, J. M. (1992). Research in probability and statistics: Reflections and directions. In Grouws, D. A. (Ed.), *Handbook of Research on Mathematics Teaching and Learning* (pp. 465–494). Macmillan.

Shaughnessy, J. M. (2003). Research on students' understandings of probability. In Kilpatrick, J., Martin, W. G., & Schifter, D. (Eds.), *A Research Companion to Principles and Standards for School Mathematics* (pp. 216–226). National Council of Teachers of Mathematics.

Shaughnessy, J. M. (2007). Research on statistics learning and reasoning. *Second handbook of research on mathematics teaching and learning, 2*, 957-1010.

Shaw, B. A. (2004). *The relationship of teacher quality variables to technology integration in K–12 schools*. The University of Texas at El Paso.

Short, D. (2002). *Computer-based mathematics instruction: Automated remedial branching* [Unpublished doctoral dissertation, California State University].

Siddiqui, U., & Khatoon, T. (2013). Teaching physical science: Should we implement teacher-centered CAI or student-centered CAI at secondary school level in India? *European Scientific Journal, ESJ*, 9(10), 136–149.

Siemens, G. (2004). *A learning theory for the digital age*. eLearning Space. https://www.elearnspace.org/Articles/connectivism.htm

Sivin-Kachala, J. (1998). *Report on the effectiveness of technology in schools, 1990-1997*. Software Publishers Association.

Skinner, B. F. (1954). *Teaching Machines and Programmed Learning*. Department of Audiovisual Education, National Education Association.

Smaldino, D. E. (2002). The forum report: E-learning adoption rates and barriers. In Rossett, A. (Ed.), *The ASTD e-learning handbook* (pp. 19–23). McGraw-Hill.

Smarkola, C. (2007). Technology acceptance predictors among student teachers and experienced classroom teachers. *Journal of Educational Computing Research*, 37(1), 65–82. 10.2190/J3GM-3RK1-2907-7U03

Smetana, L. K., & Bell, R. L. (2012). Computer simulations to support science instruction and learning: A critical review of the literature. *International Journal of Science Education*, 34(9), 1337–1370. 10.1080/09500693.2011.605182

Snee, R. D. (1993). What's missing in statistical education? *The American Statistician*, 47(2), 149–154. 10.1080/00031305.1993.10475964

Steele, H. (2006, March). The "gender gap" in computing: Implications for teacher education. In *Society for Information Technology & Teacher Education International Conference* (pp. 811-816). Association for the Advancement of Computing in Education (AACE).

Steinberg, E. R. (1984). *Teaching computers to teach*. Lawrence Erlbaum Associates.

Sternberg, R. J. (2004). *Psychology*. Wadsworth.

Stohl, H. (2005). Probability in teacher education and development. In Jones, G. A. (Ed.), *Exploring probability in school: challenges for teaching and learning* (pp. 345–366). Springer. 10.1007/0-387-24530-8_15

Stratham, D. S., & Torell, C. R. (1996). *Computers in the classroom: The impact of technology on student learning* [Doctoral dissertation, Army Research Institute]. http://www.temple.edu/lss/htmlpublications/spotlights /200/spot206.htm

Tabrani, Z. A., & Masbur, M. (2016). Islamic Perspectives On The Existence Of Soul And Its Influence In Human Learning (A Philosophical Analysis Of The Classical And Modern Learning Theories). *JURNAL EDUKASI: Jurnal Bimbingan Konseling*, 1(2), 99–112.

Talanquer, V. (2014). Threshold concepts in chemistry: The critical role of implicit schemas. *Journal of Chemical Education*, 92(1), 3–9. 10.1021/ed500679k

Tapia, M., & Marsh, G. (2004). An instrument to measure mathematics attitudes. *Academic Exchange Quarterly*, 8, 1–8.

Tapscott, D. (1998). *Growing up digital: The rise of the Net generation*. McGraw-Hill.

Taub, G. E., Benson, N., & Szente, J. (2014). Improving Mathematics: An Examination of the Effects of Specific Cognitive Abilities on College-age Students' Mathematics Achievement. *International Journal for the Scholarship of Teaching and Learning*, 8(2), 8–15. 10.20429/ijsotl.2014.080208

UNESCO. (2002). *Information and Communication Technology in Education. A Curriculum for Schools and Programme of Teacher Development*. UNESCO.

UNESCO and Commonwealth of Learning (COL). (2004). *Schoolnet Toolkit*. Bangkok: UNESCO; Vancouver: Commonwealth of Learning.

Upadyaya, K., & Eccles, J. S. (2014). Gender differences in teachers' perceptions and children's ability self-concepts. InI. Schoon & J. S. Eccles (Eds.), *Gender differences in aspirations and attainment: A life course perspective* (pp. 79–100). Cambridge, UK: Cambridge University Press.

Ursini, S., & Sánchez, G. (2008). Gender, technology and attitude towards mathematics: A comparative longitudinal study with Mexican students. *ZDM Mathematics Education*, 40(4), 559–577. 10.1007/s11858-008-0120-1

Van Mier, H. I., Schleepen, T. M., & Van den Berg, F. C. (2019). Gender differences regarding the impact of math anxiety on arithmetic performance in second and fourth graders. *Frontiers in Psychology*, 9, 2690. 10.3389/fpsyg.2018.0269030713516

Vanderlinde, R., van Braak, J., & Hermans, R. (2009). Educational technology on a turning point: Curriculum implementation in Flanders and challenges for schools. *Educational Technology Research and Development*, 57(4), 573–584. 10.1007/s11423-009-9126-9

Von Glasersfeld, E. (1995). *Radical Constructivism: A Way of Knowing and Learning. Studies in Mathematics Education Series: 6*. Falmer Press, Taylor & Francis Inc.

Wangid, M. N., & Widyastuti, P. (2021). Effect of Gender and Department on Motivation, Self-Efficacy, And Perception as Teacher of Pre-Service Teacher Students. *Multicultural Education*, 7(6), 6–8.

Wanjala, M. (2005). *Effectiveness of computer-based instruction versus conventional methods on students' attitude, motivation and achievement in the teaching of mathematics in secondary schools of Bungoma District.* [Unpublished Master's Thesis, Moi University, Eldoret, Kenya].

Wanyonyi, P. W. (2013). *An Assessment of Factors Affecting Students' Performance in Mathematics at KCSE Level in Secondary Schools in Kakamega County, Kenya* [Doctoral dissertation].

Wekesa, D. W., Wekesa, E. W., & Amadalo, M. M. (2013). A computer mediated simulation module for teaching cell division in secondary school biology. *International Journal of Educational Research and Development.*, 2(5), 114–130.

World Bank. (2007). Developing Science and Mathematics and ICT education in SSA. World Bank.

Yaghi, H. M., & Ghaith, G. M. (2002). Correlates of computing confidence among teachers in an international setting. *Computers in the Schools*, 19(1-2), 81–94. 10.1300/J025v19n01_07

Yilmaz, N. P. (2011). Evaluation of the technology integration process in the Turkish education system. *Contemporary Educational Technology*, 2(1), 37–54.

ADDITIONAL READING

Anthony, L. M., Clarke, M. C., & Anderson, S. J. (2000). Technophobia and personality subtypes in a sample of South African university students. *Computers in Human Behavior*, 16(1), 31–44. 10.1016/S0747-5632(99)00050-3

Buabeng-Andoh, C. (2012). An exploration of teachers' skills, perceptions and practices of ICT in teaching and learning in the Ghanaian second-cycle schools. *Contemporary Educational Technology*, 3(1), 36–49. 10.30935/cedtech/6066

Carmona, J., Martínez, R. J., & Sánchez, M. (2005). Mathematical background and attitudes toward statistics in a sample of Spanish college students. *Psychological Reports*, 97(1), 53–62. 10.2466/pr0.97.1.53-6216279305

Chang, K., Chen, I., & Sung, Y. (2002). The effect of concept mapping to enhance text comprehension and summarization. *Journal of Experimental Education*, 71(1), 5–23. 10.1080/00220970209602054

Charagu, S. N. (2015). *Effects of computer assisted learning on secondary school students' achievement in Chemistry in Murang'a South Sub-county, Murang'a County, Kenya*. Unpublished MED thesis, Kenyatta University, Nairobi.

Lipnevich, A. A., MacCann, C., Krumm, S., Burrus, J., & Roberts, R. D. (2011). Mathematics attitudes and mathematics outcomes of US and Belarusian middle school students. *Journal of Educational Psychology*, 103(1), 105–118. 10.1037/a0021949

Liu, Y., & Thompson, P. (2007). Teachers' understanding of probability. *Cognition and Instruction*, 25(2-3), 113–160. 10.1080/07370000701301117

Mahmud, R., & Ismail, M. A. (2010). Impact of training and experience in using ICT on in-service teachers' basic ICT literacy. *Malaysian Journal of Educational Technology*, 10(2), 5–10.

Norris, C., Sullivan, J., Poirot, T., & Soloway, E. (2003). No access, no use, no impact: Snapshot surveys of educational technology in K-12. *Journal of Research on Technology in Education*, 36(1), 15–27. 10.1080/15391523.2003.10782400

Nur, M. A. (2010). *Factors that Influence Secondary School Students 'Performance in Mathematics in Banadir Region, Somalia*. [Unpublished Master 's thesis, Kenyatta University, Nairobi, Kenya].

Nurhikmah, H. N. H., Febriati, F., & Ervianti, E. (2021). The Impact of Computer-based Test and Students' Ability in Computer Self-Efficacy on Mathematics Learning Outcomes. *Journal of Education Technology*, 5(4), 603–610. 10.23887/jet.v5i4.34942

Ogembo, J. O. (2012). *Determinants of Students' Poor Performance in Chemistry in Public Schools of Kwale County, Kenya* [Unpublished MED thesis in the School of Education of Kenyatta University, Nairobi].

Sifuna, D. N. (2004). The illusion of universal free primary education in Kenya. *Wajibu*, 19(2), 5–8.

Silverman, D. (2000). *Doing qualitative research: A practical handbook.* Sage.

Taub, G. E., Keith, T. Z., Floyd, R. G., & McGrew, K. S. (2008). Effects of general and broad cognitive abilities on mathematics achievement. *School Psychology Quarterly*, 23(2), 187–198. 10.1037/1045-3830.23.2.187

Tezci, E. (2009). Teachers' effect on ICT use in education: The Turkey sample. *Procedia: Social and Behavioral Sciences*, 1(1), 1285–1294. 10.1016/j.sbspro.2009.01.228

Tsai, M., & Tsai, C. (2003). Students computer achievement, attitude and anxiety, the role of learning strategies. *Journal of Educational Computing Research*, 28(1), 47–61. 10.2190/PL27-TC1Q-08B2-RMCL

Turner, J. C., & Meyer, D. K. (2009). Understanding motivation in mathematics: what is happening in classrooms? In *Handbook of motivation at school* (pp. 541–566). Routledge.

Vekiri, I., & Chronaki, A. (2008). Gender issues in technology use: Perceived social support, computer self-efficacy and value beliefs, and computer use beyond school. *Computers & Education*, 51(3), 1392–1404. 10.1016/j.compedu.2008.01.003

Von Glaserfeld, E. (1991). Knowing without metaphysics: Aspects of the radical constructivist position. In Steier, F. (Ed.), *Research and reflexivity* (pp. 553–571). Sage.

World Bank. (2008). *The road not traveled (MENA Development Report).* The International Bank for Reconstruction and Development/ The World Bank.

Yadavalli, R., & Manda, M. S. L. (2014). *The Effect of Gender, Medium of instruction, Parent's Educational and Socio-Economic background on Children's Conceptions and Abilities in Graphical Representation and Interpretation of Geometrical Concepts.*

KEY TERMS AND DEFINITIONS

Chance experiment: An activity with several conceivable outcomes for which we have no way of knowing which one will occur.

Computer Assisted Learning: Incorporation of computers and computer assisted learning applications into instruction.

Computer Simulation: Computer simulation of the consequences of a mathematical model connected with a system in order to reproduce its behavior.

Conventional instruction methods: A teaching method that emphasizes classroom instruction supplemented by other resources rather than computer and computer-assisted educational materials.

Event: A subset of all conceivable outcomes that has been chosen.

High Achievement: A good level of mathematics achievement equivalent to more than 40% on a standardized test.

ICT integration: Classroom learning is aided by the use of ICT tools.

Low achievement: A arithmetic achievement level that is less than 30% on a standardized test is considered unsatisfactory.

Outcome: The outcome of a trial. When a six-sided die is rolled, for example, the result might be any number between 1 and 6.

Probability: A mathematical topic that deals with the probability of an event occurring, as assessed by the ratio of favorable cases to the total number of possible cases.

Public Schools: A government-funded and operated tuition-free school.

Technology: The theory and practice of learning process and resource design, development, utilization, management, and evaluation

Trial: Within an experiment, there is only one performance. Rolling a die a few times, for example. Each time the die is rolled, it is a test.

Section 4
Equity

Chapter 9
Equity and Inclusion in Mathematics Education in India:
Strategies for Supporting Diverse Learners

Sanjeev Kumar Jha
https://orcid.org/0000-0001-5015-038X
PM Shri KV AFS Arjangarh, New Delhi, India

ABSTRACT

Mathematics education is essential for improving students' analytical thinking, problem-solving abilities, and cognitive growth. It poses several challenges for ensuring equity and inclusion for all, particularly in a multicultural nation like India. This chapter aims to examine the disparities in mathematics performance based on gender and social categories, utilising data from the National Achievement Survey. It explores the significance of equity and inclusion in mathematics education and offers strategies to support learners from diverse backgrounds. The chapter also explores pedagogical approaches aimed at fostering equity and inclusivity. The study's findings help educators, policymakers, and educational administrators to establish an equitable and inclusive mathematics education. It concludes equity and inclusion in mathematics education necessitate a comprehensive strategy that encompasses policy reforms, allocation of resources and infrastructure, teacher professional development, cultural sensitivity, and the establishment of inclusive learning environments.

DOI: 10.4018/979-8-3693-2873-6.ch009

INTRODUCTION

Equity and inclusion in education are crucial principles that aim to ensure that all students have equal access to quality education and opportunities for success, regardless of their personal characteristics. Furthermore, the 'Organization for Economic Cooperation and Development' [OECD], (2012) emphasizes "reducing school failure pays off for both society and individuals. It can also contribute to economic growth and social development. Indeed the highest performing education systems across OECD countries are those that combine quality with equity (p. 38)." These principles recognize and address the systemic barriers and disparities that exist in education systems and strive to create an inclusive learning environment where every student feels valued, supported, and empowered to reach their full potential. The Right to Education Act [RTE] (2009) in India empowers all children in age group of 6-14 years with education (equitable quality of education to all) as fundamental right.

Equity in education refers to the equal opportunities, and essential support to meet the diverse needs of all students. OECD (2012) defines "equity in education means that personal or social circumstances such as gender, ethnic origin or family background, are not obstacles to achieving educational potential (fairness) and that that all individuals reach at least a basic minimum level of skills (inclusion)" (p. 9). Therefore, it requires acknowledging and addressing the disparities that exist among different student groups, such as those based on socioeconomic status, caste, gender, disability, or language proficiency. By actively working to reduce these disparities, education systems can strive to provide an equitable education for all students. This may involve allocating additional resources to schools or communities with higher needs, implementing targeted support programs, or adapting teaching methods to accommodate different learning styles and abilities.

Inclusion, on the other hand, focuses on creating an educational environment that embraces and values diversity. It involves actively promoting the participation and engagement of all students, regardless of their individual differences. National Curriculum Framework [NCF] (2005) highlights "inclusive education is about embracing all" (p 82.). Inclusive education recognizes that every student has unique strengths, challenges, and perspectives, and strives to create a supportive and respectful learning environment that accommodates and celebrates these differences. According to United Nationals [UN] (2006) inclusion is defined as processes to overcome barriers hindering students from receiving teaching and participation in learning experiences. Inclusive practices may include adapting teaching strategies and materials to meet diverse learning needs, fostering positive relationships and peer interactions, and providing appropriate support and accommodations for students with disabilities or special educational needs.

Both equity and inclusion are interconnected and mutually reinforcing. Achieving equity requires creating an inclusive education system that welcomes and supports all students, while inclusion cannot be fully realized without addressing and eliminating systemic inequities. By prioritizing equity and inclusion in education, societies can work towards breaking down barriers, fostering social cohesion, and ensuring that every student has an equal opportunity to succeed academically, socially, and personally.

It is important to note that achieving equity and inclusion in education is an ongoing and complex process that requires collaboration among various stakeholders, including policymakers, educators, parents, communities, and students themselves. It involves continuous reflection, dialogue, and action to identify and dismantle discriminatory practices and to create an inclusive and empowering educational experience for all.

In India, the availability of high-quality mathematics education is frequently uneven, resulting in the exclusion of marginalized groups such as girls, social groups, students from low-income families, individuals with disabilities, and those residing in rural areas. Mathematics is a universally understood language that goes beyond cultural and linguistic boundaries. It offers a structure for logical thinking and problem-solving ability. Therefore, it has a pivotal role in influencing the cognitive growth and future prospects of students. Nevertheless, equity and inclusion is important in mathematics.

Mathematics Education: Importance of Equity and Inclusion

National Policy on Education [NPE] (1986) emphasizes mathematics should be visualized as the vehicle to train a child to think, reason, analyze, and particulate logically. Apart from being a specific subject it should be treated as a concomitant to any subject involving analysis and meaning. According to Jha (2014) "mathematics is in the root of development of all the civilizations (p. 3). Thus, ensuring equity and inclusion in mathematics education is crucial. It is essential for establishing a balanced and impartial learning environment that addresses the requirements of all students. In a heterogeneous society such as India, where students come from diverse socio-economic backgrounds, genders, and social categories, it is imperative to acknowledge the inequalities and establish an all-encompassing educational setting. Moreover, an equitable quality of mathematics education cultivates a feeling of inclusion, self-assurance, and drive among students, which has a beneficial effect on their overall educational journey.

Mathematics education serves as a vital foundation for individuals to develop critical thinking, problem-solving skills, and logical reasoning abilities. However, according to Jha and Tomar (2010) many individual developed some amount of phobia with mathematics. Nevertheless, mathematics not only equips students with mathematical knowledge but also provides them with essential skills for success in various academic and professional domains. However, achieving equity and inclusion in mathematics education is a pressing concern, particularly in a diverse country like India.

Thus, equity in mathematics education refers to providing fair and opportunities for all students, irrespective of their social, gender, etc. background, to access and excel in mathematics learning. Inclusion, on the other hand, emphasizes creating a supportive and welcoming learning environment that respects and values the diverse needs, experiences, and identities of students.

Promoting Equity and Inclusion: Strategies for Supporting Culturally Diverse Learners

In the pursuit of equity and inclusion in mathematics education, it is essential to prioritize cultural responsiveness. It recognizes and values the diverse cultural backgrounds of students, incorporating their experiences and perspectives into instruction. In the Indian context, where cultural diversity is rich and varied, implementing strategies that promote cultural responsiveness can significantly enhance the learning experiences of diverse learners.

One crucial aspect of promoting equity and inclusion in mathematics education is recognizing and valuing the diverse cultural backgrounds of students. India is a country with a mosaic of cultures, languages, and traditions. Teachers must acknowledge and respect the cultural diversity present in their classrooms. This can be achieved by fostering an inclusive and accepting classroom environment, where students feel safe to express their cultural identities and experiences. Creating opportunities for students to share and celebrate their cultural backgrounds can build a sense of belonging and promote a positive learning environment.

Incorporating students' cultural experiences into mathematics instruction is another effective strategy for promoting equity and inclusion. Teachers can draw on students' cultural contexts and experiences to make mathematical concepts more relatable and meaningful. For example, when teaching measurement, teachers can use examples that align with students' cultural practices, such as measuring ingredients in cooking or mapping distances to cultural landmarks. By connecting mathematics to students' everyday lives, teachers can enhance engagement and understanding, making the subject more accessible to diverse learners.

Using culturally relevant examples and contexts in problem-solving is a valuable strategy to promote equity and inclusion in mathematics education. Mathematics problems that reflect students' cultural backgrounds and experiences can foster a deeper understanding of concepts and increase student engagement. For instance, when introducing fractions, teachers can use examples related to traditional Indian art or music to illustrate the concept. By incorporating culturally relevant contexts, teachers can create a bridge between students' cultural knowledge and mathematical concepts, making learning more relatable and meaningful.

Furthermore, it is important to provide opportunities for students to explore and investigate mathematical concepts within their own cultural contexts. Encouraging students to explore their cultural heritage through mathematics can help them develop a sense of pride and ownership over their learning. Teachers can assign projects or investigations that allow students to apply mathematical concepts to explore cultural phenomena or solve real-world problems specific to their communities. This approach not only enhances students' mathematical skills but also fosters cultural pride and self-confidence.

Promoting equity and inclusion in mathematics education in the Indian context requires a focus on cultural responsiveness. Recognizing and valuing students' diverse cultural backgrounds, incorporating their cultural experiences into instruction, and using culturally relevant examples and contexts in problem-solving are strategies that can support diverse learners. By embracing cultural responsiveness, teachers can create a more inclusive and engaging mathematics classroom, where all students feel valued, respected, and empowered. Through these strategies, India can foster equity and inclusion in mathematics education, celebrating the rich cultural diversity that characterizes the nation.

According to Tabassum, and Nayak (2021) gender stereotypes in India still continue to exist and are transmitted through media, and through social, educational, and recreational socialization, which promote gender prejudice and discrimination. Furthermore, UNESCO (2022) highlights that "gender norms in society are often replicated within education, and reflected in gender disparities and inequalities in access, participation, completion, and achievement" in general.

Thus, it can be understood that gender stereotypes persist in India, perpetuated through media, social, educational, and recreational factors, leading to prejudice and discrimination in education and overall gender disparities. So, gender disparity, particularly in context of mathematics achievement exists.

Objectives of the Study

1. To compare the differences in mathematics achievement in relation to gender of the students.
2. To compare the differences in mathematics achievement in relation to social categories of the students.
3. To analyse challenges in access and participation in mathematics education, particularly in terms of equity and inclusion.

4. To analyse innovative teaching methods and instructional strategies in mathematics education.
5. To examine educational policies promoting equity and inclusion.

METHODOLOGY OF THE STUDY

For Objective Number 1 and 2

The data sources for this research include existing literature, national-level surveys –National Achievement Surveys (NAS). NAS is a significant undertaking conducted in the Government of India by the National Council of Educational Research and Training (NCERT). The survey assesses academic accomplishments of students across the country. This nationwide survey encompasses students in Class III, Class V, Class VIII, and Class X. It covers a variety of disciplines, such as mathematics, science, social science, and languages.

Through the analysis of performance data obtained from the NAS, policymakers, educators, and stakeholders acquire valuable insights into the strengths and weaknesses of the education system at various levels. Furthermore, the utilization of NAS data enables the detection of discrepancies in educational accomplishments among various demographic categories, thereby assisting in the development of focused interventions to narrow these disparities and foster inclusive education. In addition, NAS data functions as a standard for evaluating progress over a period of time and comparing the achievements of states, districts, and schools. This promotes a positive and competitive environment, motivating those involved to implement creative methods in order to enhance the quality and fairness of education. Thus, NAS data is obtained through a representative sample survey conducted nationwide. It is a periodic endeavor. It reports performance of states of India. The data is publicly available at https://ncert.nic.in/ and https://nas.gov.in/report-card/2021.

For Objective Number 3 to 5

For these objectives an extensive review of existing literature and policies was done.

The current study conducted an analysis of the NAS data for classes III, V, and VIII, and classified the states/UTs into three groups based on their mathematics achievement compared to the national level. These groups were labelled as "Significantly lower than boys," "Significantly higher than boys," and "No significant difference." Furthermore, this categorization was extended to encompass social groups such as Scheduled Castes (SC), Scheduled Tribes (ST), Other Backward Classes (OBC), and general category.

Schedule Castes (SC) refer to the designated name given to the lowest castes in society, who are currently under the protection of the government and provided with specific privileges. The Indian constitution (under Article 341 of Indian Constitution) acknowledges the SC as one of the most marginalized socio-economic groups in India.

Schedule Tribes (ST) refer to the Indigenous peoples or tribal communities who are not part of the caste system. Their primary habitat is the forest. The Indian Constitution (under Article 342 of Indian Constitution) acknowledges ST as one of the most marginalized socio-economic groups in India.

Other Backward Castes (OBC) refer to a group of castes that are considered educationally and socially disadvantaged, excluding the SC and ST, as defined by the Constitution of India (under Article 340 of Indian Constitution).General Category (in relation to caste) refers to a group of castes that are considered educationally and socially advantaged group. Commonly, all the people except SC, ST, and OBC are categorized in 'General Category' in India.

This comprehensive methodology offers a detailed comprehension of the disparity in mathematics achievement state wise, providing valuable insights into the varied educational environment across different states and social demographics in India.

Mathematics Achievement and Disparities

In the Indian context, there have been historical disparities in mathematics achievement based on gender and social categories. However, it is important to note that these disparities are not inherent to any gender or social category, but rather influenced by various societal factors.

Mathematics Achievement in Relation to Gender of the Students

Gender-based disparities in mathematics achievement have been a significant concern across the States/UTs in India. It is important to understand the performance of the students in mathematics in relation to their gender. The table-1 depicts number of states/ UTs, in which girls performed 'significantly lower', 'significantly higher' and 'no significant difference' that of boys. A total 36 States/ UTs participated in NAS 2017 whereas a total 37 States/ UTs participated in NAS 2021.

Table 1. Number of states having girl's mathematics achievement lower, higher, and no significant difference with boys

	Number of States/UTs in which girl's Mathematics Achievement is									Number of States/UTs
	Significantly lower than boys			Significantly higher than boys			No difference			
Class	III	V	VIII	III	V	VIII	III	V	VIII	
NAS 2021	9	17	14	16	14	14	12	6	9	37
NAS 2017	5	1	3	6	16	11	25	19	23	36

Class III Mathematics Achievement

- In 2021, the number of states/UTs where girls' mathematics achievement was significantly lower than boys' achievement was 9 states/UTs. This indicates a concerning trend of gender disparity in mathematics achievement at this grade level.
- Conversely, girls outperformed boys significantly in mathematics achievement in 16 states/UTs, suggesting a near positive trend towards gender equity in certain regions.
- In 12 states/UTs, there was no significant difference between girls' and boys' mathematics achievement, reflecting a balanced performance.

Class V Mathematics Achievement

- Similar to Class III, there remained a notable gender gap in mathematics achievement, with 17 states/UTs reporting significantly lower achievement for girls compared to boys.
- The gender gap seems to be decreasing, as the number of states/UTs with significantly higher achievement for girls had increased from 1 in 2017 to 14 in 2021.
- There were still 6 states/UTs where no significant difference in mathematics achievement between genders was observed.

Class VIII Mathematics Achievement

- The pattern observed in Class VIII is consistent with that of Class V, with a considerable gender gap persisting, as evidenced by 14 states/UTs reporting significantly lower achievement for girls.
- There was a notable improvement from 2017, with 14 states/UTs reporting significantly higher mathematics achievement for girls in 2021.
- Only 9 states/UTs showed no significant difference in mathematics achievement between girls and boys at this grade level.

Overall, while there are still significant gender disparities in mathematics achievement across different grade levels and states/UTs, there are promising signs of improvement, particularly in the increased number of regions where girls outperform boys. However, efforts to address and eliminate gender disparities in mathematics achievement should remain a priority in educational policy and practice.

Furthermore, traditional gender roles and stereotypes exist in India (Tabassum, & Nayak, 2021), it have often led to the perception that mathematics is more suited for boys (Paudel, 2020). These stereotypes, coupled with societal expectations and biases, have contributed to lower participation and achievement levels for girls in mathematics (Wolff, 2021; Xie & Liu, 2023). However, it is essential to recognize that such disparities are not due to any innate differences in mathematical abilities between genders but are a result of societal factors. Efforts are being made to challenge these stereotypes and promote gender equality in mathematics education, such as implementing gender-inclusive teaching practices, providing role models, and creating supportive learning environments (Nimely & Nyamu, 2023).

Mathematics Achievement in Relation to Social Categories

Social category-based disparities particularly related to socioeconomic status and caste, also impact mathematics achievement in India. Students from disadvantaged socioeconomic backgrounds often face challenges in accessing quality education, including mathematics instruction (White, Ruther, Kahn, & Dong, 2016; Juang & Schachner, 2020). Limited resources, inadequate infrastructure, and lack of educational support contribute to lower achievement levels for these students. Additionally, students from marginalized castes may face social discrimination and biases that affect their educational opportunities and outcomes. Addressing these disparities requires targeted interventions, such as providing equal access to quality mathematics education, offering financial support, and implementing affirmative action policies (Shang, 2022; Abdulkadri, 2022).

Table 2 discusses about the social disparity in mathematics achievement across the classes in Indian states/ UTs. It depicts number of states/ UTs having mathematics achievement significantly lower than general category students.

Table 2. Mathematics achievement in states/UTs in relation to social group

Class	NAS	Mathematics Achievement Significantly lower than general category			Number of States/UTs
		SC	ST	OBC	
Class III	NAS 2021	20	13	13	37
	NAS 2017	16	11	3	25
Class V	NAS 2021	21	31	22	37
	NAS 2017	7	3	5	24
Class VIII	NAS 2021	30	34	29	37
	NAS 2017	7	8	2	22

Class III Mathematics Achievement

- In 2021, there were 20 states/UTs where Scheduled Caste (SC) students' mathematics achievement was significantly lower than the general category. Similarly, 13 states/UTs reported significantly lower achievement for Scheduled Tribe (ST) and Other Backward Classes (OBC) students each.
- Compared to 2017, there had been an increase in the number of states/UTs reporting significantly lower mathematics achievement for SC, ST, and OBC students.

Class V Mathematics Achievement

- In 2021, the disparities in mathematics achievement persisted, with 21 states/ UTs reporting significantly lower achievement for SC students, 31 for ST students, and 22 for OBC students when compared to the general category.
- There had been a substantial increase in the number of states/UTs reporting significantly lower mathematics achievement for ST students from 2017 to 2021.

Class VIII Mathematics Achievement

- The disparities in mathematics achievement are most pronounced at Class 8 level, with 30 states/UTs reporting significantly lower achievement for SC students, 34 for ST students, and 29 for OBC students compared to the general category in 2021.
- Similar to Class V, there had been a notable increase in the number of states/ UTs reporting significantly lower mathematics achievement for ST students from 2017 to 2021.

Overall, the data underscores the persistent disparities in mathematics achievement among different social categories across various grade levels. While efforts to address to challenge stereotypes, promote gender equality, and address socioeconomic and caste-based inequalities such as *'Beti Bachao Beti Padhao'* initiated in 2015 (in English Save girl, Teach girls), Kasturba Gandhi Balika Vidyalaya (in English Kasturba Gandhi Girls School) started in 2004, Right to Education Act 2009, etc. still, there remains a pressing need for targeted interventions and policies to ensure equitable access to quality education and improve learning outcomes for all students, regardless of their social background.

The gender and social category-based disparities in mathematics achievement exist in the Indian context, but they are influenced by societal factors rather than innate differences in abilities (Juang & Schachner, 2020). . Creating an inclusive and supportive mathematics education system that provides equal opportunities and resources for all students is crucial to bridging these disparities and ensuring equitable mathematics achievement in India.

Challenges in Access and Participation in Mathematics Education, Particularly in Terms of Equity and Inclusion

Inclusion is a complex process of participation where both ideological and societal issues, as well as individual and subject-specific issues, must be considered in the educational endeavour (Kishore & Sandeep, 2022; Roos, 2023). In India, with its vast population and diverse cultural, linguistic, and socioeconomic backgrounds, faces unique challenges in ensuring equity and inclusion in mathematics education. Historically, certain groups, such as girls, students from marginalized communities, and those with disabilities, have faced systemic barriers and discrimination in accessing quality mathematics education (Shakunthala, 2014; Bani & Lach, 2024). Furthermore, NCF (2005) suggests "disability is a social responsibility –accept it (p. 82) "

Disparities in socioeconomic status, cultural biases, insufficient infrastructure, and inadequate resources all contribute to the difficulty of mathematics education in India. Students from low-income backgrounds often lack the resources necessary to improve their academic performance due to these factors. Inclusive practices are necessary due to the persistence of gender disparities as well as learning disabilities and special educational needs. Tackling these issues requires a thorough plan. The following section discusses specific challenges associated with inclusion and equity in terms of access and participation.

1. **Socioeconomic Disparities:** Socioeconomic disparities pose a significant challenge to equity and inclusion in mathematics education. According to Das and Sinha (2017) the socioeconomic status influences the performance in mathematics of the students. Many students from economically disadvantaged backgrounds lack resources such as textbooks, educational technology, and adequate educational support. Disparities in infrastructure, lack in educational support and teaching quality between urban and rural areas further exacerbate the issue. Limited access to quality mathematics education perpetuates educational inequity and hampers the potential of students from marginalized communities. According to Jha and Tomar (2010) the students with low socioeconomic status have higher mathematics phobia.
2. **Cultural Biases and Stereotypes:** Cultural biases and stereotypes surrounding mathematics can impact access and participation for certain groups of students. In some cases, there is a perception that mathematics is only for students who are academically inclined or from specific communities (Juang & Schachner, 2020). This can create a sense of exclusion and discourage students from diverse backgrounds from pursuing mathematics education. Addressing these biases and promoting a more inclusive and culturally responsive environment is crucial for ensuring equitable access and participation.
3. **Inadequate Infrastructure and Resources:** In many parts of India, particularly in rural and remote areas, there is a lack of adequate infrastructure and resources for mathematics education. Schools may lack well-equipped mathematics labs, libraries, or access to educational technology. Limited access to textbooks, teaching materials, and internet connectivity further hampers students' learning opportunities (Shakunthala, 2014; Kishore & Kumar, 2022). The lack of essential resources affects the quality of mathematics education and creates disparities in access and participation.
4. **Teacher Quality and Training:** The quality of mathematics teachers and their training also presents a challenge to equity and inclusion. In some cases, teachers may not have received sufficient training in inclusive teaching practices or addressing the diverse learning needs of students (Barrett, 2018; Juang & Schachner, 2020; Kishore & Kumar, 2022). Unconscious biases and stereotypes

among teachers can inadvertently hinder students' learning experiences. To promote equity and inclusion, it is essential to invest in teacher professional development programs that emphasize inclusive teaching practices, cultural competence, and strategies for addressing diverse student needs.

5. **Linguistic Diversity:** India is a linguistically diverse country with numerous regional languages. Indian Constitution recognizes 22 languages (Eight Schedule Constitution of India). Moreover, there are a number of regional language and dialects. The language barrier can pose challenges for students whose first language is different from the language of instruction in mathematics. In some cases, students may struggle to understand mathematical concepts due to language limitations. Ensuring access to mathematics education in students' native languages, providing language support, and promoting bilingual instructional approaches can help address this challenge.
6. **Gender Disparities:** Gender disparities persist in mathematics education in India, with lower participation and achievement levels for girls compared to boys. Societal stereotypes, biases, and cultural norms contribute to these disparities (Juang & Schachner, 2020). Girls may face discouragement or limited opportunities to pursue mathematics due to perceived gender roles. Addressing gender biases, promoting gender-inclusive teaching practices, and providing supportive learning environments are essential for ensuring equitable access and participation for all students. Biswas and Paul (2018) found that "males are generally finds to hold more positive attitude towards mathematics than females" (p. 1614). This adverse attitude of girls towards mathematics influences their performance and choice in this subject.
7. **Learning disabilities and special educational needs:** these are additional barriers to equity and inclusion in mathematics education. Students with learning disabilities face challenges in comprehending mathematical concepts or applying problem-solving strategies. Special educational needs require tailored instructional approaches, accommodations, and support to ensure equitable access to the mathematics curriculum. Inclusive practices, such as differentiated instruction, individualized support, and assistive technologies, are crucial to address the diverse learning needs of these students (Juang & Schachner, 2020; Kishore & Kumar, 2022).
8. **Teacher biases and lack of inclusive teaching practices:** These also impede equity and inclusion in mathematics education. Unconscious biases, stereotypes, and limited cultural competence among teachers can inadvertently hinder students' learning experiences (Kishore & Kumar, 2022). Biased expectations, differential treatment, and a lack of representation in instructional materials can marginalize certain student groups. Addressing these biases requires comprehensive teacher professional development programs that promote awareness,

cultural competence, and inclusive instructional strategies. Teachers must be equipped with the knowledge and skills to create inclusive classrooms that value diversity and support the learning needs of all students.

In the Indian context, there are various challenges to achieving equity and inclusion in mathematics education. These challenges include socioeconomic disparities, cultural biases, inadequate infrastructure, limited resources, teacher quality, language barriers, and gender disparities (Juang & Schachner, 2020; Kishore & Kumar, 2022). To tackle these challenges, a comprehensive strategy is needed, which includes implementing policy changes, allocating resources and improving infrastructure for education, providing training for teachers, promoting cultural sensitivity, and establishing inclusive learning environments. India can strive to achieve equitable access and participation for all students in mathematics education by tackling these challenges.

Innovation in Teaching-Learning Mathematics: Strategies for Supporting Diverse Learners

Innovative teaching methods and instructional strategies play a vital role in promoting equity and inclusion in mathematics education (Roos, 2018). Culturally responsive teaching practices, differentiated instruction, and inclusive pedagogical approaches can help meet the diverse learning needs of students. Integrating real-life examples and contexts relevant to students' experiences can enhance their interest and engagement in mathematics.

The following section discusses the innovation in teaching-learning mathematics to support diverse learner.

1. **Differentiated Instruction:** Implementing differentiated instruction allows teachers to tailor their teaching methods and materials to meet the individual needs of students. As Indian mathematics classrooms, having diversity and disparities such as socio-economic, linguistic disparity, etc. thus, differentiated instructions are essential. This approach recognizes that students have different learning styles, abilities, and interests (Romanelli, Bird, & Ryan, 2009). By providing varied instructional activities, resources, and assessments, teachers can ensure that all students, regardless of their backgrounds such as gender, socioeconomic status, or learning disabilities abilities, have opportunities to engage with and succeed in mathematics.
2. **Problem-Based Learning:** Problem-based learning (PBL) involves presenting students with real-world problems or scenarios that require them to apply mathematical concepts and skills to find solutions. PBL promotes critical thinking, problem-solving skills, and collaboration (Romanelli, Bird, & Ryan, 2009). By incorporating culturally relevant and authentic problems, teachers can make

mathematics more relatable and meaningful so the students from diverse background and need can connect to the mathematics class. .

3. **Collaborative Learning:** Collaborative learning encourages students to work together in groups or teams to solve mathematical problems or complete tasks. This strategy fosters peer interaction, communication skills, and cooperation (Mandhu & Bhattacharyya, 2023). By forming diverse groups that include students with different abilities, backgrounds, and perspectives, collaborative learning promotes inclusive learning environments where students learn from and support one another.
4. **Technology Integration:** The classrooms are suffering with inadequate infrastructure and resources. It is essential to integrating technology tools and resources into mathematics instruction to enhance equity and inclusion. Furthermore, digital platforms, interactive simulations, and educational apps can provide students with additional opportunities for practice, exploration, and self-paced learning (Romanelli, Bird, & Ryan, 2009). Technology can also support accessibility features, such as text-to-speech, visual aids, and adaptive interfaces, ensuring that students with diverse learning needs can engage with mathematics content effectively.
5. **Culturally Responsive Teaching:** Indian classrooms are culturally diverse so its teaching must be responsive to culture. Culturally responsive teaching recognizes and values students' cultural backgrounds, incorporating their experiences, perspectives, and examples into mathematics instruction (Romanelli, Bird, & Ryan, 2009). This approach involves using culturally relevant contexts, examples, and materials to make mathematics more relatable and meaningful for students. It also promotes the appreciation of diverse cultural contributions to mathematics and counters stereotypes or biases that may exist.
6. **Formative Assessment and Feedback:** Implementing formative assessment strategies, such as regular checks for understanding, provides teachers with valuable insights into students' learning progress and needs. By providing timely and specific feedback, teachers can address individual learning gaps and guide students towards improvement. Formative assessment helps ensure that all students receive the necessary support and resources to succeed in mathematics education.
7. **Self-Reflection and Metacognition:** Encouraging students to reflect on their learning processes and develop metacognitive skills promotes equity and inclusion (Mandhu & Bhattacharyya, 2023). By helping students become aware of their own thinking and learning strategies, teachers can support them in identifying and overcoming challenges. Metacognition also empowers students to take ownership of their learning and develop strategies for independent problem-solving.

Thus, innovative teaching methods and instructional strategies play a crucial role in promoting equity and inclusion in mathematics education. Differentiated instructions consider background and learning disabilities of the individual and with problem based learning students can get connected with the class. Furthermore, by implementing differentiated instruction, problem-based learning, collaborative learning, technology integration, culturally responsive teaching, formative assessment, and fostering metacognition, teachers can create inclusive learning environments where all students can thrive and succeed in mathematics. One of the most important aspects is teachers training is essential for inclusion and biases free teaching-learning in the classroom (Kishore and Kumar, 2022).

Educational Policies Promoting Equity and Inclusion

Examining existing educational policies and initiatives that promote equity and inclusion in mathematics education is crucial. Analyzing their effectiveness and identifying areas for improvement can guide policymakers in developing evidence-based policies. For example, policies that encourage teacher professional development in inclusive practices and provide resources to schools in marginalized areas can significantly impact equitable mathematics education.

India has implemented several educational policies and initiatives to promote equity and inclusion in the education system. These policies aim to address disparities based on gender, socioeconomic status, caste, and other factors that hinder access to quality education. Let's analyze some of the key educational policies in India that promote equity and inclusion:

1. **Right to Education Act (RTE) 2009:** The RTE Act is a landmark legislation that mandates free and compulsory education for children between the ages of 6 and 14 in India. The Act prohibits discrimination based on gender, caste, religion, or socio-economic background. It emphasizes the inclusion of marginalized and disadvantaged groups, ensuring their access to quality education. According to Jha (2018) "the enactment of RTE has a positive impact on the quantitative growth of the elementary education system (p. 48).
2. **Sarva Shiksha Abhiyan (SSA):** SSA is a flagship program launched in 2001 to achieve universal elementary education. It focuses on improving access, retention, and quality of education for all children, especially those from marginalized communities. SSA provides infrastructure development, teacher recruitment and training, free textbooks, and scholarships to promote equity and inclusion.
3. **Gender-Specific Policies:** The government has implemented various policies to address gender disparities in education. The '*Beti Bachao, Beti Padhao*' (it means 'Save the Girl Child, Educate the Girl Child') initiated in 2015 campaign aims to improve the status of girls and ensure their education. Scholarships and

incentives are provided to encourage girls' enrollment and retention in schools, especially at the secondary and higher education levels.

4. **National Program for Education of Girls at Elementary Level (NPEGEL) (2003)**: NPEGEL focuses on addressing gender disparities in education by providing additional support to girls from disadvantaged backgrounds. It includes interventions like bridge courses, residential schools, and life skills education to enhance girls' participation, retention, and learning outcomes.
5. **Mid-Day Meal Scheme (MDM) (1995):** MDM is a centrally sponsored program that provides free meals to students in government and government-aided schools. This initiative aims to improve attendance and retention rates, especially among children from marginalized communities. By addressing the issue of malnutrition, MDM contributes to creating a conducive learning environment for all students.
6. **Reservation Policies:** Reservation policies (these are part of Indian Constitution since its implementation in 1950) are aimed at addressing social category-based disparities in education. These policies provide reserved seats in educational institutions for students from Scheduled Castes (SC), Scheduled Tribes (ST), and Other Backward Classes (OBC). Reservation policies help create opportunities for historically marginalized communities and promote social inclusion.
7. **Inclusive Education:** The government has emphasized inclusive education through various policies and programs. The Inclusive Education for Disabled at Secondary Stage (IEDSS) (2009-10) scheme supports the education of students with disabilities by providing aids and appliances, special educators, and inclusive infrastructure. The Rashtriya Madhyamik Shiksha Abhiyan (RMSA) focuses on secondary education and promotes inclusive practices for students with disabilities.
8. **Digital Initiatives:** The government has launched digital initiatives like Digital India and e-PATHSHALA (initiated in 2015) to promote access to quality educational resources and technology-enabled learning. These initiatives aim to bridge the digital divide and provide equal opportunities for students, regardless of their geographic location or socioeconomic background.

While these policies and initiatives have made significant strides in promoting equity and inclusion in education, challenges remain in their effective implementation. Adequate funding, teacher training, monitoring mechanisms, and community participation are crucial for ensuring the success and impact of these policies. Continuous evaluation and improvement of policies, along with collaborative efforts from all stakeholders, are essential to achieve the goal of equitable and inclusive education in India.

DISCUSSION

Ramanujam and Subramaniam (2012) state that India exhibits diversity and cultural wealth, alongside persistent poverty and social fragmentation, which is also evident in the field of mathematics education. The findings offer valuable understanding of the persistent disparities in mathematics achievement in India, particularly in relation to gender and social classifications. Analyzing these findings in the context of previous research leads to a deeper understanding of the underlying factors that contribute to these disparities and suggests possibilities for further examination and intervention.

Gender Disparities

This chapter analyses gender disparities in mathematics achievement in the National Assessment of Students (NAS) in India. It highlights that girls tend to exhibit lower performance than boys in certain regions and grade levels. This finding aligns with prior research that highlights the extensive influence of societal norms, stereotypes, and biases on the disparities in mathematics achievement between genders (Kumar & Choudhury, 2022; Das & Singhal, 2021; Pandey, 2017; White et al., 2016; Bhagat & Vijayaraghavan, 2019; Ramanujan, 2012; Else-Quest et al., 2010). Furthermore, societal pressures and cultural biases can discourage girls who aspire to pursue disciplines that demand a solid grasp of mathematics, thus contributing to the gender disparities in academic achievement (Tessa & Mahzarin, 2019).

Disparities Based on Social Categories

Ramanujan (2012) argues that teacher preconceptions, bias, and behavior contribute to the discrimination faced by children belonging to low socio-economic status groups, commonly referred to as "Scheduled Castes" (SC) and "Scheduled Tribes" (ST). Furthermore, the chapter emphasizes notable disparities in mathematics achievement that are influenced by social factors such as socioeconomic status and caste, in addition to gender disparities. This discovery reflects the existing body of literature that highlights the significant influence of socioeconomic inequalities and discrimination based on caste on educational achievements in India. The findings are congruent with the previous research conducted by Sirin (2005), Thorat and Newman (2007), Jha and Tomar (2010), Bose and Kantha (2014), and Sreekanth (2016).

Nevertheless, Jha's (2014) study found that female students display reduced levels of mathematics phobia in comparison to their male counterparts. While the chapter acknowledges some advancements in attaining gender equality over the years, such as an increase in regions where girls outperform boys, there are still notable disparities that persist. To address gender disparities in mathematics education, it is crucial to move beyond mere numerical representation of women and instead prioritize the comprehension and resolution of sociocultural factors that contribute to these inequalities. Attempts to question gender stereotypes, provide female role models in mathematics, and create inclusive learning environments have shown promise in decreasing gender inequalities in mathematics performance (Stoet & Geary, 2018).

Challenges in Access and Participation

The problem of inclusion in mathematics education is multifaceted and calls for analysis of personal, societal, ideological, and topical aspects. Ensuring equity and inclusion presents unique challenges in India due to its diverse population and their cultural, linguistic, and socioeconomic backgrounds. Mathematics education in India is challenging due to socioeconomic inequality, cultural biases, a lack of infrastructure, and insufficient resources. Because of differences in gender, learning disabilities supported by Kishore and Kumar (2022), inclusive practices are essential. Students in rural and outlying areas, in particular, have fewer educational opportunities due to a lack of resources and infrastructure. Another challenge to inclusion and equity is the quality of teachers and their training. Teacher professional development programmes should prioritise cultural competency, inclusive teaching practices, and strategies for meeting the diverse needs of students if we are serious about promoting equity and inclusion in the classroom.

Innovation in Teaching-Learning Mathematics

To ensure that all students have equal access to mathematics education, new approaches to classroom instruction are necessary. Among these, you can find strategies for promoting metacognition, culturally responsive teaching, technology integration, problem-based learning, collaborative learning, and differentiated instruction. In order to guarantee that all students succeed in mathematics, teachers can use differentiated instruction to tailor lessons to each student's unique needs. Collaborative learning emphasises group projects, while problem-based learning emphasizes analysis and evaluation the finding is supported by Mandhu and Bhattacharyya (2023). Classrooms with limited resources must prioritise technology integration because it provides additional practice opportunities. Teaching that is culturally responsive recognises and makes use of students' cultural identities and experiences. The findings is in corroboration with Romanelli, Bird, and Ryan (2009). In order to fill in specific knowledge gaps and direct growth, formative assessment is essential. Students can take charge of their own learning and come up with creative solutions to problems when teachers promote metacognition and self-reflection.

Educational Policies

For mathematics education to be more accessible and equitable, India has launched a number of programmes and policies. Some examples of these policies and programmes are the Right to Education Act of 2009, the Sarva Shiksha Abhiyan program, policies pertaining to gender and reservation, and digital initiatives such as Digital India and e-PATHSHALA. These effectively integrated technology with the teaching-learning and inclusion. Nevertheless, there are still obstacles to overcome in order to effectively implement them, and stakeholders must continuously evaluate and collaborate in order for education to be equitable and inclusive.

CONCLUSION

Looking at mathematics achievement differences by gender and social class in India shows that these differences are caused by societal factors rather than natural abilities. Gender-based disparities stem from traditional stereotypes and biases, while social category-based disparities, such as those related to socioeconomic status and caste, reflect broader inequalities in access to quality education. Efforts to address these disparities include government initiatives, reservation policies, and the promotion of inclusive teaching practices.

Despite these efforts, challenges in access and participation persist due to socioeconomic disparities, cultural biases, inadequate infrastructure, teacher quality, language barriers, and gender disparities. However, innovative teaching methods and instructional strategies offer promising avenues for promoting equity and inclusion in mathematics education. Differentiated instruction, problem-based learning, collaborative learning, technology integration, culturally responsive teaching, formative assessment, and fostering metacognition can create inclusive learning environments where all students can thrive.

Analyzing educational policies reveals a range of initiatives aimed at promoting equity and inclusion in education, including the Right to Education Act, Sarva Shiksha Abhiyan, gender-specific policies, reservation policies, inclusive education programs, and digital initiatives. While these policies have made significant strides, their effective implementation requires adequate funding, teacher training, monitoring mechanisms, and community participation.

In conclusion, addressing disparities in mathematics achievement requires a multifaceted approach involving policy reforms, investment in resources and infrastructure, teacher training, cultural responsiveness, and creating inclusive learning environments. By addressing these challenges and building on existing initiatives,

India can work towards ensuring equitable access and participation for all students in mathematics education, thus fostering a more inclusive and equitable society.

Recommendations and Further Research

Moving forward, several key strategies can be implemented to further address disparities in mathematics achievement based on gender and social category in the Indian context:

1. **Digital interventions:** It is crucial at this point to employ digital infrastructure for the process of teaching and learning mathematics. In addition, it is imperative to incorporate governmental initiatives into the classroom.
2. **Targeted Interventions:** Develop targeted interventions specifically aimed at addressing the needs of marginalized groups, including girls, students from low socioeconomic backgrounds, and those from marginalized castes. These interventions could include additional academic support, mentorship programs, and access to resources tailored to the unique challenges faced by these groups.
3. **Teacher Training and Support:** Provide comprehensive training and support for mathematics teachers to equip them with the skills and knowledge needed to implement inclusive teaching practices. Furthermore, a training programme designed to provide teachers with the necessary skills to effectively manage a classroom that is culturally and linguistically diverse. This could involve workshops, professional development programs, and ongoing mentoring to help teachers create supportive and inclusive learning environments.
4. **Curriculum Reform:** Evaluate and revise the mathematics curriculum to ensure that it is culturally relevant, inclusive, and accessible to all students. Incorporate diverse perspectives, examples, and contexts into the curriculum to make mathematics more relatable and meaningful for students from different backgrounds.
5. **Community Engagement:** Foster collaboration and engagement with parents, communities, and local stakeholders to promote equity and inclusion in mathematics education. Encourage community involvement in schools, parent-teacher associations, and educational decision-making processes to ensure that the needs and perspectives of all students are represented.
6. **Research and Data Collection:** Conduct further research and data collection to better understand the underlying factors contributing to disparities in mathematics achievement. This could involve longitudinal studies, surveys, and qualitative research to identify barriers, challenges, and promising practices for promoting equity and inclusion in mathematics education.

7. **Policy Advocacy:** Advocate for policy reforms and initiatives that prioritize equity and inclusion in mathematics education. This could include lobbying for increased funding for education, advocating for the implementation of evidence-based practices, and pushing for policy changes that address systemic inequalities in the education system.
8. **Monitoring and Evaluation:** Establish robust monitoring and evaluation mechanisms to track progress and outcomes related to equity and inclusion in mathematics education. Regularly assess the effectiveness of interventions, policies, and programs, and use data to inform decision-making and continuous improvement efforts.
9. **Collaboration and Partnerships:** Foster collaboration and partnerships between government agencies, non-profit organizations, educational institutions, and other stakeholders to leverage resources, expertise, and best practices for promoting equity and inclusion in mathematics education. By working together, stakeholders can amplify their impact and achieve greater outcomes.

By implementing these strategies in a coordinated and holistic manner, India can continue to make strides towards ensuring equitable access and participation for all students in mathematics education. By addressing disparities based on gender and social category, India can foster a more inclusive and equitable society where all students have the opportunity to succeed in mathematics and beyond.

REFERENCES

Abdulkadri, A. (2022). Addressing gender disparities in education and employment A necessary step for achieving sustainable development in the Caribbean. United Nation, Economic Commission for Latin America and the Caribbean. https://repositorio.cepal.org/server/api/core/bitstreams/617fc7ac-82b0-4b2f-be83-5a14d825caf4/content

Bani, O. K., & Lach, L. M. (2024). Barriers to, and facilitators of, education for children with disabilities worldwide: A descriptive review. *Frontiers in Public Health*, 11, 1294849. 10.3389/fpubh.2023.129484938292375

Barrett, M. (2018). How schools can promote the intercultural competence of young people. *European Psychologist*, 23(1), 93–104. 10.1027/1016-9040/a000308

Bhagat, A., & Vijayaraghavan, R. (2019). *Gender disparity in STEM: Evidence from India*. ACER. https://www.acer.org/in/discover/article/gender-disparity-in-stem-evidence-from-india

Biswas, A., & Paul, A. (2018). GENDER DIFFERENCES IN MATHEMATICAL ACHIEVEMENTS OF HONORS STUDENTS IN WEST BENGAL. *International Journal of Creative Research Thoughts*, 6(1), 1612–1615. https://ijcrt.org/papers/IJCRT1802652.pdf

Bose, A., & Kantha, V. K. (2014). Influence of socio-economic background and cultural practices on mathematics education in India: A contemporary overview in historical perspective. *ZDM Mathematics Education*, 46(7), 1073–1084. 10.1007/s11858-014-0607-x

Das, G. C., & Sinha, S. (2017). Effect of Socioeconomic status on performance in Mathematics among students of secondary schools of Guwahati city. *IOSR Journal of Mathematics, 13*(1), 26-33. https://iosrjournals.org/iosr-jm/papers/Vol13-issue1/Version-1/F1301012633.pdf

Das, U., & Singhal, K. (2021). *Solving it correctly: Prevalence and Persistence of Gender Gap in Basic Mathematics in rural India*. https://arxiv.org/ftp/arxiv/papers/2110/2110.15312.pdf

Else-Quest, N. M., Hyde, J. S., & Linn, M. C. (2010). Cross-national patterns of gender differences in mathematics: A meta-analysis. *Psychological Bulletin*, 136(1), 103–127. 10.1037/a001805320063928

Jha, S. K. (2014). *Mathematics Phobia*. Lambert Academic Publisher.

Jha, S. K. (2018). Distributive Responses to a Collective Responsibility – Right to Education in India. *Journal of Educational Planning and Administration*, 32(1), 35–50.

Jha, S. K., & Tomar, D. P. (2010). Mathematical Phobia of Government and Private Secondary School Students in Relation to Certain Demographical Variables. *Journal of Community Guidance & Research*, 27(3), 316–323. 10.5281/zenodo.8044469

Kishore, R., & Kumar, S. (2022). A Study on the effectiveness of pedagogy used by teachers for the Inclusion of students with Disabilities in the Schools of Delhi. District Institute of Education and Training, Daryaganj, New Delhi (India).

Kumar, D., & Choudhury, P. K. (2022). Do parental resources reduce the gender gap in math for primary school-going children? Evidence from India. *Education 3*, 1–18. 10.1080/03004279.2022.2133546

Mandhu, S., & Bhattacharyya, D. (2023). Learning Style Preferences Among the Students. *International Journal of Multidisciplinary Research*, 5(1), 1–12.

National Council of Educational Research and Training [NCERT]. (2017). *National Achievement Survey [NAS] 2017: Class III.* Ministry of Human Resource and Development, Government of India.

National Council of Educational Research and Training [NCERT]. (2017). National Achievement Survey [NAS] 2017: Class V. Ministry of Human Resource and Development, Government of India.

National Council of Educational Research and Training [NCERT]. (2017). *National Achievement Survey [NAS] 2017: Class VIII.* Ministry of Human Resource and Development, Government of India.

National Council of Educational Research and Training [NCERT]. (2021). *National Achievement Survey [NAS] National Report 2021 Class III, V, VIII & X. Ministry of Education, Government of India*. NAS. https://nas.gov.in/report-card/2021

NCF. (2005). National Curriculum Framework. National Council of Educational Research and Training, New Delhi, India NEP. (1986). National Policy on Education [NEP]. Ministry of Human Resource and Development, Government of India.

Nimely, D. R., & Nyamu, F. K. (2023). Strategies for Inculcating Gender Equity in Mathematics Classrooms for Enhancement of Learners' Problem-Solving Skills. *International Journal of Research and Innovation in Social Science*, 7(7), 1446–1456. 10.47772/IJRISS.2023.70814

OECD (2012). *Equity and Quality in Education: Supporting Disadvantaged Students and Schools*. OECD Publishing. 10.1787/9789264130852-en

Paudel, T. (2020). Gender Issue in Teaching and Learning Mathematics. *Proceedings of the Tenth International Mathematics Education and Society*. Research Gate. https://www.researchgate.net/publication/342437701

Ramanujam, R., & Subramaniam, K. (2012). *Mathematics Education in India Status and Outlook. Homi Bhabha Centre for Science Education Tata Institute of Fundamental Research.* https://mathedu.hbcse.tifr.res.in/wp-content/uploads/2014/01/INP-Book_Mathematics-Edu-in-India_2012_KS-RR.pdf

Ramanujan, R. (2012). Mathematics Education in India: An Overview. In Ramanujam, R., & Subramaniam, K. (Eds.), *Mathematics Education in India Status and Outlook* (pp. 1–11). Homi Bhabha Centre for Science Education Tata Institute of Fundamental Research. https://mathedu.hbcse.tifr.res.in/wp-content/uploads/2014/01/INP-Book_Mathematics-Edu-in-India_2012_KS-RR.pdf

Right to Education. (2009). *The Right of Children to Free and Compulsory Education Act, 2009*. Right to Education. https://www.education.gov.in/sites/upload_files/mhrd/files/upload_document/rte.pdf

Romanelli, F., Bird, E., & Ryan, M. (2009). Learning styles: A review of theory, application, and best practices. *American Journal of Pharmaceutical Education*, 73(1), 9. 10.5688/aj73010919513146

Roos, H. (2023). Students' voices of inclusion in mathematics education. *Educational Studies in Mathematics*, 113(2), 229–249. 10.1007/s10649-023-10213-4

Shakunthala. (2014). Social Inequality and Access to Education A Sociological Study. *The International Journal of Creative Research Thoughts, 2*(1), 732-737.

Shang, B. (2016). *Tackling Gender Inequality: Definitions, Trends, and Policy Designs. WP/22/ 232*. International Monetary Fund.

Sirin, S. R. (2005). Socioeconomic status and academic achievement: A meta-analytic review of research. *Review of Educational Research*, 75(3), 417–453. 10.3102/00346543075003417

Sreekanth, Y. (2016). *Students Educational Achievement at School Level: An Analysis across Social Categories.* A paper presented in the International Conference on Social Statistics in India, Patna, Bihar, India.

Stoet, G., & Geary, D. C. (2018). The gender-equality paradox in science, technology, engineering, and mathematics education. *Psychological Science*, 29(4), 581–593. 10.1177/095679761774171929442575

Tabassum, N., & Nayak, B. S. (2021). Gender Stereotypes and Their Impact on Women's Career Progressions from a Managerial Perspective. *IIM Kozhikode Society & Management Review*, 10(2), 192–208. 10.1177/2277975220975513

Tessa, E. S. C., & Mahzarin, R. B. (2019). Gender in Science, Technology, Engineering, and Mathematics: Issues, Causes, Solutions. *The Journal of Neuroscience : The Official Journal of the Society for Neuroscience*, 39(37), 7228–7243. 10.1523/JNEUROSCI.0475-18.201931371423

Thorat, S., & Newman, K. (2007). Caste and economic discrimination: Causes, consequences, and remedies. *Economic and Political Weekly*, 42(41), 4121–4124. 10.1523/JNEUROSCI.0475-18.2019

UNESCO. (2022). *Challenging gender bias and stereotypes in and through education*. UNESCO. https://unesdoc.unesco.org/ark:/48223/pf0000380827

United Nations. (2006). *United Nations Convention on the Rights of Persons with Disabilities and Optional Protocol*. UN. https://www.un.org/disabilities/documents/convention/convoptprot-e.pdf

White, G., Ruther, M., Kahn, J. R., & Dong, D. (2016). Gender inequality amid educational expansion in India: an analysis of gender differences in the attainment of reading and mathematics skills. *Journal of Research in Gender Studies, 6*(2), 153+. https://link.gale.com/apps/doc/A478698526/AONE?u=anon~bec3535f&sid=googleScholar&xid=757f355b10.3389/fpsyg.2021.599199

Xie, G., & Liu, X. (2023). Gender in mathematics: How gender role perception influences mathematical capability in junior high school. *The Journal of Chinese Sociology*, 10(10), 10. 10.1186/s40711-023-00188-3

Chapter 10
Math Achievement:
How Does Income Influence Children's Third Grade Math Achievement?

Emaj Uddin
https://orcid.org/0000-0002-6721-6856
University of Rajshahi, Bangladesh

ABSTRACT

Drawing from family investment models, this study examined parental monetary investment in essential math materials and math tutoring as mediators of the association between family income and math achievement during third grade. Data were collected from third graders (N=1150, Mage= 8.01 at baseline, 52.00% boys) and mothers in Bangladesh during third-grade entry through final exam. Results from SEM suggested that there were significantly positive and direct relations (r= .36, p< .05) between low-income, fewer monetary investments, and poor math outcomes during third grade, after accounting for child and maternal characteristics. When mediators were included in the model, the estimated significant pathways from low-income to poor math outcomes were decreased by 24%, but the associations were still significant. The results suggested that parental monetary investment in math tutoring and number word was more important than other materials in children's early primary math achievement. Future directions for research and policy implications are discussed.

DOI: 10.4018/979-8-3693-2873-6.ch010

INTRODUCTION

Decades of research in the US have shown that lower-income children significantly have poorer math scores (e.g., standardized test or GPA) than their higher income peers during kindergarten through early primary grades (Duncan & Brooks-Gunn, 1997; Jordan et al., 1992, 1994; Jordan, Levine, & Huttenlocher, 1994; Rittle-Johnson et al., 2017). Despite large-scale efforts and early intervention (Dearing, McCartney, & Taylor, 2001; Dahl & Lochner, 2008; Votruba-Drzal, 2003), these poorer scores in mathematics of the lower-income children have remain unchanged over the decades (Hanushek, Peterson, Talpey, & Woessmann, 2019; Reardon, 2016; Sirin, 2005). In Bangladesh, about 40%-50% of the children perform significantly worse than their middle-income peers in math problem-solving and counting during school entry through early primary grades (Aboud, 2006; Aboud & Hossain, 2011; Haider, Islam, Mia & Sorcar, 2005; Uddin, 2017).

Developmental psychologists, economists, and policy-makers long observe that lower-income children from early childhood are deprived from parental monetary investments in education that have lasting negative effects on academic achievement (Foster, 2002; Kalil & DeLeire, 2004; Mayer, 1997; McLoyd, 1998). Growing work reveals that about 15 million low-income children in the US have no school supplies or backpack to go to school for learning (Kaufhold, Alverez, & Arnold, 2006). Despite public efforts (e.g., textbooks and some cash benefit (100 TK.) to each needy child per month) in Bangladesh, growing research reveals that 40-50% of the poor children go to primary school without school supplies and equipment (Hossain & Zeitlyn, 2010, Nath, 2007). Although lower-income children's educational disadvantages-related to school supplies and special services are rising in the US and internationally including Bangladesh, little research has specifically focused on how fewer monetary investments in essential learning materials and special services for underachievers affect their early primary math achievement. Drawing from family investment model (FIM), present study examined parental monetary investments in essential math learning materials and special math services through which lower-income might affect math achievement during early primary grade in Bangladesh. These potential mediators will help understanding pathways by which lower-income might affect poor math achievement during early primary grade in Bangladesh.

Parental Monetary Investments by Which Family Income Links to Math Achievement

Parental monetary investment founded in economics (Becker, 1993; Becker & Tomes, 1979) is one of the most fundamental aspects of the FIM to understand pathways by which family income influences children's academic achievement (Conger & Donnelan, 2007; Haveman & Wolfe, 1994; Kalil et al., 2004). The FIM from social causation view asserts that fewer parental monetary investments have poorer academic achievement of children (Conger et al., 2007; Foster, 2002; Hamilton, Cheng, & Powell, 2007; Hamilton, 2012). Conger et al. (2007) and their colleagues (Mayer, 1997; McLoyd, 1998) that lower-income parents cannot afford to buy and provide goods and services (e.g., food, dress, housing, toy, medicine, care, learning materials etc.) to their children. Children whose basic needs are not met have a hard time to develop properly and also acquire early educational skills that might help them to succeed in school. Particularly, children whose parents cannot afford to purchase school supplies (e.g., school dress, number-word books, paper-books, note-books, pen-pencil box, geometry box, calculator, and mini-computer) and special services (e.g., math coaching, math tutoring) that might enhance early primary math skills are at greater educational disadvantages that in turn affect academic achievement, including mathematics (Hamilton et al., 2007; Kaufhold et al., 2006). Theoretically, lower-income constraints parental abilities to purchase children's math related books and school supplies-related to math practice that in turn might affect math achievement during early primary grades.

Although a substantial body of empirical research from the field of home environment in the US and in some low- and middle-income countries (LMICs) have found parents' fewer investments in cognitively stimulating materials and specialized tutoring by which lower-income explains 30-50% effects on children's poor cognitive skills (Guo & Harris, 2000; Linver, Brooks-Gunn, & Kohen, 2002; Yeung, Linver, & Brooks-Gunn, 2002) and math school readiness (Lombardi & Dearing, 2020; Wolf & McCoy, 2019) during early childhood, little is known about how parental monetary investments in essential materials (e.g., different number word-books, pen-pencils, paper-books, calculators) and special services (math tutoring) that are specific to children's math story-problem-solving and math counting during early primary grades. Complimentary line of descriptive research suggests that *children's number-word books* with different language versions (e.g., Arabic, English, Hindi, Spanish etc.) enhance children's early conceptual and symbolic knowledge on number system or simply number names and later math achievement (Casey et al., 2018; Lombardi et al., 2020; National Research Council, 2001). Evidence demonstrates that lower-income parents have limited abilities to purchase quality and quantity of number-word books with different languages that, in turn, might affect their early symbolic math knowledge development and later math story-problem solving and math counting during early primary grades (Casey et al., 2018; National Research Council, 2001).

Like children's number word books, *essential materials (e.g.,* paper-books, note-books, pen-pencil box etc.) are important for children's math practice and math achievement during early primary school years. Although a great deal of research have found that lower-income parents after expenses for family living are less able than middle-income parents to purchase learning materials (e.g., calculator, tape or record player, computer, board game, card game etc.) for children that foster early cognitive and mathematical skills (Coley, Sims, & Votruba-Drzal, 2016; Ramani & Siegler, 2008; Scalise, Daubert, & Ramani, 2020; Schneider et al., 2018; Votruba-Drzal, 2003; Galindo & Sonnenschein, 2015; Wolf et al., 2019), research on lower-income parents' spending on the specific domains of children's essential materials described above is very limited. Extant research demonstrates fewer monetary investments in essential materials by which lower-income might affect children's poor math performance during early primary grades (Elliott et al., 2018).

Types of calculator use (e.g., basic, scientific, & graphing) that fosters children's conceptual, operational and problem-solving in mathematics throughout the schooling may mediate the relation between income and early primary math achievement (Close, Shiel, Oldham, & O'Leary, 2012; Ellington, 2003). Although a large body of experimental research and meta-analysis has found significantly positive effects (e.g., 0.20-0.50) of the different types of calculator use on math achievement during high school and college education (Ellington, 2003), limited research has examined on how lower-income via fewer parental monetary investments in different types of calculator affect early primary math counting skill development and math achievement (Edens, 1983; Starr, 1989; Stacey & Groves, 1994).

Special math services (e.g., math coaching, private tutoring) also positively enhances math achievement throughout the schooling (Buchman, Condron, & Roscigno, 2010; Park, Byun, & Kim, 2011). Although a large body of research in the high and low-income countries have found significantly positive relations between SES (e.g., education and income) and private tutoring (Atalmis, Yilmaz, Saatcioglu, 2016) and private tutoring and academic performance (e.g., mathematics, English) throughout the schooling (Berberoğlu & Tansel, 2014; Dawson, 2010; Lee, 2013; Park et al., 2011; Stevenson & Baker, 1992; Sohn, Lee, Jang, & Kim, 2010), little research has focused directly on how parental fewer monetary investment in private math tutoring as a mediator through which lower-income affects math achievement during early primary grades (Atalmis et al., 2016; Buchman et al., 2010; Zhang & Xie, 2016). Existing research demonstrates that lower-income children have fewer opportunities than their middle-income peers to attend different types of tutoring (e.g., one-to-one tutoring, center-base tutoring) that, in turn, affect their English and mathematics performance during high school and college education (Dang, 2007; Kim & Lee, 2010; Stevenson et al., 1992). In the US, Buchman et al. (2010) found that lower SES children than their higher SES peers had fewer opportunities to attend private tutoring such as one-to-one private tutoring than center-based private tutoring that, in turn, affected their academic math and English achievement among high school students. Particularly, Atalmis et al. (2016) examined relations between SES, private tutoring, and math achievement among seventh graders in Turkey. Applying Baron and Kenny's (1986) approach, they found significantly positive and direct relations ($_{SES}$= .202, SE_{SES}= .029, $p<0.001$) between family income and private tutoring (measured as number of private tutoring, private tutoring centers, & one-to-one tutoring) and math achievement among seven graders ($_{SES}$= .927, *SE*= .301, $p<0.01$). They explained that every one-point increase in a student's income level affects his or her private tutoring attendance by 20%, accounting for students' previous math scores and school variables. In testing mediation, they further found significant relations between family income and math achievement, but effect size decreased by 15% ($_{SES}$=.788, SE_{SES}=.307, $p<0.05$) when compared with the original model. Over all, they found private tutoring partially mediates the relations between family income and math achievement among seventh graders in the country.

Parental Monetary Investments as Contributors to Disparities in Early Primary Math Achievement

While parental monetary investments in home-enriching materials such as number books, toy, calculator, tape or record player, computer, board game, card game etc. and private tutoring in the US and internationally are important mediators by which lower-income significantly effects on younger children's poor cognitive and math achievement (Atalmis et al., 2016; Elliott et al, 2018; Galindo et al., 2015; Linver et al., 2002; Lombardi et al., 2020; Siegler & Ramani, 2008Yeung et al., 2002; Wolf et al., 2019), very little research has focused directly on fewer monetary investments in children's essential materials (e.g., number book, paper book, pen/pencil box, geometry box) and math services (e.g., math tutoring) for children's early math learning and practices that could explain lower-income-related differences in early primary math achievement. In the present study, we based upon FIM (Conger et al., 2007) and previous research (Linver et al., 2002; Lombardi et al., 2020; Yeung et al., 2002; Wolf et al., 2019) contribute to family investment literature to examine parental monetary investments in the essential materials and math tutoring through which lower-income might affect poorer math achievement during third grade among Bangladeshi children.

Primary Math Learning in Bangladesh Poverty Context

Bangladesh is a low-middle income country with 165 million people. Despite promising history of poverty reduction in the past decades, about 1 in 4 Bangladeshi were living in poor families in 2016 (BBS, 2016). According to Household Income and Expenditure Survey (HIES) 2010, 24.5% of the population were relative poor and 13% were chronic poor. In child poverty and deprivation, research demonstrates that 46% of the children under 16 live below the national poverty line; 59% live in 1.08$ (PPP) a day and 23% live in persistent poverty (BBS, 2016).

Bangladeshi poor children usually start compulsory primary schooling (equivalent to grade 1) at the age of five to six (e.g., the first of January of the calendar year). The aim of compulsory primary education in Bangladesh is to ensure optimal development of all children and to prepare them to learn basic knowledge and skills in mathematics and other courses, such as English & Bengali Language (Khan, Motin, Begum, & Uddin, 2012; Rahman & Islam, 2009). According to primary math curriculum for grade 1, children should become familiar with basic number sense, counting, comparing, and grouping objects, adding and subtracting objects from 1 to 10, and recognizing number signs (Khan et al., 2012). When they enter grade 2, the primary math curriculum focuses on recognizing and writing numbers and making simple calculations with adding and subtracting quantities (Khan, Motin, Begum, & Uddin, 2013). When they enter grade 3, children solve simple math story-problems, implying addition, subtraction, multiplication and division, including early geometry. Children at grade 3 are encouraged and motivated by several tasks at multiple levels of difficulties to learn mathematics at primary school (Khan, Motin, Begum, & Uddin, 2014). Although primary school enrolment and class attendance during primary grades are successful, literature reveals that children growing up in the low-income families have poorer math GPA during early primary grades (Aboud, 2006; Aboud et al., 2011; Rahman et al., 2009; Uddin, 2017). Previous results (2014-2016) also indicate that about 40- 50% of the poor students achieved poorer math scores (e.g., C grade) during first grade and that persists during second through third grade (BBS, 2010, 2016).

Although essential materials and private tutoring are the most important for underachievers' early primary mathematics development, several descriptive studies have found that parents with daily or monthly low-income have limited monetary abilities to purchase and provide essential materials, such as number word books, paper books (white paper books), pen/pencil box, geometry box, calculator and special services (e.g., math tutoring) for day-to-day math learning and practice (Aboud, 2006; Aboud et al., 2011; Hossain et al., 2010; Haider et al., 2005; HIES, 2010; Uddin, 2017). For example, Hossain et al. (2010) found that out of 6768 sample, most of the low-income children had no school bag (62.5%), geometry box

(55%), extra number word books (65%), and pen and pencil (23%). Other research demonstrates that low-income parents cannot provide special care and services, such as private math tutoring or center-based math care for their children's extra-math care within and outside the home (Hossain et al., 2010; Nath, 2007). In turn, fewer provisions of the essential materials by low-income parents prohibit their children's math development and math achievement during early primary grades (Hossain et al., 2010; Hossain, 2016; Kabeer & Mahmud, 2009). As a result, low-income children with fewer essential materials and private tutoring perform worse than the middle-income children in mathematics during early primary grades. Although family chronic poverty or family low income, fewer monetary investments in essential materials and special services, and children's poorer math achievement during third grade are widely reported in previous research, the extent to which these family risk factors during childhood are bidirectionally related with each other remains unknown in Bangladesh context (Uddin, 2017).

The Present Study

Therefore, the goal of the present study was to explore the extent to which parental monetary investments in the items of essential materials (e.g., number-word books, paper-books, pen-pencil box, calculator, geometry box) and math tutoring of children's math learning and practice explain the relation between lower-income and poorer math achievement during third grade final exam. Specifically, we tested the following hypotheses in Bangladesh context: 1) family lower-income during third-grade entry (January of the calendar year) is directly and significantly associated with the poorer math achievement during third-grade final exam (e.g., 12 months later, December) with control variables; 2) family lower-income during third-grade entry is directly and significantly associated with fewer parental monetary investments in essential materials and math tutoring during 12 months period (e.g., January to December) with control variables; 3) Based upon the FIM (Conger et al., 2007, Haveman et al., 1994; Kalil et al., 2004; Mayer, 1997; McLoyd, 1998) and previous research (Atalmis et al., 2016; Guo & Harris, 2000; Linver et al., 2002; Yeung et al., 2002; Coley et al., 2020) we expected that as lower-income parents would have limited monetary abilities (after expenses for family living) to invest in essential materials and special services for children, so fewer parental monetary investments would partially mediate the associations between family lower-income and children's poorer math achievement during third grade final exam.

Figure 1. Hypothesized model

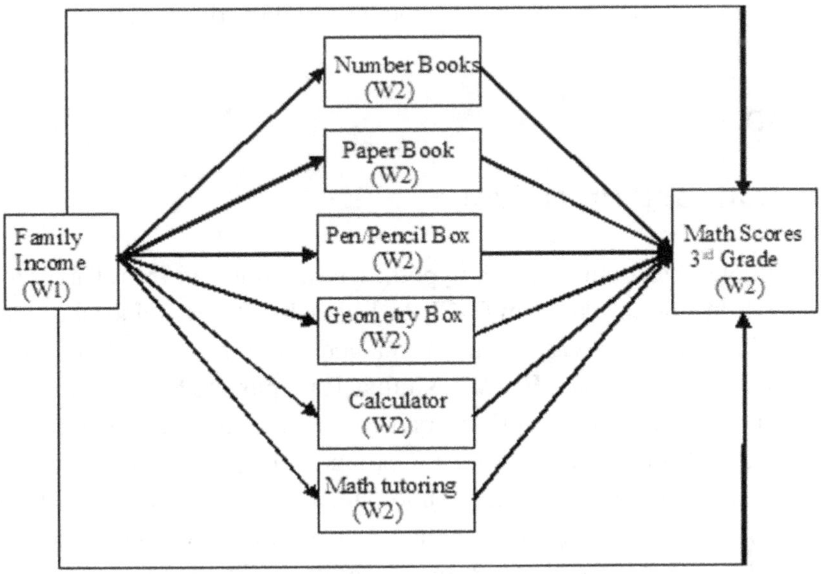

This is the first study to our knowledge to test the FIM in Bangladesh context. Our conceptual model shown in Figure 1 contributes to the family investment literature, focusing on family lower-income, parental monetary investments in essential materials and services for children's math learning and practice, and their math achievement during third grade. In order to provide empirical evidence on the contribution of parental monetary investments in essential materials and math services to the association between lower-income and poorer math achievement during third grade, the present study examines frequency of parental monetary investments across six conceptual domains, such as number-word books, number-paper books pen-pencil box, calculator, geometry box and math tutoring underlying the relation between lower-income and poorer math achievement during third grade. We simultaneously test the multiple hypotheses regarding the six mechanisms that could account for strong association between lower-income and poorer math score during third grade. Instead of taking the more common piecemeal approach that considers each mediator individually, our analysis estimates the relative contribution of each of the six mediators controlling for others- a strategy that should reduce potential bias in the estimated role of each mediator. We expect that these six mediators will account for a substantial portion of the association between lower-income and poorer math achievement during third grade. Finally, we control several background characteristics, including mother's age, education, occupation, family size, child's age and sex. Inclusion of these control variables should reduce bias in the estimation of the unique contributions that each proposed mediator makes to the relation between lower-income and poorer math achievement during third grade.

METHOD

Participants and Procedure

The sample to test our research hypotheses was selected from Underprivileged Children's Educational Program (UCEP), Rajshahi city, Bangladesh. Since 1972, UCEP has been working with the disadvantaged children in the urban slum areas of Dhaka, Gazipur, Chittagong, Barisal, Khulna, Rangpur and Rajshahi city to provide quality of primary education during first grade to eighth grade, (Rahman, Yasmin, Begum, Ara, & Nath, 2010). The UCEP expanded its programs at Rajshahi city in 1994 and established five primary schools located in Tikapara, Chhotabongram, Darikharbona, Bolonpur & Mahisbathan where skillful teachers were providing primary education to the disadvantaged children. To examine primary school achievement of disadvantaged children, a first comprehensive longitudinal survey study covering wide varieties of family characteristics and children's educational achievement in English, Math, Bengali, science and other courses during third to fifth grade was conducted in 2014-2016 approved by our university (Hossain, 2016). The researcher using simple random sampling (e.g., lottery method) selected 1150 third graders (Mage= 8.01, SD = .78) out of 2298 students from five schools and their mothers (Mage = 28.00, SD = 10.12) in 2014 and continued fourth grade to fifth grade. As the study had a rich sampling frame and good research outcome to study disadvantaged families and their children's primary school education at the city, so all third graders and their mothers (age range 20-50 years) were selected for this study purpose. The subjects selected for this study were fully matched with our research goal.

The Wave 1 (W1) data that mainly covered family income and socio-demographic characteristics were collected from the selected subjects (N=1150) in January, 2014 when the subjects enrolled third grade in the primary schools. To analyze mediating effects, we further collected data on parental monetary investments in essential materials and special services for children at the end of third grade (December, 2014). In so doing, relevant items from HOME were used for parental monetary investments in children (Caldwell & Bradley, 1984; Zadeh et al. (2010) and math tutoring (Atalmis et al., 2016). Data on third grade math scores were collected from the subjects in primary school setting. In so doing, we used Chen, Chang, Liu, & He's (2008) approach for data collection from school record. Although a total of

1150 participants and also their mothers actively participated in the research at W1, 5 subjects were absent at W2 data collection, because their parents were reluctant to participate in the study process. As results of *t-test* (not found) suggested that the absent families did not significantly affect the predictor variable (family income) and other focal variables, so we finally used 1145 subjects to analyze the hypothesized pathways model proposed in Figure 1.

We appointed two trained interviewers (e.g., one male, one female) for data collection. Before data collection, two-day workshop was arranged for the research orientation and data collection with proper instruments. All instruments were pre-tested in the sample, consulting with head teacher and necessary adjustments of the instruments were made in Bangladesh context. Interviewers built up rapport with the mothers and focal children and each head teacher of the primary schools. Interviewers also maintained some ethical considerations, including confidentiality, privacy, mutual benefit, and reciprocity. Particularly, data collectors in primary school setting worked with the head teacher to designate a quiet class room where assessment of math proficiency during third grade was conducted in school setting. Data collectors spent several minutes informally chatting or sang a song, gossiped a cricket game with children prior to begin assessments. Then, data collectors administered assessment in Bengali language. Interrater reliability between pair data collectors was calculated with Cohen's kappa and ranged from 0.57 to 0.81, with an average kappa value of 0.79 (Wolf et al., 2019). In the entire process of data collection, the interviewers conversed in Bengali language with the participants in both family and school settings and then the responses were converted into English by the present researcher (Uddin, 2015a).The present researcher supervised the entire research process.

Measures

Family Lower-Income. Family income is the sum of income from all sources received by all members of the family over some time period, typically a calendar year or month (Duncan, Magnuson, & Votruba-Drzal, 2015). In this study, family income was assessed in one point in time, when children entered third grade in 2014. Each mother of the families was asked for annual family income from all sources (e.g., job, small business, social assistance; Lugo-Gil & Tamis-LeMonda, 2008), excluding tax if any and then annual family income of the parents (e.g., father, mother) was averaged (Yeung et al., 2002) and presented in thousand Taka (M = 38.00, SD = 12.83). In this study, the raw data on annual income of the families were used as a predictor variable.

Parental Monetary Investment. Parents' monetary investments in children's developmentally age-appropriate essential materials and special services that foster children's cognitive development and math skills throughout the schooling were measured covering number word books, paper books, pen-pencil box, geometry box, and calculator adapted version from HOME (Caldwell & Bradley, 1984), United Nations Children's Fund (2006) and Zadeh et al. (2010) and math tutoring (Atalmis et al., 2016). First of all, the *number word-books* that enhance math knowledge and skills were measured with 3 items: English, Bengali, and Arabic version of number word books, following National Research Council (2001). This measure differs from the past research (Wolf et al., 2019) and the MICS survey (UNICEF, 2014) that used whether or not caregivers provided number of word-books or picture-word books in low-resourced context. *Paper-books* on which children practice math at home and at school or they submit home-work at school were measured with 3 items, including white paper book, dotted book, and note book. *Pen-pencil-box* covers different types of pens and pencils with shape, size and color that enhance children's mathematical information processing, thinking, and reasoning through longhand writing and drawing (Mueller & Oppenheimer, 2014). In the current study, pen-pencil box that differs from earlier research in conceptual content: Pen is mightier than the laptop note taking (Mueller et al., 2014) was measured with 2 items such as pen-box and pencil box. Like pen-pencil box, *geometry box* includes mathematical instruments (e.g., compass, divider, ruler, protractor, set square, & sharpener) that also enhance mathematical learning (e.g., space and shape, size, angles & distances of different objects) through drawing. In the current study, geometry box was measured with 2 items pen box and pencil box. A *calculator* is a machine that helps the students to do math operations (e.g., add, subtract, multiply, divide, square roots, calculus and graph, depending on the types of calculator use: Simple or pocket calculator or scientific calculator. In this study, we included 2 items such as simple and scientific calculator. At last, *math tutoring* was measured with 2 items: private tutoring at home and center-based tutoring adapted from Atalmis's et al. (2016) instrument. We asked the mothers whether or not you purchased and provided number word-books (3 items), paper-books (3 items), pen-pencil boxes (2 items), calculator (2 items) and math tutoring (2 items) for your focal child in the last twelve months. Each item of the six components was coded as yes =1 and no = 0 and summery measure (%) of the six components included as an observed indicators of the parental monetary investments in essential materials and special services for math practice (Black, 2003). Items showed positive bivariate correlations with one another (*rs* = .15-.61) and internal consistency of alpha = .65. These observed characteristics of the parental monetary investments were separately used as mediators in the analysis.

Third Grade Math Achievement. Primary math achievement (PMA) refers to how students (e.g., male, female) do well in math examinations in the domains of basic rules of operation and its related story-problems, fraction & decimal, average & percentage and primary geometry, as well as their behavior at school over time (Uddin, 2017). The data on math achievement during third grade were obtained from the primary schools. The scores (also marks) of mathematics during third grade were based on objective examinations conducted and evaluated by school teachers over the years (Chen et al., 2008). The total marks for math course were 100; a test score of 60 is usually considered the cutoff point between a pass and a failure in the course. In the present study, math achievement scores ($M = 55.79$, $SD = 13.26$) during third grade were used as an observed outcome variable.

Covariates. We included child and mother characteristics as control variables. Mothers reported their age (in years; $M_{age} = 30.02$, $SD = 7.22$), family size (number of members living together; $M = 4.51$, $SD = 1.87$), education in years ($M = 10.26$, $SD = 3.89$) and occupation (House wife = 1 and job involvement= 0) at W1 data collection. Regarding child characteristics, they also reported their focal child's date of birth from age (in years) at baseline, child gender (0 = female, 1 = male), and religion (0 = Muslim, 1 = other). In addition, they reported whether or not focal child's previous math score was good = 1 and bad = 0.

Analytic Plan

We began our analysis with descriptive and bivariate correlation among predictor, mediator, and outcome variables and then we tested mediating pathways from W1 average annual family income through W2 parental monetary investments to W2 third grade math achievement. Researchers argue that if there is no simultaneous relation between predictor and mediator and then between mediator and outcome variables with control extraneous factors, there is no possibility to test mediating effects (Baron et al., 1986; Cole & Maxwell, 2003). In so doing, we used structural equation modeling (SEM) with Amos (Analysis of Moment Structure, version 20.0) to conduct pathway analysis (Byrne, 2016; Kline, 1998). Annual family income and third grade math GPA represents single observed indicator, while parental monetary investments were independent latent mediators. Using latent variable modeling relations, the path analysis can decompose the extent to which annual family income and parental monetary investments have direct effects on third grade math scores (e.g., because interaction between lower annual family income and parental monetary investments). In addition, the latent variable analysis also has indirect effects of annual lower-income via parental monetary investments on later poor math scores during third grade (Linver et al., 2002).

The present SEM analysis has two components: 1) measurement model and 2) structural model (Byrne, 2016). While our measurement model describes model identification and the factor loadings of the six latent constructs such as number-word book, pen-pencil box, paper-book, calculator, geometry box and math tutoring, our structural model describes the coefficients of latent relations between lower-income and several mediators and then between mediators and third grade math score. In the structural model shown in Figure 1, we outlined both direct and indirect pathways through which W1 annual lower-income relates to W2 third graders' poorer math scores. In the model, all mediators of parental monetary investments are dependent variables of W1 annual lower-income, but independent variables of W2 third graders' poorer math scores. Before testing mediation model, we conducted confirmatory factor analysis to identify the measurement model and to evaluate factor loadings of the six mediators (see, Table 3). At last, we tested full mediation model to identify simultaneous mediating effects of the six latent variables on the strong association between lower-income and third graders' math outcome (see, Figure 2). In the model, control variables (e.g., mothers' and children's age, education, family size, previous math scores) were included to know whether or not these control variables effect on the focal model (Linver et al., 2002). Finally, we evaluated model fit, using chi-square, goodness-of-fit index (GFI), comparative fit index (CFI), and the root mean square error of approximation (RMSEA; Hu & Bentler, 1999).

RESULTS

Bivariate Correlation

Table 1. Sample descriptive characteristics (N = 1145)

	Mean or %	SD	Range
1. W1 Average annual family income*	38.00	12.83	30-50
Parental monetary investment (yes %)			
2. W2 Number word book	42.23		
3. W2 Number paper book	55.21		
4. W2 pen/pencil box	32.24		
5. W2 Geometry box	45.64		
6. W2 Calculator	55.12		
7. W2 Math tutoring	48.11		
8. W2 Third grade math scores	55.79	13.26	24-86
9. W1 Mother's age	30.02	7.22	25.40
10. W1 Mother's education in year	10.26	3.89	00-12
11. Mother's occupation (housewife %)	58.26		
12. W1Children's age in year	8.21	.78	8.02-8.44
13. W1 Child gender (male %)	52.00		
14. Previous math score (good %)	48.25		

Results of descriptive analyses can be found in Table 1. Results reveal that there were substantial differences in annual family income (range 30-50 thousand taka) within the present sample, with an average income of 38.00 (SD = 12.83). Indicators of parental monetary investments in essential materials and math tutoring for children also were different in the sample, with the lowest frequency of parental monetary investments in pen/pencil box (32.24%) to the highest frequency in number paper books (55.21%). Results of bivariate correlation are shown in Table 2. Results suggest that low-level income of the families was significantly and moderately related to fewer parental monetary investments in essential materials and math services for children ($r = 0.34 - 0.65, p < 0.05$). The indicators of parental monetary investments in turn were significantly and moderately associated with poorer math scores during third-grade ($r = .39, p < 0.05$). As the pattern of associations among predictor, mediators, and outcome variables ($r = 0.34 - 0.65$) included in bivariate correlation analysis with control variables were consistent with our theoretical model in Figure 1 as were expected in hypothesis 1 and 2 (Baron et al., 1986).

*Table 2. Bivariate correlation coefficients for all variables**

	1	2	3	4	5	6	7	8	9	10	11
1. W1Family income	1										
2. W2 Number word books	.48*	1									
3. W2 Number paper books	.22*	.34*	1								
4. W2 Pen-pencil box	.24*	.43*	.42 *	1							
5. W2 Geometry box	.24*	.30*	.51*	.45*	1						
6. W2 Calculator	.23*	.51*	.41*	.66*	.65*	1					
7. W2 Math tutoring	.23*	.33**	.23*	.52*	.26*	.35*	1				
8. W2 3rd grade math score	.19*	.61*	.24*	.31*	.41*	.31*	.32*	1			
9. W1 2nd grade math score	.10*	.31*	.17**	.14*	.23*	.34*	.35*	.18*	1		
10. W1 Mother's education	.14**	.22*	.26*	.44*	.33**	.21*	.29*	.15*	.23*	1	
11. W1 Child sex	.12*	.34**	.36*	.14*	.23*	.22*	.19*	.16*	.29*	.28*	1
12. W1 child age	.21*	.11*	.20**	.35*	.20*	.17**	.37*	.18*	.09**	.18*	.19*

*$p<.05$, **$p<.01$

Confirmatory Factor Analysis

Table 3. Factor loadings of parental monetary investments and services

Variable	Standardized loading	*SE*
Parental monetary investments in essential materials and services		
Number word books	.792	.022
Number paper books	.681	.031
Pen or pencil box	.802	.022
Geometry box	.751	.031
Calculator	.680	.034
Math tutoring	.823	.013

Standardized factor loadings from the CFA conducted for the indicators of parental monetary investments are shown in Table 3. Results indicated that the measurement model was fully identified. The convergent validity for all indicators onto their respective factors (the standardized factor loadings of the mediators) was relatively high from .680 to .823. The CFA model fitting with the data was adequate at χ^2 (215, N = 1145, df = 28) = .224.57, $p < .05$. The other fit indices were well (GFI = .972; TLI = .975; CFI = .988; RMSEA = .042).

Direct Effects

Figure 2. Full mediating model for family low-income predicting math scores during third-grade

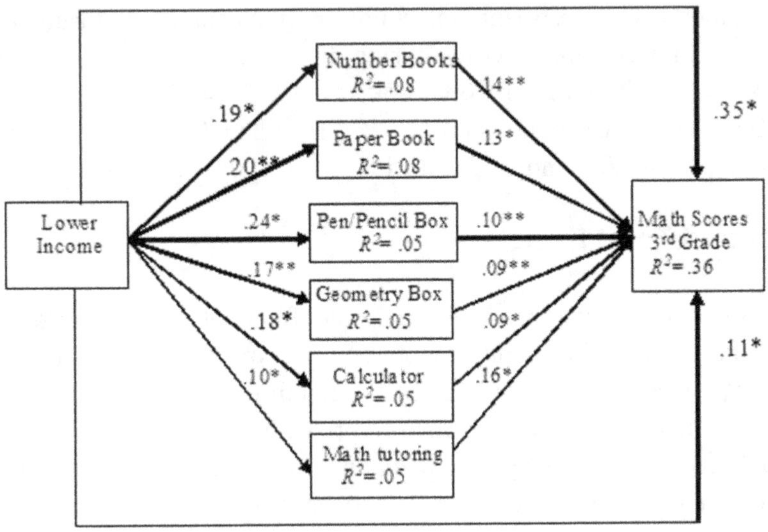

Standardized path coefficients are shown. Standardized path coefficients presented in the upper line represent parameter estimates for the direct relation of family low-income with math scores without the mediators, but those in the lower line represent for the pathways with the mediators. The mediating pathways with control variables are significant at p< .05.

χ^2 (159), N=1145, *df* = 23) = 218.11, p< .05; GFI =.974; TLI = .963; CFI = .961; RMSEA = .044

The SEM model examining the associations between family income, parental monetary investments in essential materials and math tutoring, and children's math scores during third-grade in the full sample with control for maternal and child characteristics was found to have adequate model fit at χ^2 (159), N = 1145, *df* = 23) = 218.11, *p*< .05; GFI = .974; TLI = .963; CFI = .961; RMSEA = .044. Standardized coefficients (*s*) of direct effects of the model are shown in Figure 2 and also present in text. Results of the SEM model that explained 28% variance in the children's math outcomes indicate positive and significant direct relations between family low-income, parental monetary investments, and children's math outcomes, after accounting for child and maternal characteristics. Overall, we found direct and positive associations between family low-income and children's poor math scores during third grade = .35, *SE* = .03, *p* < .01. Particularly, low-income of the families was significantly associated with whether or not number word book = .19, *SE* = .02, *p* < .01, paper book = .20, *SE* = .01, *p* < .05, pen/pencil box = .24, *SE* = .02, *p* < .01, geometry box = .17, *SE* = .03, *p* < .05, and math tutoring = .18, *SE* = .03, *p* < .01 provided. We also found significantly positive and direct relations between all parental monetary investments and children's math outcomes during third-grade. Specifically, children's math scores during third grade was positively and significantly associated with number word book, = .14, *SE* = .01, *p* < .05, paper book, = .13, SE = .03, *p* < .01, pen/pencil box = .10, *SE* = .03, *p* < .05, geometry box = .09, *SE* = .03, *p* < .05, and math tutoring = .16, *SE* = .03, *p* < .01.

Indirect and Total Effects

Like direct effects, we also found indirect effects of the indicators of parental monetary investments in essential materials and math tutoring for children that partially explained 09-24% variance in the relations between family low-income and children's math outcomes during third grade. In particular, number word book, paper book, and math tutoring than pen/pencil box and geometry box for children were potential mediators of the relations between family low-income and children's poorer math outcomes during third grade. Specifically, when we included all mediators in the baseline model, direct and positive relations between family low-income and child outcomes were substantially reduced by 24%, = .11, *SE* = .03, *p* < .001. Overall findings suggest that our mediators were partially explained the relations between family lower-income and young children's poor math outcomes during third grade with control variables in Bangladesh. Particularly, the total effects of lower-income had a marginal total effects on math scores during third grade (= .46, *SE* = .02, *p* < 0.001)

DISCUSSION

The present study examined the relevance of family investment model within a sample of Bangladeshi third graders. Specifically, we aimed to understand whether parental monetary investments in essential materials- number word books, paper books, pen/pencil box, geometry box, calculator and math services (e.g., math tutoring) for children-might partially explain the relations between family low-income and math achievement during third grade in Bangladesh urban poverty context. The results from structural equation modeling suggest that a parental monetary investment model well-developed in the high-income context (Conger et al., 2007; Linver et al., 2002; Yeung et al., 2002) and operationalized for use in low-income households (Wolf et al., 2019), particularly in Bangladesh can partially explain relations between family low-income and young children's poor math outcomes during third-grade in this lower middle-income country setting. In this particular setting, we explore several new findings for the field.

Like in the US, young children's poor math achievement during third grade is significantly associated with family low-income in Bangladesh urban poverty context. In terms of total effects, we found that a 1 *SD* increase in family low-income was associated with a .09-.28 *SD* increase in young children's early primary math outcomes, depending on providing age-appropriate essential materials and extra-math services for the children. The findings in the low-resourced context of Bangladesh confirm that family low-income is important, direct predictor of young children's poor math outcomes during third-grade. We also observe low-income of the families is also strongly and directly related to fewer parental monetary investments in essential materials and math services for the children. These findings consistent with prior research work from the US (Casey et al., 2018; Gershoff et al., 2007), Ghana (Wolf et al., 2019) and sub-Saharan Africa (McCoy et al., 2015) and Turkey (Atalmis et al., 2016) suggest social policy implication to improve family low-income that has independent- and probably additive- effects on improvement of family income, parental monetary investments, and positive children's math outcomes.

Although we found strong relations between family low-income and parental monetary investments, all investments low-income parents make in their children's math practices during third grade were marginally related to their math outcomes during third-grade, as hypothesized based on previous research (Mistry, Biesanz, Chien, Howes, & Benner, 2008; Mistry, Benner, Biesanz, & Clark, 2010). For example, parental monetary investments in geometry box and calculator were poorly related to children's early primary math outcomes. Despite this, parental monetary investments in essential materials, particularly number word-books, paper books and math tutoring are critical for children's early math development (Elliott et al., 2018; Wolf et al., 2019). Research in the low-resourced context has shown that parental monetary investments in home-enriching materials for young children were important mediators in explaining family income differences in young children's early primary math outcomes (Casey et al., 2018; Wolf et al., 2019).

We first examined adapted version of parental monetary investment model to examine relations between family low-income and poor math achievement during third-grade in Bangladesh poverty context. In so doing, we only included fundamental aspects of parental monetary investments in essential materials and math services and ignored several important dimensions of the broader framework of a parental investment model: (1) provision of adequate standard of living, (2) suitable home environment, (3) sufficient childcare and child healthcare, (4) positive parenting behavior, and (5) residing in a good location that may affect primary math outcomes over time (Conger et al., 2007; Gershoff et al., 2007; Haveman et al., 1994; Kalil et al., 2004). Although the fundamental aspects of the parental monetary investment model were positively predictive of young children's math outcomes during third-grade in this low-resourced context, the measure used to operationalize parental monetary investments in essential materials and math tutoring for math practice included limited set of items adapted version from HOME over twelve months. Other types of parent-child interactions and parental involvement in home-numeracy activities such as frequency of math talking, reading, writing and counting that underlie the relations between family income and children's math outcomes over time are not accounted for the present study. For example, family stress processes (Yeung et al., 2002) and mother's stress and mental health in low-resourced context that affect maternal support for children's math learning within and outside the home (LaForett & Mendez, 2010). Additional research is needed to understand how family income links young children's early primary math outcomes via these parental investments. Such factual data are valuable to national and international community's efforts to eliminate family income disparities in young children's early primary math outcomes (United Nations, 2015; Wolf et al., 2019).

Limitations and Direction for Future Research

Despite several strengths of this research, the current study is no exception without limitations. First, the results of this work are not causal, because we used nonexperimental nature of data. Although we accounted for several background characteristics (e.g., mother's age, mother's education and occupation, family size, children's age, sex, and previous math scores) in our bivariate correlation and SEM analyses, we are unable to determine the direction of the relations between family income, parental monetary investments, and math outcome during third grade followed in this study. For example, it is probably that parental monetary investments in essential materials and math services may change in direct response to how young children are doing well or worse in primary mathematics over time as a function of changes in family income. Additional two or more rounds of longitudinal data are needed for better understanding of how parental monetary investments relate to children's developmental trajectories of math achievement in the primary school years (Watts et al., 2015).

Second, the measures used to capture parental monetary investments in essential materials and math services reported by mothers only did not assess the quality of mother-child interactions and that were not fully representative of all the possible ways that mothers invest in young children's math learning. Particularly, the adapted version measures of the parental monetary investments and math services developed in the US and other context were primarily used in this study were not validated in this cultural setting (Iruka, LaForett, & Odom, 2012). In the future, additional research work is needed for objective metrics of the parental monetary investments (e.g., the HOME Inventory) in essential materials and math tutoring for children to ensure their relevance, validity, and comprehensiveness in this particular cultural and educational setting (Pisani, Borisova, & Dowd, 2018; Wolf et al., 2019).

Third, about 90% of the participants at District level of Bangladesh were ethnically Muslims. Although Bangladesh educational policy places value on educational achievement, irrespective of gender (e.g., male, female), religion (e.g., Muslim, Hindu, Christian, and Buddhists) and ethnic groups (e.g. 50-55 groups), the findings of the study cannot generalize beyond this specific study context. Like studies in US, additional research is needed to test similar models of parental monetary investments in diverse cultural, linguistic, and geographical context of this country (Tudge & Doucet, 2004; Wolf et al., 2019). As all caregivers (e.g., fathers, guardians) were not included due to increase cost and collaboration, there may be possible bias in the external validity of the findings within this particular context. Furthermore, although nutrition and physical development in low-resourced context are important markers of early child development and later school achievement, these variables were not included in this study (Larson et al., 2018). Similarly, cultural relevance of children's healthy development, creativity, and achievement also has been shown important to provide educational services also were not included (Huntsinger, Jose, & Luo, 2016; Iruka et al., 2012). Future longitudinal research examining children's primary math trajectories should include parental investments in math expectation, parental math support, and parental involvement in numeracy activities to understand complexity of young children's primary math achievement over time (Kornrich & Furstenberg, 2013; Lombardi et al., 2020).

CONCLUSION AND IMPLICATIONS

Despite young children exposing to a variety of family risks in the LMICs (Walker et al., 2007; Wolf et al., 2019), results of the study clearly indicate the strong parental monetary investment pathways through which low-income of the families affects young children's poor math outcomes during third grade in Bangladesh. The findings will contribute not only to enrich family investment model developed in the US but also to new theory-building by providing evidence in an understudied context. Because this context is characterized by high levels of family and child poverty, food insecurity, child and maternal malnutrition, and low levels of women's literacy and employment (BBS, 2016; Bornstein & Putnick, 2012; Bornstein, Putnick, Bradley, Lansford, & Deater-Deckard, 2015). Child development economists emphasize the importance of direct monetary investments that increases earnings, food security, childcare, and children's math essential materials and educational resources in the low-income families that in turn ultimately foster children's literacy and numeracy development (Duncan & Magnuson, 2013; Duncan, Magnuson, Votruba-Drzal, 2014; Foster, 2002; Hoddinott et al., 2011). For example, the Dayton Foundation in the US provides essential school supplies such as paper, pens and pencils, note books, arts and craft materials for low-income children for their classroom success. Like material support, developmental psychologists and sociologists target multipronged interventions to improve mother-child interactions, maternal numeracy skills, and maternal involvement in children's early math learning within and outside the home that promote children's math development over time (Wolf et al., 2019).

Both of the intervention approaches (e.g., economic & psychosocial) focus on the potential utilities of two-generation policy and programs to simultaneously identify economic, social, cultural, and psychological needs of both the parties- mother and children- intended to improve early primary math skills development and achievement in high-risk context (Duncan et al., 2014; Engle et al., 2011; Neville, Pakulak, & Stevens, 2015). Indeed, evidence in interventional and experimental studies also clearly demonstrates that positive changes in low-income families and maternal education make parents able to purchase and provide more math learning resources for their children. Educational training and literacy development for mothers make them able to involve more in their children's home numeracy activities (Huntsinger, Jose, & Luo, 2016; LeFevre, Skwarchuk, Smith-Chant, Fast, Kamawar, & Bisanz, 2009). All these ultimately and positively enhance children's early primary math outcomes (Dearing et al., 2001; Votruba-Drzal, 2003). Overall, our findings consistent with these perspectives highlight that low-levels of family income and fewer parental monetary investments are the main targets of interventions to mitigate children's poor math outcomes during early primary grades in this country (Uddin, 2015b). Public and voluntary agencies in the low-middle income country may provide material and psychosocial supports to enhance low-income children's healthy development and educational achievement.

REFERENCES

Aboud, F. E. (2006). Evaluation of an early childhood preschool program in rural Bangladesh. *Early Childhood Research Quarterly*, 21(1), 46–60. 10.1016/j.ecresq.2006.01.008

Aboud, F. E., & Hossain, K. (2011). The impact of preprimary school on primary school achievement in Bangladesh. *Early Childhood Research Quarterly*, 261(2), 237–246. 10.1016/j.ecresq.2010.07.001

Atalmis, E. H., Yilmaz, M., & Saatcioglu, A. (2016). How does private tutoring mediate the effects of socioeconomic status on mathematics performance? Evidence from Turkey. *Policy Futures in Education*, 14(8), 1135–1152. 10.1177/1478210316674159

Bangladesh Bureau of Statistics. (2010). *Household income and expenditure survey(HIES) 2010*. Bangladesh Bureau of Statistics.

Bangladesh Bureau of Statistics. (2016). *Bangladesh demographic profile 2016*. Bangladesh Bureau of Statistics.

Baron, R. M., & Kenny, D. A. (1986). The mediator-moderator variable distinction in social psychological research: Conceptual, strategic, and statistical considerations. *Journal of Personality and Social Psychology*, 51(6), 1173–1182. 10.1037/0022-3514.51.6.11733806354

Becker, G. S. (1993). *Human capital: A theoretical and empirical analysis, with special reference to education* (3rd ed.). University of Chicago Press. 10.7208/chicago/9780226041223.001.0001

Becker, G. S., & Tomes, N. (1979). An equilibrium theory of the distribution of income and intergenerational mobility. *Journal of Political Economy*, 87(6), 1153–1189. 10.1086/260831

Berberoğlu, G., & Tansel, A. (2014). Does private tutoring increase students' academic performance? Evidence from Turkey. *International Review of Education*, 60(5), 683–701. 10.1007/s11159-014-9436-y

Bornstein, M. H., & Putnick, D. L. (2012). Cognitive and socioemotional caregiving in developing countries. *Child Development*, 83(1), 46–61. 10.1111/j.1467-8624.2011.01673.x22277006

Bornstein, M. H., Putnick, D. L., Bradley, R. H., Lansford, J. E., & Deater-Deckard, K. (2015). Pathways among caregiver education, household resources, and infant growth in 39 low-and middle-income countries. *Infancy*, 20(4), 353–376. 10.1111/infa.1208626273231

Buchman, C., Condron, D. J., & Roscigno, V. J. (2010). Shadow education, American Style: Test preparation, the SAT and college enrollment. *Social Forces*, 89(2), 435–461. 10.1353/sof.2010.0105

Byrne, B. B. (2016). *Structural equation modeling with Amos: Basic concepts, applications, and programming* (3rd ed.). Rutledge. 10.4324/9781315757421

Caldwell, B. M., & Bradley, R. H. (1984). *Home observation for measurement of the environment*. University of Arkansas at Little Rock.

Chen, X., Chang, L., Liu, H., & He, Y. (2008). Effects of the peer group on the development of social functioning and academic achievement: A longitudinal study in Chinese children. *Child Development*, 79(2), 235–251. 10.1111/j.1467-8624.2007.01123.x18366421

Cole, D. A., & Maxwell, S. E. (2003). Testing mediational models with longitudinal data: Questions and tips in the use of structural equation modeling. *Journal of Abnormal Psychology*, 112(4), 558–577. 10.1037/0021-843X.112.4.55814674869

Coley, R. L., Kruzik, C., & Votruba-Drzal, E. (2020). Do family investments explain growing socioeconomic disparities in children's reading, math, and science achievement during school versus summer months? *Journal of Educational Psychology*, 112(6), 1183–1196. 10.1037/edu0000427

Coley, R. L., Sims, J., & Votruba-Drzal, E. (2016). Family expenditures supporting children across income and urbanicity strata. *Children and Youth Services Review*, 70, 129–142. 10.1016/j.childyouth.2016.09.017

Conger, R. D., & Donnellan, M. B. (2007). An interactionist perspective on the socioeconomic context of human development. *Annual Review of Psychology*, 58(1), 175–199. 10.1146/annurev.psych.58.110405.08555116903807

Dahl, G. B., & Lochner, L. (2008). The impact of family income on the child achievement: Evidence from the Earned Income Tax Credit. *Cambridge, M. A.: National Bureau of Economic Research, NBER Working Paper No. 14599.*

Dang, H. A. (2007). The determinants and impact of private tutoring classes in Vietnam. *Economics of Education Review*, 26(6), 684–699. 10.1016/j.econedurev.2007.10.003

Davis-Kean, P. E. (2005). The influence of parent education and family income on child achievement: The indirect role of parental expectations and home environment. *Journal of Family Psychology*, 19(2), 294–304. 10.1037/0893-3200.19.2.29415982107

Dawson, W. (2010). Private tutoring and mass schooling in East Asia: Reflections in inequality in Japan, South Korea and Cambodia. *Asia Pacific Education Review*, 11(1), 14–24. 10.1007/s12564-009-9058-4

Duncan, G. J., Dowsett, C. J., Claessens, A., Magnuson, K., Huston, A. C., Klebanov, P., Pagani, L. S., Feinstein, L., Engel, M., Brooks-Gunn, J., Sexton, H., Duckworth, K., & Japel, C. (2007). School readiness and later achievement. *Developmental Psychology*, 43(6), 1428–1446. 10.1037/0012-1649.43.6.142818020822

Duncan, G. J., Magnuson, K., & Votruba-Drzal, E. (2014). Boosting family income to promote child development. *The Future of Children*, 24(1), 99–120. 10.1353/foc.2014.000825518705

Duncan, J. G., & Brooks-Gunn, J. (1997). *Consequences of growing up poor*. Russell Sage Foundation.

Elliot, L., & Bachman, H. J. (2018). How do parents foster young children's math skills? *Child Development Perspectives*, 12(1), 16–21. 10.1111/cdep.12249

Engle, P. L., Fernald, L. C., Alderman, H., Behrman, J., O'Gara, C., Yousafzai, A., de Mello, M. C., Hidrobo, M., Ulkuer, N., Ertem, I., & Iltus, S. (2011). Strategies for reducing inequalities and improving developmental outcomes for young children in low-income and middle-income countries. *Lancet*, 378(9799), 1339–1353. 10.1016/S0140-6736(11)60889-121944378

Foster, E. M. (2002). How economists think about family resources and child development. *Child Development*, 73(6), 1904–1914. 10.1111/1467-8624.0051312487501

Galindo, C., & Sonnenschein, S. (2015). Decreasing the SES math achievement gap: Initial math proficiency and home learning environments. *Contemporary Educational Psychology*, 43, 25–38. 10.1016/j.cedpsych.2015.08.003

Gershoff, E. T., Aber, J. L., Raver, C. C., & Lennon, M. C. (2007). Income is not enough: Incorporating material hardship into models of income associations with parenting and child development. *Child Development*, 78(1), 70–95. 10.1111/j.1467-8624.2007.00986.x17328694

Guo, G., & Harris, K. M. (2000). The mechanisms mediating the effects of poverty on children's intellectual development. *Demography*, 37(4), 431–447. 10.1353/dem.2000.000511086569

Haider, M., Islam, M. R., Mia, A., & Sorcar, N. R. (2005). Income generation perspective in non-formal education NGO initiatives in Bangladesh. *Asia Pacific Journal of Social Work*, 15(1), 17–29. 10.1080/21650993.2005.9755966

Hamadani, J. D., Tofail, F., Hilaly, A., Huda, S. N., Engle, P., & Grantham-McGregor, S. M. (2010). Use of family care indicators and their relationship with child development in Bangladesh. *Journal of Health, Population and Nutrition*, 28(1), 23–33. 10.3329/jhpn.v28i1.452020214083

Haveman, R., & Wolfe, B. (1994). *Succeeding generations: On the effects of investments in children*. Russell Sage Foundation.

Hossain, A., & Zeitlyn, B. (2010). *Poverty, Equity and access to education in Bangladesh (Research Monograph No. 51)*. Consortium for Research on Educational Access, Transition and Equity.

Hu, L., & Bentler, P. M. (1999). Cutoff criteria for fit indexes in covariance structure analysis. *Structural Equation Modeling*, 6(1), 1–55. 10.1080/10705519909540118

Hussain, K. G. (2016). *Achievement of non-formal primary education among the working children at Rajshahi city, Bangladesh* [Unpublished M. Phil Thesis, Bangladesh: University of Rajshahi].

Iruka, I. U., LaForett, D. R., & Odom, E. C. (2012). Examining the validity of the family investment and stress models and relationship to children's school readiness across five cultural groups. *Journal of Family Psychology*, 26(3), 359–370. 10.1037/a002829022545934

Jordan, N. C., Huttenlocher, J., & Levine, S. C. (1992). Differential calculation abilities in young children from middle- and low-income families. *Developmental Psychology*, 28(4), 644–653. 10.1037/0012-1649.28.4.644

Jordan, N. C., Huttenlocher, J., & Levine, S. C. (1994). Assessing early arithmetic abilities: Effects of verbal and nonverbal response types on the calculation performance of middle- and low-income children. *Learning and Individual Differences*, 6(4), 413–432. 10.1016/1041-6080(94)90003-5

Jordan, N. C., Levine, S. C., & Huttenlocher, J. (1994). Development of calculation abilities in middle- and low-income children after formal instruction in school. *Journal of Applied Developmental Psychology*, 15(2), 223–240. 10.1016/0193-3973(94)90014-0

Kabeer, N., & Mahmud, S. (2009). Imagining the future: Children, education and intergenerational transmission of poverty in urban Bangladesh. *IDS Bulletin*, 40(1), 10–21. 10.1111/j.1759-5436.2009.00003.x

Kalil, A., & DeLeire, T. (Eds.). (2004). *Family investments in children's potential: Resources and parenting behaviors that promote success*. Lawrence Erlbaum Associates. 10.4324/9781410610874

Kaufhold, J. A., Alverez, V. G., & Arnold, M. (2006). Lack of school supplies, materials and resources as an elementary cause of frustration and burnout in South Texas special education teachers. *Journal of Instructional Psychology*, 33(3), 159–161.

Khan, A. F. M. K., Motin, S., Begum, H. B., & Uddin, M. M. (2012). *Elementary mathematics: Class one*. National Textbook Board, Bangladesh.

Kim, S., & Lee, J. H. (2010). Private tutoring and demand for education in South Korea. *Economic Development and Cultural Change*, 58(2), 259–296. 10.1086/648186

Kline, R. B. (1998). *Principles and practice of structural equation modeling*. New York: NY: Guildford.

Kornrich, S., & Furstenberg, F. (2013). Investing in children: Changes in parental spending on children: 1972-2007. *Demography*, 50(1), 1–23. 10.1007/s13524-012-0146-422987208

Lee, J. Y. (2013). *Private tutoring and its impact on students' academic achievement, formal schooling, and educational inequality in Korea*. [Unpublished Ph. D Dissertation, The US: Columbia University].

Lee, K. S. (2010). Parental educational investments and aspirations in Japan. *Journal of Family Issues*, 31(12), 1579–1603. 10.1177/0192513X10368268

Linver, M. R., Brooks-Gunn, J., & Kohen, D. E. (2002). Family processes as pathways from income to young children's development. *Developmental Psychology*, 38(5), 719–734. 10.1037/0012-1649.38.5.71912220050

Lombardi, C. M., & Dearing, E. (2020). Maternal support of children's math learning in associations between family income and math school readiness. *Child Development*, 92(1), 39–55. 10.1111/cdev.1343632797635

Lugo-Gil, J., & Tamis-Lemonda, C. S. (2008). Family resources and parenting quality: Links to children's cognitive development across the first 3 years. *Child Development*, 79(4), 1065–1085. 10.1111/j.1467-8624.2008.01176.x18717907

Mayer, S. E. (1997). *What money can't buy: Family income and children's life chances*. Harvard University Press.

McCoy, D. C., Zuilkowski, S. S., & Fink, G. (2015). Poverty, physical stature, and cognitive skills: Mechanisms underlying children's school enrollment in Zambia. *Developmental Psychology*, 51(5), 600–614. 10.1037/a003892425844851

McLoyd, V. C. (1998). Socioeconomic disadvantage and child development. *The American Psychologist*, 53(2), 185–204. 10.1037/0003-066X.53.2.1859491747

Mistry, R. S., Benner, A. D., Biesanz, J. C., Clark, S. L., & Howes, C. (2010). Family and social risk, and parental investments during the early childhood years as predictors of low-income children's readiness outcomes. *Early Childhood Research Quarterly*, 25(4), 432–449. 10.1016/j.ecresq.2010.01.002

Mistry, R. S., Biesanz, J. C., Chien, N., Howes, C., & Benner, A. D. (2008). Socioeconomic status, parental investments, and cognitive and behavioral outcomes of low-income children from immigrant and native households. *Early Childhood Research Quarterly*, 23(2), 193–212. 10.1016/j.ecresq.2008.01.002

Mueller, P. A., & Oppenheimer, D. M. (2014). The pen is mightier than the keyboard: Advantages of longhand over laptop note taking. *Psychological Science*, 25(6), 1159–1168. 10.1177/095679761452458124760141

Nath, S. R. (2007). Remunerated supplementary tutoring in out-of-school study in rural Bangladesh. BRAC Research Report, Research and Education Division. Dhaka, 1212, Bangladesh.

National Research Council. (2001). *Adding it up: Helping children learn mathematics*. National Academy Press.

Neville, H., Pakulak, E., & Stevens, C. (2015). Family-based training to improve cognitive outcomes for children from lower socioeconomic status backgrounds: Emerging themes and challenges. *Current Opinion in Behavioral Sciences*, 4, 166–170. 10.1016/j.cobeha.2015.05.006

Park, H., Byun, S., & Kim, K. (2011). Parental involvement and students' cognitive outcomes in Korea: Focusing on private tutoring. *Sociology of Education*, 84(1), 3–22. 10.1177/0038040710392719

Rahman, M. O., & Islam, M. T. (2009). Universal primary education for all towards millennium development goal 2: Bangladesh perspective. *Journal of Health Management*, 11(2), 405–418. 10.1177/097206340901100209

Ramani, G. B., & Siegler, R. S. (2008). Promoting board and stable improvements in low-income children's numerical knowledge through playing number games. *Child Development*, 79(2), 375–394. 10.1111/j.1467-8624.2007.01131.x18366429

Reardon, S. F., & Portilla, X. A. (2016). Recent trends in income, racial, and ethnic school readiness gaps at kindergarten entry. *AERA Open*, 2(3), 1–18. 10.1177/2332858416657343

Rittle-Johnson, B., Fyfe, E. R., Hofer, K. G., & Farran, D. C. (2017). Early math trajectories: Low-income children's mathematics knowledge from ages 4 to 11. *Child Development*, 88(5), 1727–1742. 10.1111/cdev.1266227921305

Scalise, N. R., Daubert, E. N., & Ramani, G. B. (2020). Benefits of playing numerical card games on head start children's mathematical skills. *Journal of Experimental Education*, 88(2), 200–220. 10.1080/00220973.2019.158172133100376

Schneider, D., Hastings, O. P., & LaBriola, J. (2018). Income inequality and class divides in parental investments. *American Sociological Review*, 83(3), 475–507. 10.1177/0003122418772034

Siegler, R. S., & Ramani, G. B. (2008). Playing numerical board games promotes low-income children's numerical development. *Developmental Science*, 11(5), 655–661. 10.1111/j.1467-7687.2008.00714.x18801120

Sirin, S. R. (2005). Socioeconomic status and academic achievement: A meta-analytic review of research. *Review of Educational Research*, 75(3), 417–453. 10.3102/00346543075003417

Sohn, H., Lee, D., Jang, S., & Kim, T. K. (2010). Longitudinal relationship among private tutoring, student-parent conversation, and student achievement. *KEDI Journal of Educational Policy*, 7(1), 23–41.

Stevenson, D. L., & Baker, D. F. (1992). Shadow education and allocation in formal schooling: Transition to university in Japan. *American Journal of Sociology*, 97(6), 1639–1657. https://www.jstor.org/stable/2781551. 10.1086/229942

Tudge, J. R. H., & Doucet, F. (2004). Early mathematical experiences: Observing young black and white children's everyday activities. *Early Childhood Research Quarterly*, 91(1), 21–39. 10.1016/j.ecresq.2004.01.007

Uddin, M. E. (2015a). Family socio-cultural values affecting early marriage between Muslim and Santal communities in rural Bangladesh. *The International Journal of Sociology and Social Policy*, 35(3/4), 141–164. 10.1108/IJSSP-06-2014-0046

Uddin, M. E. (2015b). Ethnic disparity in family socio-economic status in Bangladesh: Implication for family welfare policy-practice. *Global Social Welfare : Research, Policy & Practice*, 2(2), 29–38. 10.1007/s40609-014-0021-x

Uddin, M. E. (2017). Ecological framework for primary school attainment in the tri-ethnic communities in rural Bangladesh. *Child Indicators Research*, 10(3), 693–713. 10.1007/s12187-016-9401-3

United Nations Children's Fund. (2006). *Multiple indicator cluster survey manual 2005: Monitoring the situation of children and women*. UNICEF.

United Nations Children's Fund. (2015). *Transforming our world: The 2030 agenda for sustainable development*. UNICEF.

Vandermaas-Peeler, M., Nelson, J., Bumpass, C., & Sassine, B. (2009). Numeracy-related exchanges in joint story-book reading and play. *International Journal of Early Years Education*, 17(1), 67–84. 10.1080/09669760802699910

Vasilyeva, M., Dearing, E., Ivanova, A., Shen, C., & Kardanova, E. (2018). Testing the family investment model in Russia: Estimating indirect effects of SES and parental beliefs on the literacy skills of first graders. *Early Childhood Research Quarterly*, 42, 11–20. 10.1016/j.ecresq.2017.08.003

Votruba-Drzal, E. (2003). Income changes and cognitive stimulation in young children's home learning environments. *Journal of Marriage and Family*, 65(2), 341–355. 10.1111/j.1741-3737.2003.00341.x

Walker, S. P., Wachs, T. D., Gardner, J. M., Wasserman, G. A., Pollitt, E., & Carter, J. A. (2007). Child development in developing countries 2. Child development and risk factors for adverse outcomes in developing countries. *Lancet*, 369(9556), 145–157. 10.1016/S0140-6736(07)60076-217223478

Watts, T. W., Duncan, G. J., Chen, M., Claessens, A., Davis-Kean, P. E., Duckworth, K., & Susperreguy, M. I. (2015). The role of mediators in the development of longitudinal mathematics achievement associations. *Child Development*, 86(6), 1892–1907. 10.1111/cdev.1241626332124

Watts, T. W., Duncan, G. J., Clements, D. H., & Sarama, J. (2018). What is the long-run impact of learning mathematics during preschool. *Child Development*, 89(2), 539–555. 10.1111/cdev.1271328105650

Watts, T. W., Duncan, G. J., Siegler, R. S., & Davis-Kean, P. E. (2014). What's past is prologue: Relations between early mathematics knowledge and high school achievement. *Educational Researcher*, 43(7), 352–360. 10.3102/0013189X14 55366026806961

Wolf, S., & McCoy, D. C. (2019). Household socioeconomic status and parental investments: Direct and indirect relations with school readiness in Ghana. *Child Development*, 90(1), 260–278. 10.1111/cdev.1289928708296

Yeung, W. J., Linver, M. R., & Brooks-Gunn, J. (2002). How money matters for young children's development: Parental investment and family processes. *Child Development*, 73(6), 1861–1879. 10.1111/1467-8624.t01-1-0051112487499

Zadeh, Z. Y., Farnia, F., & Ungerleider, C. (2010). How home enrichment mediates the relationship between maternal education and children's achievement in reading and math. *Early Education and Development*, 21(4), 568–594. 10.1080/10409280903118424

Math Achievement

Zhang, Y., & Xie, Y. (2016). Family background, private tutoring, and children's educational performance in contemporary China. *Chinese Sociological Review*, 48(1), 64–82. 10.1080/21620555.2015.109619330416869

Chapter 11
Foundational-Level Mathematics Teaching Using Learners' Mother Tongue

Tebogo J. Mampane
University of South Africa, South Africa

ABSTRACT

This study investigated the use of mother tongue teaching in South African township schools for Black people. It hoped to contribute to the limited literature on mathematics education using mother tongue. The literature review used a descriptive qualitative approach. Data were collected, analysed, and interpreted from semi-structured interviews. The study investigated mathematical concepts in the learners' native language. Inadequate standards, legislation, rules for implementation, and inadequate teacher training, were barriers to instruction in mother language. The constitution mandated the use of 11 official languages as outlined in the Languages in Education Policy. The study advocates for trans-languaging in educating in mother tongue. The results of mother tongue teaching were illuminating and reaffirmed the value of exposing learners to more than one language in primary schools. Effective learning and instruction of mathematics in mother language require policies, methods, and future research sensitive to dynamic and distinctive qualities of several school environments.

DOI: 10.4018/979-8-3693-2873-6.ch011

INTRODUCTION

This article provides a summary and interpretation of study findings on mother-tongue (MT) teaching of mathematics. For a long time, children in township schools in South Africa went to class learning a language they were unable to comprehend (Thuketana & Makgabo, 2022; Fleisch, 2008). As a result, this article focuses on how to use MT in a math classroom. The first debate concentrated on a working definition of MT. The topic then shifted to whether or not math educators had a sufficient grasp of mathematical language to adequately instruct their learners. The issue surfaced when attempting to use MT for mathematical education. The report also discusses the challenges that educators experience while using MT to teach mathematical concepts. Language of Learning and Teaching (LoLT) policy developments are discussed as a conclusion. Existing research (Njoroge, 2017; Mashige, Cekiso, and Meyiwa, 2019) has shown that LoLT is important for learning mathematics (Phakeng, Planas, Bose & Njurai, 2018). The researchers found that using the mother tongue as the language of learning and teaching (LoLT) in mathematics led to higher-quality instruction and improved learner performance. Hafiz and Farik (2016) offered an argument in support of this claim, arguing that it was easier for teachers to teach mathematics in their mother tongue. Further research, however, revealed that most educators struggled with the transition from English to their mother tongue (MT), or what is now known as LoLT in mathematics (Makeleni & Sethusha, 2014:107).

There is a plethora of research on educational and policy shifts, but surprisingly little on the experiences of math educators who employ MT in their classrooms. Teachers, according to the literature (Makeleni & Sethusha, 2014; Muthivhi & Broom, 2008), need to be fluent not only in the language of instruction but also in the process of teaching and learning through a mother tongue. The achievement of both the educators and the learners will suffer if the medium of instruction is not used to its full potential. The fundamental reason why students fail to learn English as a second language (ESL) and why they perform poorly in school is that they are not fluent in their mother tongue. However, the effectiveness of MT instruction is contested by certain South African scholars. These assumptions, along with the goal to gain insight into mathematics education in MT, led to the selection of a qualitative descriptive method for data collection, analysis, and interpretation.

LITERATURE REVIEW

The problem-plagued area of many South African schools is teaching mathematics through mother tongue (Monyooe, 2017). Learners in South Africa have the right to receive instruction in their mother tongue wherever possible, as per Section 29 of the Constitution (Republic of South Africa, 1996). As a result, most public schools wisely switched the LoLT from English to MT. Provincial structures manage educational matters differently depending on where they have jurisdiction and what they must accomplish, despite the national government setting policy criteria. Use of the 11 official languages is promoted by the CRSA (RSA, 1996) and the Language in Education Policy (LiEP) of 1997 (Plüddemann, 2015). The Constitution changed official school language guidelines. The eleven official teaching languages encourage learners to learn a second language. However, not all elementary school children in townships speak the LoLT taught there, despite the strategy's goal of facilitating learners' access to MT teaching. Blacks, Coloureds, and Indians have historically, and still live in impoverished urban areas known as "townships" (Monyooe, 2017). Lower math achievement can be observed in South African township and rural schools. Teachers' inexperience and a lack of materials in the subject contribute to students' low grades. The research took place at Mamelodi, a township in Gauteng Province. The local languages—are Sepedi, Setswana, Sesotho, Tsonga Venda, Zulu, Swazi, Xhosa, and Ndebele, and Afrikaans and English spoken by Coloureds, Whites and Indians.

The GDE is in charge of Tshwane South schools in Mamelodi. In these classrooms, MT is used to help students, yet students' mathematical abilities still suffer as a result. Inconsistent policy directives and curricular practice; learner quality; performance and monitoring; school culture and educational ethos; and inadequate infrastructure are all examples of operational challenges (Monyooe, 2017). Underperforming schools and learners who struggle across grade levels are the direct result of insufficient training and a lack of coordination of corrective efforts. The International Mathematics and Science Study (IMSS) results show that South African children are not doing well in mathematics. South African kids were included in the survey, and their country lags significantly below others. Makeleni and Sethusha (2014) advocate for math educators to have a solid foundation of knowledge and training. They need to understand the ideas behind mathematics. Guaranteeing that educators meet expectations and adapt to new curricula requires a system of standards for both teacher performance and student achievement (Monyooe, 2017). The growing number of students attending township schools faces a lack of resources to help them succeed in mathematics. Schools cannot function with above mentioned challenges and less funding. The Department of Education and professional learning groups should review the math mother tongue policy, as teachers with a Diploma in Teaching are respected even if they have not had additional training in mathematics.

Language of Learning in Foundation Phase

Learning mathematics in the Foundation Phase (FP) should enable learners to gain an understanding of universal truths that may be applied in any context since concepts and procedures in mathematics are introduced early on. (Republic of South Africa, 1996). All students, regardless of age, are considered "learners" under the South African Schools Act. The term "learner" refers to children in the third grade of the public school's Foundation Phase. Measurement, data management, space and shape, number and relationship operations, patterns, functions, and algebra are the five foundational ideas that must be introduced to learners in accordance with the Mathematics Curriculum and Assessment Policy Statement (CAPS) (Department of Basic Education, 2011: 10-11). The lessons must be offered in the South African spoken languages. The government recognises English, Afrikaans, and nine other Black languages. There is a mandate for primary, middle, and high school learners students in the township to take math. Locally and regionally developed literature, as well as a curriculum taught in the child's original language, have all been shown to improve early education (Mashige, Cekiso, & Meyiwa, 2019). If a child has solid literacy skills in MT, these can be easily transferred to the second language. Most MTs lack a thorough understanding of mathematical terminology, making it difficult for English-to-MT translations by math teachers. Therefore, education policies should assist in addressing such demands to raise the bar of inadequate education. Students' intellectual, social, and academic growth will be delayed if they are unable to communicate in the language of instruction (BiteneIkome, 2013). To teach and learn MT effectively, then, requires reliable information and comprehension.

The SGB is responsible for choosing the MT of each school. Teachers in schools where learners speak more than one language are expected to teach children in the MT of their choice. When children undergo schooling in their MT languages, this is called "mother tongue education" (Khosa, 2012). Most kids' first language is the one they use at home with their families. What we mean by 'mother tongue' is not necessarily the language of the mother's household. In a similar vein, Moyo (2009) defines mother tongue as the first language that a child learns at home. For the sake of this analysis, a person's mother tongue is defined as their primary language of communication within their household, as well as their preferred and most understood language outside of the household. When it comes to teaching kids to math. reading and writing, oral language skills are prioritized in primary school settings.

Kajee (2012) states that most black learners experience barriers in the language used in the classroom. The most common reason is that these learners are unable to use the language which they are familiar with, and end up not performing well in the classroom, because they are not taught the subject matter in their mother tongue. As stated in the LiEP for MT (DoE, 2010: 6), the DoE promotes the usage of the MT among primary school learners. At one time, English was the LoLT in South African primary schools. NEPA Section 3(4)(m), LiEP, Section 6(1) of the South African School Act, and The Norms and Standards, form the basis for the language policy. Schools that use a single or more languages make this decision. However, MT multilingual classrooms require teachers who are fluent in the languages being used. Teachers who were educated in English but who speak MT have difficulty making the transition to MT in Grade 3 (Makeleni & Sethusha, 2014). Some South African math textbooks are also poorly translated into MT and inhibit higher-order thinking, and the Revised National Curriculum Statement (RNCS) (DoE, 2005) does not include any MT instructional assistance resources. Professional development is necessary for teachers in MT.

Research Design and Methodology

To assess educators' knowledge and comfort with using MT in the classroom, the researcher conducted a qualitative case study. Ethics guidelines were strictly followed throughout the study (Creswell, 2014). MT mathematics instruction was measured against international standards of practice and policy. Six third-grade math teachers in Mamelodi were interviewed employing qualitative, semi-structured interviews (Creswell, 2014) to get insight into children's cultural and MT-specific beliefs, attitudes, behaviours, and social situations. The following questions were posed to math teachers: What makes you qualified to teach mathematics using MT? Where did you train as a teacher to become prepared for teaching math at a township school? 4) Which language makes it the least difficult to introduce math to third graders, and why? Additional resources included the Annual Teaching Plan (ATP), teacher lesson plans, and learners' activity books. Based on teacher qualifications and student textbooks and lesson plans, MT math programs were established. The researcher began by looking back, analysing, and clarifying MT math education. Key components of MT mathematics education were demonstrated through participants' responses, literature, and policy papers. With proper coding, triangulation, and interpretations, MT could serve as a benchmark for mathematics instruction.

RESULTS

Theme One: Conceptualising the MT Teaching Policy

In South Africa, MT is widely used as a medium of instruction in mathematics. When teaching mathematics, teachers and leaders alike require the ability to translate mathematical concepts from English to MT. Math teachers can apply the required language policy with the help of the academic research study and policy documents. The following comments were provided by teachers in township schools on their experiences with MT in the classroom:

As an MT teacher, I found it difficult to explain the concept of 2-dimensional shapes like the square, triangle, etc. to my students. This is because there are English words that simply do not exist in the MT vocabulary. So, I had to go find the English equivalents. There are some MT words that are only known in English, such as the Annual Teaching Plan (TA1). (T3B)

In addition to this, it is challenging for me because I do not use the Sepedi terminology for monetary units when I am interacting with my children at home. When we bring it up in the classroom, it comes as a surprise to them, and it does not generate any interest among the learners. It undermines the instructor's faith in their own abilities. Even at the beginning of the year, we make sure to congratulate one another on the next year no one uses that expression in MT (TC3),

I had to settle for using a dictionary to look up the MT words because I am a teacher. However, there are times when I need to look up a term in English because MT isn't providing it. (TE5)

Most Sepedi terms have no equivalent in English. It's not uncommon for me to stutter when trying to explain the line of symmetry to my students. Prisms and pyramids, for example, defy our efforts to give them suitable Sepedi names. Due to our limited linguistic abilities (we can only read and write in English), we are prohibited from doing so (TF6).

The comments from participants revealed how challenging it was to meet the LiEP's MT standards. The employment of MT in the classroom complicates the process of learning English as a second language. Umar, Muhoro, Opondo, Okumu, Kasimu, Wachira, Odoyo, Muturi, Ogari, Namikoye, and Wayong'o (2018), among others, argue that using MT to teach a child is a fantastic method since it helps build on the child's prior knowledge and experiences.

Teachers have also voiced concerns that it could be challenging to teach in a LoLT if certain learners struggle to understand it because it is not their MT. This lends credence to the statements of Kotzé, Van der Westhuizen, and Barnard (2017) and

Spaull (2015), who argue that teachers do not have the specific expertise necessary to deliver instruction in different LoLT simultaneously.

Huddlestone and Southwood's (2016) research provides supporting evidence for this claim by showing that the incapacity of instructors to convert concepts from English to the MT was preventing them from using the MT as the LoLT. The study's participants agreed with this notion, as they had personally experienced MT distortion owing to the employment of various languages in their daily lives. Some learners' parents have college degrees and communicate with their children using a mix of English and their MT. Promoting a single-language policy disproportionately harms low-income, multilingual populations due to the difficulty of teaching and learning in a community where different languages are spoken. Teachers' abilities may be questioned if there is a disagreement between the Department of Education and the schools on whether or not MT should be used as the medium of instruction (Gobana, 2013).

Theme Two: The Process of Training Teachers for LoLT

The workshops and pre-service teacher training were all conducted in English, according to the teachers interviewed. Teachers have complained that it is difficult when learners do not understand the LoLT due to the variety of MTs spoken in the classroom. According to Kotzé, Van der Westhuizen, Barnard, and Spaull (2016; 2017), educators who do not speak the MT used at school, lack the skills and abilities to effectively instruct children in mathematics. The following comments were provided by teachers in township schools regarding their training for math teaching through LoLT:

No one ever taught me MT; instead, I studied the English language. During workshops even the speakers speak English and use English topics. (TA1),

I received no Setswana instruction during my time at university. The university courses I took were all taught in English, so all I had to do was interpret. Even workshops are conducted in English, although most words in class need to be checked in a dictionary (TC3).

No preparation for teaching in Sepedi was provided. We were just informed the school's LoLT will be shifting. In MT, we were only ever taught how to switch languages; we were never given any actual training as educators. My English skills were sharpened in college, where I took English-language math classes. I also attended a yearlong course given in English by the Department of Education on teaching third-grade math. Due to the diversity of our teaching staff, the District has a hard time organising workshops in MT.(TD4)

This is because I went to school to learn the English language. Workshops are where we receive English language instruction. Even in the math workshops, we're expected to use English. The MT they insist we use is not something they employ themselves. We had no formal instruction in MT. They speak English and demand an MT translation from us. (TF6E)

According to Umar et al. (2018), teachers acquire PTD (professional teacher development) through structured fieldwork and classroom instruction. This is sometimes referred to as a "student teaching" experience. The language policy of the school can be implemented successfully if instructors are well-prepared. Universities and colleges are supposed to prepare teachers, yet they aren't preparing instructors to teach a subject like mathematics in MT. As teacher education is crucial for closing the policy-practice gap, it is important to know if the programs at teacher training institutions are in line with the language policy's implementation. Begi (2014) argues that providing educators with in-service training on the value of MT will encourage them to support the policy's implementation in their classrooms. Mata (2014) agrees, arguing that educators can better assist their students' growth and learning if they are trained to use MT in the classroom. Nonetheless, Ssentanda, Huddlestone, and Southwood (2016) discovered that in some institutions, only selected educators had received training in the use of MT.

Theme Three: Preferred MT for Mathematics Teaching

When asked in what language they felt most at ease instructing mathematics, all participants indicated English. This explanation aligns with the data presented in the second question. The educators complained that they did not have access to enough materials to teach learners with MT. Some of the teachers have offered the following feedback:

Ten years into my career as a teacher, I have never used MT. When at all feasible, English should be used. I feel more at ease communicating in English. Rather than using the MT word "lekgolo," which is more appropriate for indoor use, children outside use words like "hundred" (TA1).

Because the local languages are not mathematically pure, teaching mathematics through the English language is the most effective option (TB2).

Since all ATP professionals are expected to be fluent in English, that's the one I'll choose. The concepts have been presented in English, and the majority of the remarks have also been offered in English because even the dictionary that has been delivered to us has many words that were borrowed from other languages (TC3).

Since most resources are written in English, MT translation can be difficult. If you try to find them on Google, you will be disappointed. When we make a purchase, the vendor usually sends us paperwork in a language we don't understand. It should be written in Sepedi, however some of the words have been translated into Sesotho erroneously. (TD4)

When it comes to teaching math, there is no better language than English. Teaching in English benefits both me and my students. Our MT is unknown to them. The Tsonga spoken in our township is not the same language as the real Xitsonga, hence English is preferable. This is why we are once again communicating in English (TE5).

Most educators disagreed with the idea of using MT in math classrooms. Instead, they used English as if it were their mother tongue. Similar findings were found in Maţa's (2014) study, where participants admitted that the difficulties of teaching and learning content in MT necessitated the use of dictionaries to translate concepts into MT. This confirms the results of a separate research that found English as a second language to be vastly more popular than the MT. Despite the recommendation of national and provincial education departments that schools use the MT language in Grades 1, 2, and 3, and then switch to English as the language of instruction in Grade 4, Taylor and von Fintel (2016:76) indicate that some schools chose to go 'straight-for-English' as the LoLT from Grade 1. Others who have studied the question of which language should be used to teach mathematics have come to the same conclusion: English is preferable to MT. For instance, despite the National Policy on Education's requirement that students in Nigeria's lower primary schools be taught in their MT, Umar et al. (2018) discovered that teachers in Adamawa State typically teach Mathematics in English.

Theme Four: Challenges of Teaching Through MT

Using English vocabulary in an MT classroom is detrimental to student learning. The Malay-language study by Yunus and Sukri (2017) concluded that the materials were out of date and could not be translated from English to science, therefore they could not be used in classrooms. The participants in the study's comments align with this study when they say:

Since I only use ATPs for teaching and they are in English, I'm going to guess that English is the way to go. Most of the feedback, however, is written in English and even the given dictionary includes words that were modified from other languages. Because our suppliers only provide us with materials in a foreign language, it is difficult for us to make purchases. Several terms have been mistranslated from Sesotho to Sepedi. (TA1)

Sepedi Language is a major source of difficulty. When I use MT, learners lose interest immediately. It simply doesn't exist in their mental framework; therefore they have trouble making connections to it. I am the first person they have heard to say this. Since their parents don't speak that language, they aren't making the transition from one familiar to one unfamiliar (TB2).

These learners have a hard time using MT for math because they are accustomed to English terminology. Learners were more successful overall while using English as their language of instruction as compared to when they were using MT. This MT has poor performance. I can't tell if they just don't understand it or they just don't like the language.(TC3)

The learners like English and have fun with it; trust me on that. I've been there before. If you were an English teacher, you would adore them. They like speaking English in this context. Even the slower ones do very well when speaking English; their heads are held high with pride. However, while using Sepedi or MT, they simply appear exhausted (TD4).

From what I've seen in teaching with MT, we face several obstacles while working with these students. English is my first language of choice. From kindergarten onward, the vast majority of students are exposed to instruction in the English language. The use of MT at our school makes things more challenging for them.(TE5)

Because their foundation is in English, especially when it comes to numbers and number names, it is difficult for us teachers to teach in MT. English has fairly simple numerals and number names. It's challenging, though, because of MT. The calculations are drastically off. Also, the kids here prefer English. They find their MT annoying. The parents are English fans as well. The majority of local parents send their children to schools that offer English as a second language (LoLT), so the market is flooded with options.(TF6)

Teachers have reported that they have difficulty converting MT concepts into ideas in the classroom because they have difficulty acquiring the proper MT concepts. Some people have said, yet again, that the MT dictionary does not have all the ideas and therefore using Google is useless. Third-graders who are taught in a language they do not know, face significant difficulties in mathematics. If not addressed in a timely and effective manner, the educational barrier that arises from students' needing to learn in a LoLT that is not the same as their home language, has the potential to become a major impediment to a truly inclusive education system (MacKay, 2014). However, participant data shows that instructors find it simpler to instruct in English rather than the MT of the learners they teach. Learners in Grade 3 who started their formal education in a different language from the school's LoLT (in this case, English) have a difficult time adjusting, according to the study's participants. Since these students lack conceptual knowledge, they are unable to fully benefit from their MT (Kotzé, Van der Westhuizen, & Barnard, 2017). According to the participants, some students were not exposed to the LoLT before the Foundation Phase (Grades 1–3). The LoLT was neither their mother tongue nor their everyday spoken language, and this meant that they were unprepared to use their MT in the classroom. Maake (2014), MacKay (2014), and Kotzé, Van der Westhuizen, and Barnard (2017) all referred to this difficulty as a linguistic barrier. Learners' struggles can be traced back to a lack of fluency in the target language (Kotzé, Van der Westhuizen, & Barnard, 2017).

DISCUSSION

Many math teachers believe that, depending on the context, they must be fluent in multiple MTs to convey mathematical ideas from English into MT. Math teachers don't obey the language policy since they can't keep up with the policy's expectations. They've concluded that they need to get formal education. The problem with using MT to teach mathematics is that university training is typically given in English. They felt their voices should have been heard when crafting the math education policy. Staff morale and student assistance are two factors that should be considered while trying to improve output. For mathematical transformation, MT should be used to educate primary school learners on mathematical concepts. Multilingual workplace participants disagree, arguing that the policy's lack of clarity and accessibility negatively affects low-income communities that speak more than one language. This causes MT distortion on their part.

Despite widespread disagreement, the Department of Education insists that MT be implemented in schools. Many people, however, are sceptical of teachers because of the MT problem. It is difficult to teach pupils math when the LoLT is not their MT and teachers do not have access to sufficient resources. Even though many township teachers have advanced degrees, they believe it is unjust for the government to ignore the language barriers experienced by instructors in the early grades.

CONCLUSION

A well-considered language policy will improve the MT learning experience. Based on their knowledge of the benefits of using the mother tongue in teaching mathematics, teachers of the mother tongue should have meaningful conversations with their students. The goal of teacher preparation programs should be to prepare future educators to teach in their mother tongue. To ensure that only adequately educated teachers are hired to teach mathematics using their mother tongue, there should be a system of controlled monitoring to go along with the training. It is also suggested that future studies evaluate teachers' abilities to teach mathematics to students in their MT by comparing their native language knowledge to what they learned in teacher preparation programs. Parents should be involved and give their consent when determining a child's first language. Helpful materials for teaching mathematics through MT should be provided to schools. There is a lack of research on how math teachers teach math through MT, thus similar investigations are warranted. In a multilingual situation, the government and the school must consult with instructors and school administrations to choose the LoLT.

REFERENCES

Bitenelkome, L. (2013). *Investigating the English language proficiency of Grade 4 pupils in two contexts* [Doctoral dissertation, University of Pretoria].

Creswell, J. W. (2014). *Research design: Qualitative, quantitative, and mixed methods approaches* (4th ed.). Sage Publications.

Department of Basic Education. (2011). *Curriculum and Assessment Policy Statement: Foundation Phase Grades 1-3.* South Africa: Pretoria.

Department of Education (DoE). (1997). The language. In *Education Policy and National Curriculum Statement. Announcement by the Minister of education, 14 July*. Department of Education.

Department of Education (DoE). (2005). *Revised National Curriculum Statement Grades R-9 (Schools)*. Department of Education.

Fleisch, B. (2008). *Primary education in crisis: Why South African schoolchildren underachieve in reading and mathematics*. Juta and Company Ltd.

Hafiz, M., & Farik, M. (2016). Effectiveness of teaching and learning mathematics' using children's home language and cultural tools. *International Journal of Scientific and Technology Research*, 5(1), 123–127.

Makeleni, N. T., & Sethusha, M. J. (2014). The experiences of foundation phase teachers in implementing the curriculum. *Mediterranean Journal of Social Sciences*, 5(2), 103. 10.5901/mjss.2014.v5n2p103

Mashige, M. (2019). Foundation Phase teachers' experiences with instruction in the mother tongue in the Eastern Cape. *South African Journal of Childhood Education*, 9(1), 1–10.

Mashige, M., Cekiso, M., & Meyiwa, T. (2019). Foundation Phase teachers' experiences with instruction in the mother tongue in the Eastern Cape. *South African Journal of Childhood Education*, 9(1), 1–10.

Monyooe, L. (2017). Reclassifying Township Schools—South Africa's Educational Tinkering Expedition! *Creative Education*, 8(3), 471–485. 10.4236/ce.2017.83036

Muthivhi, E., & Broom, Y. (2008). Continuities across schooling transition: A case of classroom practices among teachers in Venda, South Africa. *Journal of Educational Studies*, 7(1), 98–121.

Njoroge, M. C. (2017). The efficacy of using mother tongue language in the teaching of mathematics and science in primary schools: Evidence from grade one classrooms in Kenya. *Journal of Applied Linguistics and Language Research*, 4(8), 133–150.

Owu-Ewie, C. (2006). The language policy of education in Ghana: A critical look at the English-only language policy of education. In *Selected proceedings of the 35th annual conference on African linguistics* (pp. 76-85). Somerville, MA: Cascadilla Proceedings Project.

Phakeng, M. S., Planas, N., Bose, A., & Njurai, E. (2018). Teaching and learning mathematics in trilingual classrooms: Learning from three different continents. In *Mathematical discourse that breaks barriers and creates space for marginalized learners* (pp. 277–293). Brill Sense.

Plüddemann, P. (2015). Unlocking the grid: Language-in-education policy realisation in post-apartheid South Africa. *Language and Education*, 29(3), 186–199. 10.1080/09500782.2014.994523

Republic of South Africa, (1996). Constitution of the Republic of South Africa. *Pretoria: Government Gazette, 378*(17678).

Republic of South Africa. (1996). *South African Schools Act, No. 84*. Government Printers.

Thuketana, N. S., & Makgabo, M. C. (2022). The use of English to offer learner support and enhance perceptual skills development in South African township schools. *The Journal for Transdisciplinary Research in Southern Africa*, 18(1), 1209. 10.4102/td.v18i1.1209

Umar, F. A., Muhoro, S. N., Opondo, F. O., Okumu, J. M., Kasimu, T. M., Wachira, D. W., Odoyo, E. O., Muturi, E. K., Ogari, C. O., Namikoye, E. S., & Wayong'o, S. N. (2018). *Teacher factors influencing the use of mother tongue in teaching of mathematics in lower primary schools in Yola South local government.* [Doctoral dissertation. School of Education. Kenya: Kenyatta University].

Yunus, M. M., & Sukri, S. I. A. (2017). The use of English in teaching mathematics and science: The PPSMI policy vis-à-vis the DLP. *Advances in Language and Literary Studies*, 8(1), 133–142. 10.7575/aiac.alls.v.8n.1p.133

Compilation of References

Abdel-Hafez Mahoud, H. (2021). Using a situated learning-based strategy to develop some critical thinking and enhance English Language Learning Interests at Preparatory Stage Pupils. *Journal of Education*, 87(87), 1–48. 10.21608/edusohag.2021.174506

Abduljabbar, R., Dia, H., Liyanage, S., & Bagloee, S. A. (2019). Applications of artificial intelligence in transport: An overview. *Sustainability (Basel)*, 11(1), 189. 10.3390/su11010189

Abdulkadri, A. (2022). Addressing gender disparities in education and employment A necessary step for achieving sustainable development in the Caribbean. United Nation, Economic Commission for Latin America and the Caribbean. https://repositorio.cepal.org/server/api/core/bitstreams/617fc7ac-82b0-4b2f-be83-5a14d825caf4/content

Aboud, F. E. (2006). Evaluation of an early childhood preschool program in rural Bangladesh. *Early Childhood Research Quarterly*, 21(1), 46–60. 10.1016/j.ecresq.2006.01.008

Aboud, F. E., & Hossain, K. (2011). The impact of preprimary school on primary school achievement in Bangladesh. *Early Childhood Research Quarterly*, 261(2), 237–246. 10.1016/j.ecresq.2010.07.001

Abraham, A., & Dingel, M. (2021). Exploring the relationship among mathematics attitude, gender, and achievement of undergraduate Health Science Students. *Mathematics teaching Research Journal, 13*.

Adelabu, F. M., & Alex, J. (2022). Online baseline assessment in mathematics: Initial teacher education entry-level student performance. *Academic Journal of Interdisciplinary Studies*, 11(1), 68–68. 10.36941/ajis-2022-0006

Afshari, M., Bakar, K. A., Luan, W. S., Samah, B. A., & Fooi, F. S. (2009). Factors affecting teachers' use of ICT. *Online Submission, 2*(1), 77-104. 201

Agbehadji, I. E., Awuzie, B. O., Ngowi, A. B., & Millham, R. C. (2020). Review of big data analytics, artificial intelligence and nature-inspired computing models towards accurate detection of COVID-19 pandemic cases and contact tracing. *International Journal of Environmental Research and Public Health*, 17(15), 5330. 10.3390/ijerph1715533032722154

Agonács, N., & Matos, J. F. (2019). Heutagogy and self-determined learning: A review of the published literature on the application and implementation of the theory. *Open Learning*, 34(3), 223–240. 10.1080/02680513.2018.1562329

Ahiatrogah, P. D., Madjoub, M. B., & Bervell, B. (2013). *Effect of Computer Assisted Instruction on the Achievement of Basic School Students in Pre- Technical Skills*. Academic Journal of Interdisciplinary. 10.5901/ajis.2013.v2n1p77

Alam, A., & Mohanty, A. (2022, December). Facial Analytics or Virtual Avatars: Competencies and Design Considerations for Student-Teacher Interaction in AI-Powered Online Education for Effective Classroom Engagement. In *International Conference on Communication, Networks and Computing* (pp. 252-265). Cham: Springer Nature Switzerland.

Alam, A. (2022). Employing adaptive learning and intelligent tutoring robots for virtual classrooms and smart campuses: Reforming education in the age of artificial intelligence. In *Advanced Computing and Intelligent Technologies* [Singapore: Springer Nature Singapore.]. *Proceedings of ICACIT*, 2022, 395–406.

Alam, A. (2023). Harnessing the Power of AI to Create Intelligent Tutoring Systems for Enhanced Classroom Experience and Improved Learning Outcomes. In *Intelligent Communication Technologies and Virtual Mobile Networks* (pp. 571–591). Springer Nature Singapore. 10.1007/978-981-99-1767-9_42

Alesina, A., Carlana, M., La Ferrara, E., & Pinotti, P. (2018). *Revealing streotypes: Evidence from immigrants in schools*. National Bureau of Economics Research. 10.3386/w25333

Alessi, S. M., & Trollip, S. R. (2001). Simulations. *Multimedia for Learning (Methods and Development)*, 213-269.

Alex, J. K., & Mukuka, A. (2024). Heutagogy in action: Unveiling the transformative power of virtual "air campus" experiences of mathematics trainee teachers. *Eurasia Journal of Mathematics, Science and Technology Education*, 20(3), em2416. 10.29333/ejmste/14320

Amsami, B. U., Mohammed, Y., & Mazila, E. A. (2015). Visual art teachers and performance assessment methods in Nigerian senior secondary schools. Mgbakoigba. *Journal of African Studies*, 4, 1–18. https://www.ajol.info/index.php/mjas/article/view/118501

Ancker, J. S., & Kaufman, D. (2007). Rethinking health numeracy: A multidisciplinary literature review. *Journal of the American Medical Informatics Association : JAMIA*, 14(6), 713–772. 10.1197/jamia.M246417712082

Anderson, J., & Kriesler, A. (2018). Making maths matter: Engaging students from low SES schools through social justice contexts. In E. Bergqvist, M. Österholm, C. Granberg, & L. Sumpter (Eds.), *Proceedings of the 42nd Conference of the International Group for the Psychology of Mathematics Education,* 2, 35–42). Umeå, Sweden: PME.

Andriani, D., Widada, W., Herawaty, D., Ardy, H., Nugroho, K. U. Z., Ma'rifah, N., & Anggoro, A. F. D. (2020). Understanding the number concepts through learning Connected Mathematics (CM): A local cultural approach. *Universal Journal of Educational Research*, 8(3), 1055–1061. 10.13189/ujer.2020.080340

Angermuller, J. (2020). *Poststructural discourse studies: From Structure to practice. The Cambridge Handbook of Discourse Studies*. Routledge.

Ansari, S., & Burns, E. (2023, November 23). *The evolution of online learning: heutagogy meets MOOCs in the digital age*. Jamk Arena Pro. https://urn.fi/urn:nbn:fi:jamk-issn-2984-0783-27

Anthony, G., Hunter, J., & Hunter, R. (2019). Working towards equity in mathematics education: Is differentiation the answer? In Hine, G., Blackley, S., & Cooke, A. (Eds.), *Proceedings of the 42nd Annual Conference of the Mathematics Education Research Group of Australasia.* MERGA.

Anthony, G., & Hunter, R. (2016). Grouping practices in New Zealand mathematics classrooms: Where are we at and where should we be? *New Zealand Journal of Educational Studies*, 52(1), 73–79. 10.1007/s40841-016-0054-z

Anthony, G., Hunter, R., & Hunter, J. (2018). Challenging teachers' perceptions of student capability through professional development: A telling case. *Professional Development in Education*, 44(5), 650–666. 10.1080/19415257.2017.1387868

Arroyo-Barrignete, J., Carabias-Lopez, S., Borras-Pala, F., & Martin-Anton, G. (2023). *Gender differences in Mathematics Achievement: The case of a business school in Spain. Sage Open, 1-14.*

Artzt, A., & Newman, C. (1990). *How to Use Cooperative Learning in the Mathematics Class*. National Council of Teachers of Mathematics.

Ashri, R. (2019). *The AI-powered workplace: how artificial intelligence, data, and messaging platforms are defining the future of work.* Apress.

Atalmis, E. H., Yilmaz, M., & Saatcioglu, A. (2016). How does private tutoring mediate the effects of socioeconomic status on mathematics performance? Evidence from Turkey. *Policy Futures in Education*, 14(8), 1135–1152. 10.1177/1478210316674159

Atkinson, L. Z., & Cipriani, A. (2018). How to carry out a literature search for a systematic review: A practical guide. *BJPsych Advances*, 24(2), 74–82. 10.1192/bja.2017.3

Attard, C., Edwards-Groves, C., & Grootenboer, P. (2018). Dialogic practices in the mathematics classroom. In Hunter, J., Perger, P., & Darragh, L. (Eds.), *Proceedings of the 41st Annual Conference of the Mathematics Education Research Group of Australasia* (pp. 122–129). Auckland: MERGA.

Attwell, P; Battle, J. (1999). Home Computers and School Performance. *The Information Society.*

Australian Curriculum, Assessment and Reporting Authority. (2015). *Australian curriculum: Mathematics, version 8.* Australia Curriculum. https://www.australiancurriculum.edu.au/Mathematics/Curriculum/F-10

Averill, R. (2018). Using rich investigative mathematics activities. In Hunter, R., Civil, M., Herbel-Eisenmann, B., Planas, N., & Wagner, D. (Eds.), *Mathematical discourse that breaks barriers and creates space for marginalized learners* (pp. 195–212). Sense Publishers. 10.1163/9789004378735_010

Baams, L., & Kaufman, T. (2023). Sexual orientation and gender identity/expression in adolescent research: Two decades in review. *Journal of Sex Research*, 60(7), 1004–1019. 10.1080/00224499.2023.221924537307300

Baas, J., Schotten, M., Plume, A., Côté, G., & Karimi, R. (2020). Scopus as a curated, high-quality bibliometric data source for academic research in quantitative science studies. *Quantitative Science Studies*, 1(1), 377–386. 10.1162/qss_a_00019

Baidoo-Anu, D., & Ansah, L. O. (2023). Education in the era of generative artificial intelligence (AI): Understanding the potential benefits of ChatGPT in promoting teaching and learning. *Journal of AI*, 7(1), 52–62. 10.61969/jai.1337500

Bandura, A. (1977). Self-efficacy: Toward a unifying theory of behavioral change. *Psychological Review*, 84(2), 191–215. 10.1037/0033-295X.84.2.191847061

Bangladesh Bureau of Statistics. (2010). *Household income and expenditure survey(HIES) 2010*. Bangladesh Bureau of Statistics.

Bangladesh Bureau of Statistics. (2016). *Bangladesh demographic profile 2016*. Bangladesh Bureau of Statistics.

Bani, O. K., & Lach, L. M. (2024). Barriers to, and facilitators of, education for children with disabilities worldwide: A descriptive review. *Frontiers in Public Health*, 11, 1294849. 10.3389/fpubh.2023.129484938292375

Banks, J. A. (2013). The nature of multicultural education. *Multicultural education: Issues and perspectives*, 3-24. Research Gate.

Baron, R. M., & Kenny, D. A. (1986). The mediator-moderator variable distinction in social psychological research: Conceptual, strategic, and statistical considerations. *Journal of Personality and Social Psychology*, 51(6), 1173–1182. 10.1037/0022-3514.51.6.11733806354

Barrett, M. (2018). How schools can promote the intercultural competence of young people. *European Psychologist*, 23(1), 93–104. 10.1027/1016-9040/a000308

Becker, G. S. (1993). *Human capital: A theoretical and empirical analysis, with special reference to education* (3rd ed.). University of Chicago Press. 10.7208/chicago/9780226041223.001.0001

Becker, G. S., & Tomes, N. (1979). An equilibrium theory of the distribution of income and intergenerational mobility. *Journal of Political Economy*, 87(6), 1153–1189. 10.1086/260831

BECTA. (2003). *ICT Research. What the Research Says about Using ICT in Maths*. British Educational Communications and Technology Agency.

Becta. (2003). What the Research Says about Using ICT in Maths. *UK:Becta ICT Research.*

Bennani, S., Maalel, A., & Ben Ghezala, H. (2022). Adaptive gamification in E-learning: A literature review and future challenges. *Computer Applications in Engineering Education*, 30(2), 628–642. 10.1002/cae.22477

Bennison, A., Goos, M., & Bielinski, D. (2018). Identifying practices that promote engagement with mathematics among students from disadvantaged backgrounds. In J. Hunter, P. Perger, & L. Darragh (Eds.), *Proceedings of the 41st Annual Conference of the Mathematics Education Research Group of Australasia*, (pp. 154–16). Auckland: MERGA.

Berberoğlu, G., & Tansel, A. (2014). Does private tutoring increase students' academic performance? Evidence from Turkey. *International Review of Education*, 60(5), 683–701. 10.1007/s11159-014-9436-y

Bernius, J. P., Krusche, S., & Bruegge, B. (2022). Machine learning based feedback on textual student answers in large courses. *Computers and Education: Artificial Intelligence*, 3, 100081. 10.1016/j.caeai.2022.100081

Bertrand, M. (2018). Course lecture-The glass ceiling. *Economica*, 85(338), 205–338. 10.1111/ecca.12264

Besar, P. (2018). The situated learning theory: The key to effective classroom teaching? *International Journal for Educational, Social, Political &. Cultural Studies*, 1(1), 49–60.

Beswick, K., Rawlings-Sanaei, F., & Tuohilampi, L. (2023). Interactive Mathematics Teacher Activities. In Manizade, A., Buchholtz, N., & Beswick, K. (Eds.), *The Evolution of Research on Teaching Mathematics. Mathematics Education in the Digital Era* (Vol. 22). Springer. 10.1007/978-3-031-31193-2_5

Bezuk, N. S., Cathcart, W. G., Vance, J. H., & Pothier, Y. M. (2001). Learning Mathematics in Elementary and Middle Schools. In *Merrill Prentice Hall* (6th ed.). Pearson Education.

Bhagat, A., & Vijayaraghavan, R. (2019). *Gender disparity in STEM: Evidence from India.* ACER. https://www.acer.org/in/discover/article/gender-disparity-in-stem-evidence-from-india

Bhandary, S. (2017). *Understanding some cultural barriers to women's access to education: A study in rural Nepal* [Madter's thesis, Norway: MF Norwegian School of Theology].

Bharadwaj, P., De Giorgi, G., Hansen, D., & Neilson, C. (2016). The gender gap in mathematics: Evidence from Chile. *Economic Development and Cultural Change*, 65(1), 1–166. 10.1086/687983

Bhattacharya, I., & Sharma, K. (2007). India in the knowledge economy-an electronic paradigm. *International Journal of Educational Management*, 21(6), 543–568. 10.1108/09513540710780055

Bicknell, B., Young-Loveridge, J., & Nguyen, N. (2016). A design study to develop young children's understanding of multiplication and division. *Mathematics Education Research Journal*, 28(4), 567–58. 10.1007/s13394-016-0180-4

Bishop, M., Fish, J. N., Hammack, P., & Russell, S. (2020). Sexual identity development milestones in three generations of sexual minority people: A national probability sample. *Developmental Psychology*, 56(11), 2177–2193. 10.1037/dev000110532833471

Biswas, A., & Paul, A. (2018). GENDER DIFFERENCES IN MATHEMATICAL ACHIEVEMENTS OF HONORS STUDENTS IN WEST BENGAL. *International Journal of Creative Research Thoughts*, 6(1), 1612–1615. https://ijcrt.org/papers/IJCRT1802652.pdf

Bitenelkome, L. (2013). *Investigating the English language proficiency of Grade 4 pupils in two contexts* [Doctoral dissertation, University of Pretoria].

Bjørkli, K.BjÃ¸rkli. (2014). The impact on learning outcomes in mathematics of mobile-enhanced, combined formative and summative assessment. *International Journal of Technology Enhanced Learning*, 6(4), 343–360. 10.1504/IJTEL.2014.069025

Blaschke, L. M. (2014). Moving students forward in the PAH continuum: Maximizing the power of the social web. In L. M. Blaschke, C. Kenyon, & S. Hase (Eds.), *Experiences in self-determined learning*. Amazon.

Blaschke, L. M., & Hase, S. (2016). Heutagogy: A holistic framework for creating twenty-first-century self-determined learners. In B. Gros, Kinshuk, & M. Maina (Eds.), *The Future of Ubiquitous Learning. Lecture Notes in Educational Technology* (pp. 25–40). Springer. https://doi.org/10.1007/978-3-662-47724-3_2

Blaschke, L. M. (2012). Heutagogy and lifelong learning: A review of heutagogical practice and self-determined learning. *International Review of Research in Open and Distance Learning*, 13(1), 56. 10.19173/irrodl.v13i1.1076

Blaschke, L. M. (2018). Self-determined learning (Heutagogy) and digital media creating integrated educational environments for developing lifelong learning skills. In Kergel, D., Heidkamp, B., Telléus, P., Rachwal, T., & Nowakowski, S. (Eds.), *The Digital Turn in Higher Education* (pp. 129–140). Springer. 10.1007/978-3-658-19925-8_10

Blaschke, L. M. (2019). The pedagogy–andragogy–heutagogy continuum and technology-supported personal learning environments. In Jung, I. (Ed.), *Open and Distance Education Theory Revisited* (pp. 75–84). Springer. 10.1007/978-981-13-7740-2_9

Blaschke, L. M., & Hase, S. (2015). Heutagogy, technology, and lifelong learning for professional and part-time learners. In Dailey-Hebert, A., & Dennis, K. (Eds.), *Transformative Perspectives and Processes in Higher Education* (Vol. 6, pp. 75–94). Springer International Publishing. 10.1007/978-3-319-09247-8_5

Blue, L. E., O'Brien, M., & Makar, K. (2018). Exploring the classroom practices that may enable a compassionate approach to financial literacy education. *Mathematics Education Research Journal*, 30(2), 143–16. 10.1007/s13394-017-0223-5

Bobis, J., & Tregoning, M. (2019). "The task is not the challenge": Changing teachers' practices to support student struggle in mathematics. In G. Hine, S. Blackley, & A. Cooke (Eds.), *Proceedings of the 42nd Annual Conference of the Mathematics Education Research Group of Australasia,* (pp. 133–139). Perth: MERGA.

Bobis, J., Way, J., Anderson, J., & Martin, A. J. (2016). Challenging teacher beliefs about student engagement in mathematics. *Journal of Mathematics Teacher Education*, 19(1), 33–55. 10.1007/s10857-015-9300-4

Boitshwarelo, B., Reedy, A. K., & Billany, T. (2017). Envisioning the use of online tests in assessing twenty-first century learning: A literature review. *Research and Practice in Technology Enhanced Learning*, 12(1), 1–16. 10.1186/s41039-017-0055-730595721

Booth, A., Cardona-Sasa, L., & Nolen, P. (2018). Do single sex classes affect academic achievement? An experiment in a coeducational university. *Journal of Public Economics*, 168, 109–126. 10.1016/j.jpubeco.2018.08.016

Bornstein, M. H., & Putnick, D. L. (2012). Cognitive and socioemotional caregiving in developing countries. *Child Development*, 83(1), 46–61. 10.1111/j.1467-8624.2011.01673.x22277006

Bornstein, M. H., Putnick, D. L., Bradley, R. H., Lansford, J. E., & Deater-Deckard, K. (2015). Pathways among caregiver education, household resources, and infant growth in 39 low-and middle-income countries. *Infancy*, 20(4), 353–376. 10.1111/infa.1208626273231

Bose, A., & Kantha, V. K. (2014). Influence of socio-economic background and cultural practices on mathematics education in India: A contemporary overview in historical perspective. *ZDM Mathematics Education*, 46(7), 1073–1084. 10.1007/s11858-014-0607-x

Bragg, L. A., Herbert, S., Loong, E., Vale, C., & Widjaja, W. (2016). Primary teachers notice the impact of language on children's mathematical reasoning. *Mathematics Education Research Journal, 28*(4). .10.1007/s13394-016-0178-y

Braun, V., & Clarke, V. (2006). Using thematic analysis in psychology. *Qualitative Research in Psychology*, 3(2), 77–101. 10.1191/1478088706qp063oa

Brokensha, S., Kotzé, E., & Senekal, B. A. (2023). *AI in and for Africa: A Humanistic Perspective*. CRC Press. 10.1201/9781003276135

Brophy, J. (1999). Toward a model of the value aspects of motivation in education: Developing appreciation for. *Educational Psychologist*, 34(2), 75–85. 10.1207/s15326985ep3402_1

Brown, R., & Redmond, T. (2017). Privileging a contextual approach to teaching mathematics: A secondary teacher's perspective. In A. Downton, S. Livy, & J. Hall (Eds.), *Proceedings of the 40th Annual Conference of the Mathematics Education Research Group of Australasia*, (pp. 109–116). Melbourne: MERGA.

Brown, J. S., Collins, A., & Duguid, P. (1989). Situated cognition and the culture of learning. *Educational Researcher*, 18(1), 32–42. 10.2307/1176008

Brown, L. I., & Kanyongo, G. Y. (2010). Gender differences in performance in mathematics in Trinidad and Tobago: Examining affective factors. *International Electronic Journal of Mathematics Education*, 5, 113–130. 10.29333/iejme/253

Brown, R. (2017). Using collective argumentation to engage students in a primary mathematics classroom. *Mathematics Education Research Journal*, 29(2), 183–199. 10.1007/s13394-017-0198-2

Bruner, J. (1966). *Towards a Theory of Instruction*. Norton.

Bryk, A. S. (2015). Accelerating how we learn to improve. *Educational Researcher*, 44(9), 467–477. 10.3102/0013189X15621543

Buchholtz, N. F., Krosanke, N., Orschulik, A. B., & Vorhölter, K. (2018). Combining and integrating formative and summative assessment in mathematics teacher education. *ZDM Mathematics Education*, 50(4), 715–728. 10.1007/s11858-018-0948-y

Buchman, C., Condron, D. J., & Roscigno, V. J. (2010). Shadow education, American Style: Test preparation, the SAT and college enrollment. *Social Forces*, 89(2), 435–461. 10.1353/sof.2010.0105

Buck Institute for Education (BIE). (2014). *Introduction to project-based learning*. BIE. http://bie.org/images/uploads/general/20fa7d42c216e2ec171a212e97fd4a9e.pdf

Bunoti, S. (2011). *The quality of higher education in developing countries needs professional support*. A paper presented in 22nd International Conference on Higher Education: Quality in Higher Education, Ankara, Turkey.

Burbules, N. C., & Torres, C. A. (2000). An introduction to globalization and education. In Burbules, N. C., & Torres, C. A. (Eds.), *Globalization and education: Critical perspectives* (pp. 1–26). Routledge.

Burbules, N. C., & Torres, C. A. (2013). Globalization and education: An introduction. In *Globalization and education* (pp. 1–26). Routledge. 10.4324/9781315022642-5

Burdman, P. (2020). *Let's Make Math Education Relevant for Real Life*. EdSurge. https://www.edsurge.com/news/2020-10-26-let-s-make-math-education- relevant-for-real-life

Burgos, M., & Godino, J. D. (2022). Assessing the epistemic analysis competence of prospective primary school teachers on proportionality tasks. *International Journal of Science and Mathematics Education*, 20(2), 367–389. 10.1007/s10763-020-10143-0

Bybee, R. (2000). Achieving technological literacy: A national perspective. *Technology Teacher*, 60(1), 23–28.

Byrne, B. B. (2016). *Structural equation modeling with Amos: Basic concepts, applications, and programming* (3rd ed.). Rutledge. 10.4324/9781315757421

Caldwell, B. M., & Bradley, R. H. (1984). *Home observation for measurement of the environment*. University of Arkansas at Little Rock.

Caprara, L., & Caprara, C. (2022). Effects of virtual learning environments: A scoping review of literature. *Education and Information Technologies*, 27(3), 3683–3722. 10.1007/s10639-021-10768-w34629934

Care, E., Scoular, C., & Griffin, P. (2016). Assessment of collaborative problem solving in education environments. *Applied Measurement in Education*, 29(4), 250–264. 10.1080/08957347.2016.1209204

Carlana, M. (2019). Implicit streotypes: Evidence from teachers' gender bias. *The Quarterly Journal of Economics*, 134(3), 1163–1224. 10.1093/qje/qjz008

Cavaglia, C., Mavhin, S., McNally, S., & Ruiz-Valenzuela, J. (2020). Gender, achievement, and achievement, and subjet choice in English education. *Oxford Review of Economic Policy*, 36(4), 816–835. 10.1093/oxrep/graa050

Celedón-Pattichis, S., Peters, S. A., Borden, L. L., Males, J. R., Pape, S. J., & Chapman, O. (2018). Asset-based approaches to equitable mathematics education research and practice. *Journal for Research in Mathematics Education*, 49(4), 373–389. 10.5951/jresematheduc.49.4.0373

Ceylan, B. (2020). Ubiquitous learning and heutagogy in teacher education. In Durak, G., & Çankaya, S. (Eds.), *Managing and Designing Online Courses in Ubiquitous Learning Environments* (pp. 279–295). IGI Global. 10.4018/978-1-5225-9779-7.ch015

Chakabwata, W. (2023). *Intersections of gender,identity and subject choice in shaping student learning experience at a technical teacher training College in Zimbabwe.* [Thesis, University of South Africa].

Chang, Z., Liu, S., Xiong, X., Cai, Z., & Tu, G. (2021). A survey of recent advances in edge-computing-powered artificial intelligence of things. *IEEE Internet of Things Journal*, 8(18), 13849–13875. 10.1109/JIOT.2021.3088875

Chan, M. C. E., Clarke, D., & Cao, Y. (2018). Erratum to: The social essentials ofq learning: An experimental investigation of collaborative problem solving andq knowledge construction in mathematics classrooms in Australia and China. *Mathematics Education Research Journal*, 30(1), 51–55. 10.1007/s13394-017-0221-7

Cheeseman, J., Downton, A., & Livy, S. (2017). Investigating teachers' perceptions of enabling and extending prompts. In A. Downton, S. Livy, & J. Hall (Eds.), *Proceedings of the 40th Annual Conference of the Mathematics Education Research Group of Australasia,* (pp. 141–148). Melbourne: MERGA.

Cheeseman, J. (2018). Teachers' perceptions of obstacles to incorporating a problem solving style of mathematics into their teaching. In Hunter, J., Perger, P., & Darragh, L. (Eds.), *Proceedings of the 41st Annual Conference of the Mathematics Education Research Group of Australasia* (pp. 210–217). Auckland: MERGA.

Cheeseman, J., & McDonough, A. (2019). Coding young learners' pictorial responses to an open-ended assessment task. In Hine, G., Blackley, S., & Cooke, A. (Eds.), *Proceedings of the 42nd Annual Conference of the Mathematics Education Research Group of Australasia,* (pp. 112–116). Perth: MERGA.

Cheng, X., Liu, S., Sun, X., Wang, Z., Zhou, H., Shao, Y., & Shen, H. (2021). Combating emerging financial risks in the big data era: A perspective review. *Fundamental Research*, 1(5), 595–606. 10.1016/j.fmre.2021.08.017

Chen, I.-H., Gamble, J. H., Lee, Z.-H., & Fu, Q.-L. (2020). Formative assessment with interactive whiteboards: A one-year longitudinal study of primary students' mathematical performance. *Computers & Education*, 150, 103833. 10.1016/j.compedu.2020.103833

Chen, L., Chen, P., & Lin, Z. (2020). Artificial intelligence in education: A review. *IEEE Access : Practical Innovations, Open Solutions*, 8, 75264–75278. 10.1109/ACCESS.2020.2988510

Chen, X., Chang, L., Liu, H., & He, Y. (2008). Effects of the peer group on the development of social functioning and academic achievement: A longitudinal study in Chinese children. *Child Development*, 79(2), 235–251. 10.1111/j.1467-8624.2007.01123.x18366421

Chen, X., Zou, D., Xie, H., Cheng, G., & Liu, C. (2022). Two decades of artificial intelligence in education. *Journal of Educational Technology & Society*, 25(1), 28–47.

Cheung, A., & Slavin, R. (2011). *The effectiveness of education technology for enhancing reading achievement: A meta-analysis.* Best Evidence. https://www.bestevidence.org/reading/tech/tech.html

Chimpololo, A. (2020). An analysis of heutagogical practices through mobile device usage in a teacher training programme in Malawi. *Journal of Learning for Development*, 7(2), 190–203. 10.56059/jl4d.v7i2.391

Chipman, S. F. (2005). Research on women and mathematical issue. In Gallagher, A., & Kaufman, J. (Eds.), *Gender differences in mathematics: An intergrative psychological approach* (pp. 1–24). Cambridge University Press.

Cholin, V. S. (2005). Study of the application of information technology for effective access to resources in Indian university libraries. *The International Information & Library Review, 37*(3).

Choy, B. H., & Dindyal, D. (2017a). Noticing affordances of a typical problem. In B. Kaur, W. K. Ho, T. L. Toh, & B. H. Choy (Eds.), *Proceedings of the 41st Conference of the International Group for the Psychology of Mathematics Education* (pp. 249–256). Singapore: PME.

Choy, B. H., & Dindyal, D. (2017b). Snapshots of productive noticing: Orchestrating learning experiences using typical problems. In A. Downton, S. Livy, & J. Hall (Eds.), *Proceedings of the 40th Annual Conference of the Mathematics Education Research Group of Australasia,* (pp. 157–16). Melbourne: MERGA.

Choy, B. H., & Dindyal, D. (2018). Orchestrating mathematics lessons: Beyond the use of a single rich task. In J. Hunter, P. Perger, & L. Darragh. (Eds.), *Making waves, opening spaces. Proceedings of the 41st annual conference of the Mathematics Education Research Group of Australasia.* Auckland: MERGA.

Choy, B. H. (2016). Snapshots of mathematics teacher noticing during task design. *Mathematics Education Research Journal*, 28(3), 421–440. 10.1007/s13394-016-0173-3

Choy, B. H., Thomas, M. O. J., & Yoon, C. (2017). The FOCUS framework: Characterising productive noticing during lesson planning, delivery and review. In Schack, E. O., Fisher, M. H., & Wilhelm, J. A. (Eds.), *Teacher noticing: Bridging and broadening perspectives, contexts, and frameworks* (pp. 445–466). Springer. 10.1007/978-3-319-46753-5_26

Christensen, R. W., & Knezek, G. A. (2009). Construct validity for the teachers' attitudes toward computers questionnaire. *Journal of Computing in Teacher Education*, 25(4), 143–155.

Christodoulou, J. (2005). *Glossary of gender-related terms.* Mediterranean Institute of Gender Studies.: European Institute for Gender Equality. https://medinstgen derstudies.org/wp-content/uploads/Gender-Glossaryupdated_final.pdf

Chubbuck, K., Curley, W., & King, T. (2016). Who's on first? Gender differences in performance on the SAT test on critical reading intems with sports and science content. *ETS Research Report Series*, 2(2), 1–116. 10.1002/ets2.12109

Chytry, V., & Kubiatko, M. (2021). Pupils' Summative Assessments in Mathematics as Dependent on Selected Factors. *Eurasia Journal of Mathematics, Science and Technology Education*, 17(8). https://eric.ed.gov/?id=EJ1309268

Cimpian, J. R., Luienski, S., Timmer, J., Makowski, M., & Miller, E. (2016). Have gender gaps in math closed? Achievement, teacher perceptions, and learnining behaviours across two ECLS-K chorts. *AERA Open*, 2(4), 1–19. 10.1177/2332858416673617

Clarkson, P. C. (2009). Globalisation and mathematics teaching: The global importance of local language contexts. In Zajda, J., & Rust, V. (Eds.), *Globalisation, Policy and Comparative Research. Globalisation, Comparative Education and Policy Research* (Vol. 5, pp. 139–156). Springer. 10.1007/978-1-4020-9547-4_10

Closs, L., Mahat, M., & Imms, W. (2022). Learning environments' influence on students' learning experience in an Australian Faculty of Business and Economics. *Learning Environments Research*, 25(1), 271–285. 10.1007/s10984-021-09361-233814969

Coates, D., & Humphreys, B. R. (2004). No Significant Distance' between Face-to-face and Online Instruction: Evidence from Principles of Economics. *Economics of Education Review., 23*(6), 533-546.

Cochrane, T., Antonczak, L., Gordon, A., Sissons, H., & Withell, A. (2012). Heutagogy and mobile social media: post Web 2.0 pedagogy. In M. Brown (Ed.), *ascilite 2012: Future challenges, sustainable futures* (pp. 204-214)4. Wellington, New Zealand: ascilite..https://bit.ly/3OHTnBM

Cockburn, I. M., Henderson, R., & Stern, S. (2018). The impact of artificial intelligence on innovation: An exploratory analysis. In *The economics of artificial intelligence: An agenda* (pp. 115–146). University of Chicago Press.

Cole, D. A., & Maxwell, S. E. (2003). Testing mediational models with longitudinal data: Questions and tips in the use of structural equation modeling. *Journal of Abnormal Psychology*, 112(4), 558–577. 10.1037/0021-843X.112.4.55814674869

Coley, R. L., Kruzik, C., & Votruba-Drzal, E. (2020). Do family investments explain growing socioeconomic disparities in children's reading, math, and science achievement during school versus summer months? *Journal of Educational Psychology*, 112(6), 1183–1196. 10.1037/edu0000427

Coley, R. L., Sims, J., & Votruba-Drzal, E. (2016). Family expenditures supporting children across income and urbanicity strata. *Children and Youth Services Review*, 70, 129–142. 10.1016/j.childyouth.2016.09.017

Colineauty, H., Soulier, A., Lepage, B., & Kelly-Irving, M. (2022). Considering sex and gender in Epidemiology: a challenge beyond terminology . From conceptual analysis to methodological strategies. *Biol Sex Differ. 13*(23), 1-10.

Collins, A. (1996). Design issues for learning environments. In Vosniadou, S. (Ed.), *International perspectives on the design of technology-supported learning environments* (pp. 347–361). Lawrence Erlbaum.

Colliver, Y. (2018). Fostering young children's interest in numeracy through demonstration of its value: The footsteps study. *Mathematics Education Research Journal*, 30(4), 407–428. 10.1007/s13394-017-0216-4

Conger, R. D., & Donnellan, M. B. (2007). An interactionist perspective on the socioeconomic context of human development. *Annual Review of Psychology*, 58(1), 175–199. 10.1146/annurev.psych.58.110405.08555116903807

Connor, C. M., Mazzocco, M. M., Kurz, T., Crowe, E. C., Tighe, E. L., Wood, T. S., & Morrison, F. J. (2018). Using assessment to individualize early mathematics instruction. *Journal of School Psychology*, 66, 97–113. 10.1016/j.jsp.2017.04.00529499792

Cosker-Rowland, R. (2023). Recent Work on gender identity and gender. *Analysis*, 83(4), 801–820. 10.1093/analys/anad027

Coviliac, A. (2022). *Improvement of AI-assessment systems in grading open questions based on the teaching assistant's view* [Bachelor's thesis, University of Twenty].

Crenshaw, K. (1991). Mapping the margins: Intersectionality politics, and violence against women of colour. *Stanford Law Review*, 43(6), 1241–1299. 10.2307/1229039

Creswell, J. W. (2009). *Research design: Qualitative, quantitative and mixed methods approaches* (3rd ed.). Sage.

Creswell, J. W. (2014). *Research design: Qualitative, quantitative, and mixed methods approaches* (4th ed.). Sage Publications.

Cross, T., & Palese, K. (2015). Increasing learning: Classroom assessment techniques in the online classroom. *American Journal of Distance Education*, 29(2), 98–108. 10.1080/08923647.2015.1023594

Dahl, G. B., & Lochner, L. (2008). The impact of family income on the child achievement: Evidence from the Earned Income Tax Credit. *Cambridge, M. A.: National Bureau of Economic Research, NBER Working Paper No. 14599.*

Damla, I. (2022). A poststructural feminist approach to the acqusition of gender roles in early childhood. *İstanbul Üniversitesi Kadın Araştırmaları Dergisi, 25,* 141-154.

Dang, H. A. (2007). The determinants and impact of private tutoring classes in Vietnam. *Economics of Education Review*, 26(6), 684–699. 10.1016/j.econedurev.2007.10.003

Das, G. C., & Sinha, S. (2017). Effect of Socioeconomic status on performance in Mathematics among students of secondary schools of Guwahati city. *IOSR Journal of Mathematics,13*(1), 26-33. https://iosrjournals.org/iosr-jm/papers/Vol13-issue1/Version-1/F1301012633.pdf

Das, U., & Singhal, K. (2021). *Solving it correctly: Prevalence and Persistence of Gender Gap in Basic Mathematics in rural India.* https://arxiv.org/ftp/arxiv/papers/2110/2110.15312.pdf

Davis, N. E., & Tearle, P. (Eds.). (1999). *A core curriculum for telematics in teacher training.* Telemetics. www.ex.ac.uk/telematics.T3/corecurr/tteach98.htm

Davis-Kean, P. E. (2005). The influence of parent education and family income on child achievement: The indirect role of parental expectations and home environment. *Journal of Family Psychology*, 19(2), 294–304. 10.1037/0893-3200.19.2.29415982107

Dawson, W. (2010). Private tutoring and mass schooling in East Asia: Reflections in inequality in Japan, South Korea and Cambodia. *Asia Pacific Education Review*, 11(1), 14–24. 10.1007/s12564-009-9058-4

Delgado, H. O. K., de Azevedo Fay, A., Sebastiany, M. J., & Silva, A. D. C. (2020). Artificial intelligence adaptive learning tools: The teaching of English in focus. *BELT-Brazilian English Language Teaching Journal*, 11(2), e38749–e38749. 10.15448/2178-3640.2020.2.38749

Deng, L. (2018). Artificial intelligence in the rising wave of deep learning: The historical path and future outlook [perspectives]. *IEEE Signal Processing Magazine*, 35(1), 180–177. 10.1109/MSP.2017.2762725

Denzin, N. K., & Lincoln, Y. S. (Eds.). (2000). *Handbook of qualitative research* (2nd ed.). Sage.

Department of Basic Education. (2011). *Curriculum and Assessment Policy Statement: Foundation Phase Grades 1-3.* South Africa: Pretoria.

Department of Education (DoE). (1997). The language. In *Education Policy and National Curriculum Statement. Announcement by the Minister of education, 14 July*. Department of Education.

Department of Education (DoE). (2005). *Revised National Curriculum Statement Grades R-9 (Schools)*. Department of Education.

Dewantara, A. H., & Zulkardi, D. (2015). Assessing seventh graders' Mathematical literacy in solving PISA-like tasks. *Indonesian Mathematical Society Journal on Mathematics Education, 6*(2), 39-49. https://eric.ed.gov/?id=EJ1079604

Dickerson, A., McIntosh, S., & Valente, C. (2015). Do the mathematics: An analysis of the gender gap in mathematics in Africa. *Economics of Education Review*, 46, 1–22. 10.1016/j.econedurev.2015.02.005

Divaharan, S., & Koh, J. H. L. (2010). Learning as students to become better teachers: Pre-service teachers' IWB learning experience. *Australasian Journal of Educational Technology*, 26(4). Advance online publication. 10.14742/ajet.1072

Dossie, G. (2021). fIGLIO, D., Giuliano, P., & Sapienza, P. (2021). The family origin of the math gender gap is a white affluent phenomenon. *AEA Papers and Proceedings. American Economic Association*, 111(May), 179–183. 10.1257/pandp.20211124

Downes, S. (2006). *An introduction to connective knowledge*. Downes. http://www.downes.ca/post/33034

Downey, G., Gray, T., Hall, T., & Singh, M. (2018). Developing global perspectives: Responding to the state of international education in Australian Universities. In T. Hall, T. Gray, Downey, G, & M. Singh's, *The Globalisation of Higher Education Developing Internationalised Education Research and Practice* (pp. 3-20). Western Sydney University: Palgrave/ MacMillan.

Duncan, G. J., Dowsett, C. J., Claessens, A., Magnuson, K., Huston, A. C., Klebanov, P., Pagani, L. S., Feinstein, L., Engel, M., Brooks-Gunn, J., Sexton, H., Duckworth, K., & Japel, C. (2007). School readiness and later achievement. *Developmental Psychology*, 43(6), 1428–1446. 10.1037/0012-1649.43.6.142818020822

Duncan, G. J., Magnuson, K., & Votruba-Drzal, E. (2014). Boosting family income to promote child development. *The Future of Children*, 24(1), 99–120. 10.1353/foc.2014.000825518705

Duncan, J. G., & Brooks-Gunn, J. (1997). *Consequences of growing up poor*. Russell Sage Foundation.

EACEA. (2011). *Science Education in Europe: National Policies*. Practices and Research.

Educational Testing Services. (2018). *A snap of the individuals who took the GRE general test 2013-2018.*

Eisenberg, M. B. (2008). Information Literacy: Essential Skills for the Information Age. *DESIDOC Journal of Library and Information Technology*, 28(2), 39–47. 10.14429/djlit.28.2.166

Elliot, L., & Bachman, H. J. (2018). How do parents foster young children's math skills? *Child Development Perspectives*, 12(1), 16–21. 10.1111/cdep.12249

Elliott, A. (2019). *The culture of AI: Everyday life and the digital revolution.* Routledge. 10.4324/9781315387185

Else-Quest, N. M., Hyde, J. S., & Linn, M. C. (2010). Cross-national patterns of gender differences in mathematics: A meta-analysis. *Psychological Bulletin*, 136(1), 103–127. 10.1037/a001805320063928

Engle, P. L., Fernald, L. C., Alderman, H., Behrman, J., O'Gara, C., Yousafzai, A., de Mello, M. C., Hidrobo, M., Ulkuer, N., Ertem, I., & Iltus, S. (2011). Strategies for reducing inequalities and improving developmental outcomes for young children in low-income and middle-income countries. *Lancet*, 378(9799), 1339–1353. 10.1016/S0140-6736(11)60889-121944378

English, L. D., & King, D. (2019). STEM integration in sixth grade: Designing and constructing paper bridges. *International Journal of Science and Mathematics Education*, 17(5), 863–888. 10.1007/s10763-018-9912-0

English, L. D., & Kirshner, D. (Eds.). (2015). *Handbook of International Research in Mathematics Education* (3rd ed.). Routledge. 10.4324/9780203448946

Erwin, J. (1993). *Friendship and peer relationship in children.* John Wiley & Sons Ltd.

Esen, Y. (2013). Making room for gender sensitivity in pre-service teacher education. *Evropejskij Issledovatel*, 61(10), 2544–2554. 10.13187/er.2013.61.2544

Eshiwani, G. S. (1982). A study of Women's Access to higher Education in Kenya with a special reference to Mathematics and Science Education. Bureau of Education research, Nairobi.

Evans, C., Kandiko Howson, C., & Forsythe, A. (2018). Making sense of learning gain in higher education. *Higher Education Pedagogies*, 3(1), 1–45. 10.1080/23752696.2018.1508360

Fan, L., & Zhu, Y. (2023). Using performance assessment in secondary school mathematics: An empirical study in a Singapore classroom. *Journal of Mathematics Education*, 1(1), 132–152. https://journalofmathed.scholasticahq.com/article/90517.pdf

FAWE. (1999). Improving Performance of Girls in Schools. A Report. Nairobi, Kenya.

Feldman, J., Monteserin, A., & Amandi, A. (2015). Automatic detection of learning styles: State of the art. *Artificial Intelligence Review*, 44(2), 157–186. 10.1007/s10462-014-9422-6

Fernandez, D., White, S., Smith, H., Conner, P., & Ryan, M. (2023)). Gender inequality in science, technology, and mathematics, gendered time disparities in perceived and actual time spent in perceived and actual in practical laboratory based activities . *Front. Educ. 8*, 1-10.

Fielding, L., Kerr, N., & Rosier, P. (2007). *Annual growth for all students, catch-up growth for those who are behind.* The New Foundation Press, Inc.

Fielding-Wells, J., & Fry, K. (2019). Introducing guided mathematical inquiry in the classroom: Complexities of developing norm of evidence. In Hine, G., Blackley, S., & Cooke, A. (Eds.), *Proceedings of the 42nd Annual Conference of the Mathematics Education Research Group of Australasia,* (pp. 268–275). Perth: MERGA.

Firdaus, F., Kailani, I., Bakar, M. N. B., & Bakry, B. (2015). Developing critical thinking skills of students in mathematics learning. [EduLearn]. *Journal of Education and Learning*, 9(3), 226–236.

Fischbein, E., & Schnarch, D. (1997). The evolution with age of probabilistic, intuitively based misconceptions. *Journal for Research in Mathematics Education*, 28(1), 96–105. 10.2307/749665

Fleisch, B. (2008). *Primary education in crisis: Why South African schoolchildren underachieve in reading and mathematics*. Juta and Company Ltd.

Foster, E. M. (2002). How economists think about family resources and child development. *Child Development*, 73(6), 1904–1914. 10.1111/1467-8624.0051312487501

Galindo, C., & Sonnenschein, S. (2015). Decreasing the SES math achievement gap: Initial math proficiency and home learning environments. *Contemporary Educational Psychology*, 43, 25–38. 10.1016/j.cedpsych.2015.08.003

Gallardo, K. (2021). The importance of assessment literacy: Formative and summative assessment instruments and techniques. *Workgroups eAssessment: Planning, Implementing and Analysing Frameworks*, 3-25. 10.1007/978-981-15-9908-8_1

Gamwell, L. (2015). Mathematics: Geometries of beauty. *Nature*, 528(7583), 476–477. 10.1038/528476a

Ganley, C., & Lubienski, S. T. (2016). Mathematics confidence, interest and performance: Examining gender patterns and reciprocal relations. *Learning and Individual Differences*, 47, 182–193. 10.1016/j.lindif.2016.01.002

Garrison, J. (1998). *Toward a pragmatic social constructivism.* In M. Larochelle, N. Bednarz, & J. Garrison (Eds.), *Constructivism and education* (pp. 3-60). Cambridge, UK: Cambridge. 10.1017/CBO9780511752865.005

Geiger, V., Goos, M., & Forgasz, H. (2015). A rich interpretation of numeracy for the 21st century: A survey of the state of the field. *ZDM Mathematics Education*, 47(4), 531–548. 10.1007/s11858-015-0708-1

Geiger, V., Mulligan, J., Date-Huxtable, L., Ahlip, R., Jones, D. H., May, E. J., Rylands, L., & Wright, I. (2018). An interdisciplinary approach to designing online learning: Fostering pre-service mathematics teachers' capabilities in mathematical modelling. *ZDM Mathematics Education*, 50(1), 217–223. 10.1007/s11858-018-0920-x

George, B., & Wooden, O. (2023). Managing the strategic transformation of higher education through artificial intelligence. *Administrative Sciences*, 13(9), 196. 10.3390/admsci13090196

Gershoff, E. T., Aber, J. L., Raver, C. C., & Lennon, M. C. (2007). Income is not enough: Incorporating material hardship into models of income associations with parenting and child development. *Child Development*, 78(1), 70–95. 10.1111/j.1467-8624.2007.00986.x17328694

Gerstein, J. (2013). *Education 3.0 and the pedagogy (andragogy, heutagogy) of mobile learning. [Blog post.] User generated education.* https://usergeneratededucation.wordpress.com/2013/05/13/

Gherasim, L. R., Butnaru, S., & Mairean, C. (2013). Classroom environment, achievement goals and maths performance: Gender differences. *Educational Studies*, 39(1), 1–12. 10.1080/03055698.2012.663480

Gibbs, B., & Hunter, B. (2018). Making mathematics accessible for all: A classroom inquiry. In J. Hunter, P. Perger, & L. Darragh (Eds.), *Proceedings of the 41st Annual Conference of the Mathematics Education Research Group of Australasia,* (pp. 330–336). Auckland: MERGA.

Ginns, P., Hu, F. T., Byrne, E., & Bobis, J. (2016). Learning by traced worked examples. *Applied Cognitive Psychology*, 30(2), 160–169. 10.1002/acp.3171

Graetz, G., & Karimi, A. (2019). *Examining gender gap variation across Forms* (Working Paper 8). Institute for Evaluation of Labour Market and Education Policy (IFAU).

Gregory, S., Bannister-Tyrrell, M., Charteris, J., & Nye, A. (2018). Heutagogy in postgraduate education: Cognitive advantages for higher degree online students. In Erwee, R., Harmes, M., Harmes, M., & Danaher, P. (Eds.), *Postgraduate Education in Higher Education. University Development and Administration* (pp. 189–209). Springer. 10.1007/978-981-10-5249-1_32

Griffin, P., & Care, E. (Eds.). (2014). *Assessment and teaching of 21st-century skills: Methods and approach.* Springer.

Grivokostopoulou, F., Perikos, I., & Hatzilygeroudis, I. (2017). An educational system for learning search algorithms and automatically assessing student performance. *International Journal of Artificial Intelligence in Education*, 27(1), 207–240. 10.1007/s40593-016-0116-x

Guo, G., & Harris, K. M. (2000). The mechanisms mediating the effects of poverty on children's intellectual development. *Demography*, 37(4), 431–447. 10.1353/dem.2000.000511086569

Gutstein, E., & Peterson, B. (Eds.). (2013). *Rethinking mathematics: Teaching social justice by the numbers*. Rethinking Schools Ltd.

Hafiz, M., & Farik, M. (2016). Effectiveness of teaching and learning mathematics' using children's home language and cultural tools. *International Journal of Scientific and Technology Research*, 5(1), 123–127.

Haider, M., Islam, M. R., Mia, A., & Sorcar, N. R. (2005). Income generation perspective in non-formal education NGO initiatives in Bangladesh. *Asia Pacific Journal of Social Work*, 15(1), 17–29. 10.1080/21650993.2005.9755966

Hall, J., Robinson, T., Flegg, J., & Wilkinson, J. (2019). Students' gendered experiences in undergraduate programs in University Mathematics Departments. In M. Graven, H. Venkat, A. Essien, & P. Vale, *Proceedings of the 43rd Conference of the International Group for the Psychology of Mathematics Education Vol* (pp. 320-327). Pretoria: PME.

Hallermann, S., Larmer, J., & Mergendoller, J. (2011). *PBL in the elementary grades: Step-by step guidance, tools, and tips for standards-focused K-5 projects*. Novato, CA: Buck Institute for Education.

Hall, J., & Noren, E. (2021). Innovations in "gender issues" research in mathematics education. *Mathematics Education Research Journal*, 33(4), 787–791. 10.1007/s13394-021-00404-8

Halupa, C. (2021). Reaching "creating" in Bloom's taxonomy: The merging of heutagogy and technology in online learning. In Information Resources Management Association (Ed.), *Research Anthology on Developing Critical Thinking Skills in Students* (pp. 15–35). IGI Global. https://doi.org/10.4018/978-1-7998-3022-1.ch002

Hamadani, J. D., Tofail, F., Hilaly, A., Huda, S. N., Engle, P., & Grantham-McGregor, S. M. (2010). Use of family care indicators and their relationship with child development in Bangladesh. *Journal of Health, Population and Nutrition*, 28(1), 23–33. 10.3329/jhpn.v28i1.452020214083

Hamdi, F. M. (2013). The impact of globalization in the developing countries. *Developing Country Studies*, 3(11), 142–144.

Hansen, B. B., & Klopfer, S. O. (2006). Optimal full matching and related designs via network flows. *Journal of Computational and Graphical Statistics*, 15(3), 609–627. 10.1198/106186006X137047

Harding, S. (1996). Gendered ways of knowing and the "epistemology crisis" of the west. In Goldberger, J., & Tarule, B. (Eds.), *C, & M. Belensky, Knowledge difference and power: Essays inspired by Women's ways of knowing* (pp. 431–454). Basic Books.

Hase, S., & Kenyon, C. (2000). From Andragogy to Heutagogy. In *ultiBASE In-Site*. RMIT. https://webarchive.nla.gov.au/awa/20010220130000/http://ultibase.rmit.edu.au/Articles/dec00/hase2.htm

Hase, S. (2009). Heutagogy and e-learning in the workplace: Some challenges and opportunities. *Impact: Journal of Applied Research in Workplace E-Learning*, 1(1), 43–52.

Hase, S., & Kenyon, C. (2013). *Self-determined learning: Heutagogy in action* (1st ed.). Bloomsbury Publishing.

Hassan, M. (2024, January 1). *Documentary analysis: A research method for analysing documents*. Scribbr. https://www.scribbr.com/methodology/documentary-analysis/

Haveman, R., & Wolfe, B. (1994). *Succeeding generations: On the effects of investments in children*. Russell Sage Foundation.

Henningsen, I. (2008). Gender mainstreaming of adult mathematics education: Opportunities and challenges. *ALM-IJ*, 3(1), 32–40.

Herbert, J., & Stipek, D. (2005). The emergence of gender differences in children's perceptions of their academic competence. *Journal of Applied Developmental Psychology*, 26(3), 276–295. 10.1016/j.appdev.2005.02.007

Herzig, A. (2010). Women belonging in the social worlds of graduate mathematics. *The Mathematics Enthusiast, 7*(2).

Hightower, A. M., Lloyd, S. C., & Swanson, C. B. (2011). *Improving student learning by supporting quality teaching : Key Issues, Effective Strategies*. Editorial Projects in Education, Inc.

Holm, M. (2011). Project-Based Instruction: A Review of the Literature on Effectiveness in Prekindergarten through 12th Grade Classrooms. *InSight: Riviera Academic Journal, 7*(2), 1-1. http://independent.academia.edu/MaggieHolm

Hossain, A., & Zeitlyn, B. (2010). *Poverty, Equity and access to education in Bangladesh (Research Monograph No. 51)*. Consortium for Research on Educational Access, Transition and Equity.

Hu, L., & Bentler, P. M. (1999). Cutoff criteria for fit indexes in covariance structure analysis. *Structural Equation Modeling*, 6(1), 1–55. 10.1080/10705519909540118

Hunter, R., & Hunter, J. (2018). Opening the space for all students to engage in mathematical talk within collaborative inquiry and argumentation. In R. Hunter, M. Civil, B. Herbel-Eisenmann, N. Planas, & D. Wagner (Eds.), *Mathematical discourse that breaks barriers and creates space for marginalized learners.* Rotterdam: Sense Publishers. 10.1163/9789463512121_001

Hunter, R., Hunter, J., & Bills, T. (2019). Enacting culturally responsive or socially response-able mathematics education. In C. Nicol, S. Dawson, J. Archibald, & F. Glanfield (Eds.), *Living culturally responsive mathematics curriculum and pedagogy: Making a difference with/in indigenous communities.* Springer: Netherlands. 10.1163/9789004415768_007

Hunter, R., Hunter, J., Anthony, G., & McChesney, K. (2018 September). Developing mathematical inquiry communities: Enacting culturally responsive, culturally sustaining ambitious teaching. *SET, 2,* 25–3

Hunter, J. (2017). Developing interactive mathematical talk: Investigating student perceptions and accounts of mathematical reasoning in a changing classroom context. *Cambridge Journal of Education*, 47(4), 475–49. 10.1080/0305764X.2016.1195789

Hunter, R., & Hunter, J. (2017). Maintaining a cultural identity while constructing a mathematical disposition as a Pāsifika learner. In McKinley, E. A., & Tuhiwai Smith, L. (Eds.), *Handbook of Indigenous education.* Springer. 10.1007/978-981-10-1839-8_14-1

Hurst, C., & Hurrell, D. (2016). Investigating children's multiplicative thinking: Implications for teaching. *European Journal of STEM Education*, 1(3), 1–1. 10.20897/lectito.201656

Hussain, K. G. (2016). *Achievement of non-formal primary education among the working children at Rajshahi city, Bangladesh* [Unpublished M. Phil Thesis, Bangladesh: University of Rajshahi].

Hykle, J. A. (2011). Interrelationships among Cognitive Controls, Gender Science Achievement, and Science Process Skills. University of Cincinnati. in Mathematics Achievement: A Comparative Study. „Social Science" 42, John, P. (2002). Quantitative methods. In Marsh, D., & Stoker, G. (Eds.), *Theory and methods in political science* (2nd ed., pp. 184–211). Palgrave Macmillan.

Iannone, P., & Miller, D. (2019). Guided notes for university mathematics and their impact on students' note-taking behaviour. *Educational Studies in Mathematics*, 101(3), 387–404. 10.1007/s10649-018-9872-x

Iruka, I. U., LaForett, D. R., & Odom, E. C. (2012). Examining the validity of the family investment and stress models and relationship to children's school readiness across five cultural groups. *Journal of Family Psychology*, 26(3), 359–370. 10.1037/a002829022545934

Ittigson, R. J., & Zewe, J. G. (2003). Technology in the mathematics classroom. In L. A. Tomei (Ed.), Challenges of teaching with technology across the curriculum: Issues and Solutions. Hershey: Information Science Publishing. 10.4018/978-1-59140-109-4.ch004

James, A. (2007). *Teaching the male brain: How boys think, feel, and learn in school*. Corwin Press.

Janiesch, C., Zschech, P., & Heinrich, K. (2021). Machine learning and deep learning. *Electronic Markets*, 31(3), 685–695. 10.1007/s12525-021-00475-2

Jazby, D., & Widjaja, W. (2019). Teacher noticing of primary students' mathematical reasoning in a problem-solving task. In Hine, G., Blackley, S., & Cooke, A. (Eds.), *Proceedings of the 42nd Annual Conference of the Mathematics Education Research Group of Australasia,* (pp. 380–387). Perth: MERGA.

Jenkins, K. (2018). Towards an account of gender identity. *Ergo (Ann Arbor, Mich.)*, 5(27), 713–744.

Jha, S. K. (2014). *Mathematics Phobia*. Lambert Academic Publisher.

Jha, S. K. (2018). Distributive Responses to a Collective Responsibility – Right to Education in India. *Journal of Educational Planning and Administration*, 32(1), 35–50.

Jha, S. K., & Tomar, D. P. (2010). Mathematical Phobia of Government and Private Secondary School Students in Relation to Certain Demographical Variables. *Journal of Community Guidance & Research*, 27(3), 316–323. 10.5281/zenodo.8044469

Jin, C., Chen, W., Cao, Y., Xu, Z., Tan, Z., Zhang, X., Deng, L., Zheng, C., Zhou, J., Shi, H., & Feng, J. (2020). Development and evaluation of an artificial intelligence system for COVID-19 diagnosis. *Nature Communications*, 11(1), 5088. 10.1038/s41467-020-18685-133037212

Johnson, L., Smith, R., Willis, H., Levine, A., & Haywood, K. (2010). *The horizon report*. Horizon.

Johnson, D. W., & Johnson, R. T. (1996). Cooperation and the use of technology. In D. H.

Johnson, R. B., & Onwuegbuzie, A. J. (2004). Mixed methods research: A research paradigm whose time has come. *Educational Researcher*, 3(7), 14–26. 10.3102/0013189X033007014

John-Steiner, V. (2014). *Sociocultural and feminist theory: Mutuality and relevance*. Research Gate. https://www.researchgate.net/publication/241344833

Jonassen, D. H., Carr, C., & Hsiu-Ping, Y. (1998). Computers As Mindtools For Engaging Learners In Critical Thinking. *TechTrends*, 43(2), 24–32. 10.1007/BF02818172

Jonassen, D., & Reeves, T. (1996). Learning with technology: Using computers as cognitive tools. In Jonassen, D. (Ed.), *Handbook of Research Educational on Educational Communications and Technology* (pp. 693–719). Macmillan.

Jones, G. A., Langrall, C. W., & Mooney, E. S. (2007). Research in probability: Responding to classroom realities. *Second handbook of research on mathematics teaching and learning, 2*, 909-955.

Jones, I., & Alcock, L. (2014). Peer assessment without assessment criteria. *Studies in Higher Education*, 39(10), 1774–1787. 10.1080/03075079.2013.821974

Jones, I., Swan, M., & Pollitt, A. (2015). Assessing mathematical problem solving using comparative judgement. *International Journal of Science and Mathematics Education*, 13(1), 151–177. 10.1007/s10763-013-9497-6

Jordan, N. C., Huttenlocher, J., & Levine, S. C. (1992). Differential calculation abilities in young children from middle- and low-income families. *Developmental Psychology*, 28(4), 644–653. 10.1037/0012-1649.28.4.644

Jordan, N. C., Huttenlocher, J., & Levine, S. C. (1994). Assessing early arithmetic abilities: Effects of verbal and nonverbal response types on the calculation performance of middle- and low-income children. *Learning and Individual Differences*, 6(4), 413–432. 10.1016/1041-6080(94)90003-5

Jordan, N. C., Levine, S. C., & Huttenlocher, J. (1994). Development of calculation abilities in middle- and low-income children after formal instruction in school. *Journal of Applied Developmental Psychology*, 15(2), 223–240. 10.1016/0193-3973(94)90014-0

Jules-Van-Belle & Soetaert. (2001). Breakdown into the virtual user- involved design and learning. *Journal of Technology and Teacher Education*, 8(2), 31–42.

Kaack, L. H., Donti, P. L., Strubell, E., Kamiya, G., Creutzig, F., & Rolnick, D. (2022). Aligning artificial intelligence with climate change mitigation. *Nature Climate Change*, 12(6), 518–527. 10.1038/s41558-022-01377-7

Kabeer, N., & Mahmud, S. (2009). Imagining the future: Children, education and intergenerational transmission of poverty in urban Bangladesh. *IDS Bulletin*, 40(1), 10–21. 10.1111/j.1759-5436.2009.00003.x

Kalil, A., & DeLeire, T. (Eds.). (2004). *Family investments in children's potential: Resources and parenting behaviors that promote success*. Lawrence Erlbaum Associates. 10.4324/9781410610874

Kalogeropoulos, P., Russo, J., Sullivan, P., Klooger, M., & Gunningham, S. (2019). Re-enfranchising mathematically-alienated Students: Teacher and tutor perceptions of the Getting Ready in Numeracy (G.R.I.N.) Program. *International Electronic Journal of Mathematics Education*, 14(1), 1–17. 10.29333/iejme/5881

Kamal, T., Khan, M. A., Hamad, Z., Illiyan, A., Das, A. K., & Alkhuraydili, A. (2023). Perception and challenges of virtual classes with gender digital divide amidst and post-COVID-19 pandemic in Iraq: An empirical analysis. *Education Research International*, 2023, 1–14. 10.1155/2023/3665768

Kaufhold, J. A., Alverez, V. G., & Arnold, M. (2006). Lack of school supplies, materials and resources as an elementary cause of frustration and burnout in South Texas special education teachers. *Journal of Instructional Psychology*, 33(3), 159–161.

Kavatsyuk, O., Ioannou, M., & Avraamidou, L. (2021). Gender-Inclusive instructional practices in university mathematics classes. *International Journal of Gender, Science and Technology*, 14(2), 1–16.

Kazima, M., Pillay, V., & Adler, J. (2008). Mathematics for teaching: Observations from two case studies. *South African Journal of Education*, *28*(2), 283–299.

Keengwe, J. (2018). Globalization, digital technology, and teacher education in the United States. In *Oxford Research Encyclopedia of Education*. Oxford University Press. 10.1093/acrefore/9780190264093.013.366

Kennewell, S., Parkinson, J., & Tanner, H. (2000). *Developing the ICT capable school*. Routledge Falmer.

Khan, A. F. M. K., Motin, S., Begum, H. B., & Uddin, M. M. (2012). *Elementary mathematics: Class one*. National Textbook Board, Bangladesh.

Khan, T., & Thomas, S. (2022). Promoting positive education through constructivist digital learning heutagogy: An intervention outcome. *Journal of Learning for Development*, 9(2), 305–316. 10.56059/jl4d.v9i2.646

Kibble, J. D. (2017). Best practices in summative assessment. *Advances in Physiology Education*, 41(1), 110–119. 10.1152/advan.00116.201628188198

Kibrislioglu, N. (2015). An investigation about 6th students' attitude towards Mathematics. *Procedia: Social and Behavioral Sciences*, 186, 64–69. 10.1016/j.sbspro.2015.04.024

Kim, H. (2019). Teacher learning opportunities provided by implementing formative assessment lessons: Becoming responsive to student mathematical thinking. *International Journal of Science and Mathematics Education*, 17(2), 341–363. 10.1007/s10763-017-9866-7

Kim, S., & Lee, J. H. (2010). Private tutoring and demand for education in South Korea. *Economic Development and Cultural Change*, 58(2), 259–296. 10.1086/648186

Kim, T. (2007). Old borrowings and new models of the university in East Asia. *Globalisation, Societies and Education*, 5(1), 39–52. 10.1080/14767720601133140

Kinshuk, C., Chen, N.-S., Cheng, I.-L., & Chew, S. W. (2016). Evolution is not enough: Revolutionizing current learning environments to smart learning environments. *International Journal of Artificial Intelligence in Education*, 26(2), 561–581. 10.1007/s40593-016-0108-x

Kishore, R., & Kumar, S. (2022). A Study on the effectiveness of pedagogy used by teachers for the Inclusion of students with Disabilities in the Schools of Delhi. District Institute of Education and Training, Daryaganj, New Delhi (India).

Kitchenham, B. (2004). *Procedures for Performing Systematic Reviews. Keele University Technical Report.* TR/SE-0401. http://artemisa.unicauca.edu.co/~ecaldon/docs/spi/kitchenham_2004.pdf

Klimova, B., Pikhart, M., & Kacetl, J. (2023). Ethical issues of the use of AI-driven mobile apps for education. *Frontiers in Public Health*, 10, 1118116. 10.3389/fpubh.2022.111811636711343

Kline, R. B. (1998). *Principles and practice of structural equation modeling*. New York: NY: Guildford.

Kornrich, S., & Furstenberg, F. (2013). Investing in children: Changes in parental spending on children: 1972-2007. *Demography*, 50(1), 1–23. 10.1007/s13524-012-0146-422987208

Korucu-Kış, S. (2021). Preparing student teachers for real classrooms through virtual vicarious experiences of critical incidents during remote practicum: A meaningful-experiential learning perspective. *Education and Information Technologies*, 26(6), 6949–6971. 10.1007/s10639-021-10555-733935573

Kulik, J. (2003). *Effects of using instructional technology in elementary and secondary schools: What controlled evaluation studies say (Final Report No. P10446.001).* Arlington, VA: SRI International.

Kulik, C., & Kulik, J. (1991). Effectiveness of computer-based instruction: An updated analysis. *Computers in Human Behavior*, 7(1-2), 75–95. 10.1016/0747-5632(91)90030-5

Kumar, D., & Choudhury, P. K. (2022). Do parental resources reduce the gender gap in math for primary school-going children? Evidence from India. *Education 3*, 1–18. 10.1080/03004279.2022.2133546

Kyaruzi, F., Strijbos, J.-W., Ufer, S., & Brown, G. T. (2019). Students' formative assessment perceptions, feedback use and mathematics performance in secondary schools in Tanzania. *Assessment in Education: Principles, Policy & Practice*, 26(3), 278–302. 10.1080/0969594X.2019.1593103

Lacárcel, A. M. (2022). Artificial intelligence in education as a means to personalize learning. In Saura, J. R., & Debasa, F. (Eds.), *Handbook of Reearch on Artificial Intelligence in Government Practices and Processes* (pp. 285–308). IGI Global., 10.4018/978-1-7998-9609-8.ch016

Lankford, B. A. (2021). Multimedia in teacher-directed and student-centered mathematics classrooms. In Y. Wen, Y. Wu, G. Qi, S.-C. Guo, J. M. Spector, S. Chelliah, Kinshuku, & Y.-J. Lan (Eds.), *Expanding Global Horizons Through Technology Enhanced Language Learning. Lecture Notes in Educational Technology* (pp. 211–227). Springer. 10.1007/978-981-15-7579-2_13

Larmer, J., Mergendoller, J., & Boss, S. (2015). *Setting the standard for project-based learning*. ASCD.

Lave, J. (2008). Situated learning and changing practice. In Amin, A., & Roberts, J. (Eds.), *Community, Economic and Organization* (pp. 283–296). Oxford University Press.

Lave, J., & Wenger, E. (1991). *Situated learning: Legimate peripheral participation*. Cambridge University Press. 10.1017/CBO9780511815355

Le, A., Bui, T., Tran, B., & Phung, T. (2023). Gender gap mathematics achievement: Vietnamese students in America mathematics competitions. *Issues in Educational Research*, 33(1), 137–138.

Leach, G. (2019). Strength-based grouping: A call for change. In G. Hine, S. Blackley, & A. Cooke (Eds.), *Mathematics Education Research: Impacting Practice, Proceedings of the 42nd Annual Conference of the Mathematics Education Research Group of Australasia*. Perth: MERGA.

Leder, G. (2015). Gender and mathematics review revisted. *The Proceedings of the 12th International Congress on Mathematical Education* (pp. 145-170). Melbourne: Monash University, Melbourne.

Lee, J. Y. (2013). *Private tutoring and its impact on students' academic achievement, formal schooling, and educational inequality in Korea*. [Unpublished Ph. D Dissertation, The US: Columbia University].

Lee, K., & Anderson, J. (2015). Gender differences in mathematics attitudes in coeducational and single sex secondary education. In M. V. Geiger, & A. Bennsion, *Mathematics education in the margins (Proceedings of the 38th Conference of the mathematics education research group of Australasia* (pp. 357-364). Sunshine Coast: MERGA.

Lee, K. S. (2010). Parental educational investments and aspirations in Japan. *Journal of Family Issues*, 31(12), 1579–1603. 10.1177/0192513X10368268

Leenknecht, M., Wijnia, L., Köhlen, M., Fryer, L., Rikers, R., & Loyens, S. (2020). Formative assessment as practice: The role of students' motivation. *Assessment & Evaluation in Higher Education*, 46(2), 236–255. 10.1080/02602938.2020.1765228

Lemke, C., & Coughlin, E. C. (1998). *Technology in American schools*. PNBS._www.mff.org/pnbs/ME158.pdf

Leong, Y. H., Cheng, L. P., Toh, W. T., Kaur, B., & Toh, T. L. (2019). Making things explicit using instructional materials: A case study of a Singapore teacher's practice. *Mathematics Education Research Journal*, 31(1), 47–66. 10.1007/s13394-018-0240-z

LeTendre, G. (2022). Globalization and the impact of ICT on teacher's work and professional status. In *The Palgrave Handbook of Teacher Education Research* (pp. 1–22). Palgrave Macmillan. 10.1007/978-3-030-59533-3_83-1

Leung, K.-C., Leung, F. K., & Zuo, H. (2014). A study of the alignment of learning targets and assessment to generic skills in the new senior secondary mathematics curriculum in Hong Kong. *Studies in Educational Evaluation*, 43, 115–132. 10.1016/j.stueduc.2014.09.002

Levitt, T. (1983). The globalization of markets. *Harvard Business Review*, 93–94.

Liang, W., Tadesse, G. A., Ho, D., Fei-Fei, L., Zaharia, M., Zhang, C., & Zou, J. (2022). Advances, challenges and opportunities in creating data for trustworthy AI. *Nature Machine Intelligence*, 4(8), 669–677. 10.1038/s42256-022-00516-1

Lim, C. P. & Chai, C.S. (2004), An activity-theoretical approach to research of ICT integration in Singapore schools: Orienting activities and learner autonomy. *Computers & Education, 43*(3).

Li, M., Zhang, Y., Liu, H., & Hao, Y. (2018). Gender differences in mathematics achievement in Beijing: A meta-analysis. *The British Journal of Educational Psychology*, 88(4), 566–583. 10.1111/bjep.1220329265190

Links, M. J. (2018). Beyond competency-based continuing professional development. *Medical Teacher*, 40(3), 253–258. 10.1080/0142159X.2017.140121929228838

Linn, M. C., Chang, H. Y., Chiu, J., Zhang, H., & McElhaney, K. (2010). Can desirable difficulties overcome deceptive clarity in scientific visualizations? *Successful remembering and successful forgetting: A Festschrift in honor of Robert A. Bjork*, 239-262.

Linver, M. R., Brooks-Gunn, J., & Kohen, D. E. (2002). Family processes as pathways from income to young children's development. *Developmental Psychology*, 38(5), 719–734. 10.1037/0012-1649.38.5.71912220050

Li, Q., & Ma, X. (2010). A meta-analysis of the effects of computer technology on school students' mathematics learning. *Educational Psychology Review*, 22(3), 215–243. 10.1007/s10648-010-9125-8

Li, S., Hong, Y.-C., & Craig, S. D. (2023). A Systematic literature review of social learning theory in Online Learning Environments. *Educational Psychology Review*, 35(4), 108. 10.1007/s10648-023-09827-0

Lissitz, R., & Schafer, W. (2002). *Assessment in Educational Reform: Both Means and Ends*. Allyn and Bacon.

Logan, T. (2015). The influence of test mode and visuospatial ability on mathematics assessment performance. *Mathematics Education Research Journal*, 27(4), 423–441. 10.1007/s13394-015-0143-1

Lombardi, C. M., & Dearing, E. (2020). Maternal support of children's math learning in associations between family income and math school readiness. *Child Development*, 92(1), 39–55. 10.1111/cdev.1343632797635

Lou, J., & Xu, Q. (2022). The development of positive education combined with online learning: Based on theories and practices. *Frontiers in Psychology*, 13, 1–10. 10.3389/fpsyg.2022.95278436118474

Lubienski, S., & Pinheiro, W. (2020). Gender and mathematics: What can other disciplines tell us. *Journal of Urban Mathematics Education*, 13(1), 1–14. 10.21423/jume-v13i1a377

Lugo-Gil, J., & Tamis-Lemonda, C. S. (2008). Family resources and parenting quality: Links to children's cognitive development across the first 3 years. *Child Development*, 79(4), 1065–1085. 10.1111/j.1467-8624.2008.01176.x18717907

Luis, A. (2017). Unpacking the male superiority myth and masculinisation of mathematics at the intersections: A review of Research on Gender in mathematics education. *Journal for Research in Mathematics Education*, 48(4), 397–433. 10.5951/jresematheduc.48.4.0397

Luneta, K. (2022a). Can the Fourth Industrial Revolution Resolve Why the Teaching of Mathematics in the Current Paradigm Continues to Be Decontextualised and Ineffective. In Chirinda, B., Luneta, K., & Uworwabayeho, A. (Eds.), *Mathematics Education in Africa* (pp. 19–38). Springer International Publishing. 10.1007/978-3-031-13927-7_2

Luneta, K. (2022b). Special challenges in mathematics education in Sub Sahara Africa. *Current Opinion in Behavioral Sciences*, 48, 101211. 10.1016/j.cobeha.2022.101211

Lusyana, E., & Wangge, M. (2016). Increasing Higher Order Thinking Skill to Build Students" s Character by Using Mathematical Reasoning. In *Proceeding of 3rd International Conference On Research, Implementation And Education Of Mathematics And Science* (pp. 119-126). IEEE.

Luz, Y., & Yerushalmy, M. (2019). Students' conceptions through the lens of a dynamic online geometry assessment platform. *The Journal of Mathematical Behavior*, 54, 100682. 10.1016/j.jmathb.2018.12.001

Lykke, N. (2010). *Feminist studies*. Routledge. 10.4324/9780203852774

Maass, K., Geiger, V., Ariza, M. R., & Goos, M. (2019). The role of mathematics in interdisciplinary STEM education. *ZDM Mathematics Education*, 51(6), 869–884. 10.1007/s11858-019-01100-5

MacCarthy, S., Saunders, C., & Elliot, M. (2020). Increased reporting of sexual minority orientation from 2009 to 2017 in England and implications for measuring sexual minority health disparities. *LGBT Health*, 7(7), 393–400. 10.1089/lgbt.2019.018132758045

MacDonald, A., & Murphy, S. (2019). Using the drawing-telling approach to reveal young children's mathematical knowledge. In Hine, G., Blackley, S., & Cooke, A. (Eds.), *Proceedings of the 42nd Annual Conference of the Mathematics Education Research Group of Australasia,* (pp. 99–10). Perth: MERGA.

Maher, M., Skow, B., & Cicognani, A. (1999). Designing the virtual campus. *Design Studies*, 20(4), 319–342. 10.1016/S0142-694X(98)00043-X

Mahlambi, S. B., van den Berg, G., & Mawela, A. S. (2022). Exploring the use of assessment for learning in the mathematics classroom. *Journal of Education (University of KwaZulu-Natal), 89*, 23-46. 10.17159/2520-9868/i89a02

Mahmud, M., Kaiser, M. S., Hussain, A., & Vassanelli, S. (2018). Applications of deep learning and reinforcement learning to biological data. *IEEE Transactions on Neural Networks and Learning Systems*, 29(6), 2063–2079. 10.1109/TNNLS.2018.279038829771663

Makeleni, N. T., & Sethusha, M. J. (2014). The experiences of foundation phase teachers in implementing the curriculum. *Mediterranean Journal of Social Sciences*, 5(2), 103. 10.5901/mjss.2014.v5n2p103

Makransky, G., Borre-Gude, S., & Mayer, R. E. (2019). Motivational and cognitive benefits of training in immersive virtual reality based on multiple assessments. *Journal of Computer Assisted Learning*, 35(6), 691–707. 10.1111/jcal.12375

Malamed, C. (2023). *Applying Richard Mayer's Cognitive Theory of Multimedia Learning*. The e-learning coach. https://theelearningcoach.com/learning/learning-theory-and-multimedia/

Maloney, E., Rmirez, G., Gunderson, E., Levine, S., & Beilock, S. (2015). Intergenerational effects of parents math anxiety on children's math achievement and anxiety. *Psychological Science*, 26(9), 1480–1488. 10.1177/0956797615592630 26253552

Mandhu, S., & Bhattacharyya, D. (2023). Learning Style Preferences Among the Students. *International Journal of Multidisciplinary Research*, 5(1), 1–12.

Mandina, S., Mashingaidze, S., & Mafuta, J. (2013). Increasing female participation in advanced level mathematics: A perspwective from teachers and students in Zimbabwe. *African Educational Research Journals*, 1(3), 183–190.

Marcus, S., & Joakim, S. (2016). Gender differences in boys and girls and girls' perception of teaching and learning mathematics. *Open Review of Educational Research*, 3(1), 18–34. 10.1080/23265507.2015.1127770

Marinho, P., Leite, C., & Fernandes, P. (2017). Mathematics summative assessment practices in schools at opposite ends of performance rankings in Portugal. *Research in Mathematics Education*, 19(2), 184–198. 10.1080/14794802.2017.1318085

Marshman, M. (2018). Identifying the mathematics middle year students use as they address a community issue. *Mathematics Education Research Journal*, 30(4), 355–38. 10.1007/s13394-017-0195-5

Martin, M. O., Mullis, S., & Foy, P. (2008). *TIMSS 2007 International Science Report: Findings from IEA's Trends in International Mathematics and Science Study at the Fourth and Eighth Grades*. TIMSS & PIRLS International Study Center, Boston College.

Marzano, R. J., Waters, J. T., & McNulty, B. A. (2005). *School leadership that works: From research to results*. Association for Supervision and Curriculum Development.

Mashige, M. (2019). Foundation Phase teachers' experiences with instruction in the mother tongue in the Eastern Cape. *South African Journal of Childhood Education*, 9(1), 1–10.

Mason, J. (2002). *Researching your own practice: The discipline of noticing*. Routledge. 10.4324/9780203471876

Mayer, R. E. (1996). Learners as information processors: Legacies and limitations of educational psychology's second metaphor. *Educational Psychologist*, 31(3/4).

Mayer, R. E. (2014). Incorporating motivation into multimedia learning. *Learning and Instruction*, 29, 171–173. 10.1016/j.learninstruc.2013.04.003

Mayer, S. E. (1997). *What money can't buy: Family income and children's life chances*. Harvard University Press.

Mazana, Y. M., Suero Montero, C., & Olifage, C. R. (2019). *Investigating students' attitude towards learning mathematics*.

McCluskey, C., Mulligan, J., & Mitchelmore, M. (2016). The role of reasoning in the Australian Curriculum: Mathematics. In B. White, M. Chinnappan, & S. Trenholm (Eds.), *Proceedings of the 39th Annual Conference of the Mathematics Education Research Group of Australasia,* (pp. 447–45). Adelaide: MERGA.

McCoy, D. C., Zuilkowski, S. S., & Fink, G. (2015). Poverty, physical stature, and cognitive skills: Mechanisms underlying children's school enrollment in Zambia. *Developmental Psychology*, 51(5), 600–614. 10.1037/a003892425844851

McDowall, T., & Jackling, B. (2010). Attitudes towards the accounting profession: An Australian perspective. *Asian Review of Accounting*, 18(1), 30–49. 10.1108/13217341011045999

McLachlan, K., & Essien, A. A. (2022). Language and multilingualism in the teaching and learning of mathematics in South Africa: A review of literature in Pythagoras from 1994 to 2021. *Pythagoras*, 43(1), a669. 10.4102/pythagoras.v43i1.669

McLeod, D. B. (1992). Research on affect in mathematics education: A reconceptualization. In Grouws, D. A. (Ed.), *Handbook of research on mathematics teaching and learning* (pp. 575–596). Macmillan.

McLoyd, V. C. (1998). Socioeconomic disadvantage and child development. *The American Psychologist*, 53(2), 185–204. 10.1037/0003-066X.53.2.1859491747

McMillan, J. (2001). *Essential Assessment Concepts for Teachers and Administrators*. Corwin Press, Inc.

McNabb, D. E. (2004). *Research methods for political science: Quantitative and qualitative methods*. M.E. Sharpe, Inc.

McTighe, J., & Wiggins, G. (2012). *Understanding by Design* (2nd ed.). Association for Supervision of Curriculum Development.

Mejia-Rodriguez, A., Luyten, H., & Meelisen, M. (2021). Across the world: An exploration of student and parental data of TIMSS. *International Journal of Science and Mathematics Education*, 19(1), 1229–1250.

Mejias, P., McAllister, D., Diaz, K. G., & Ravest, J. (2021). A longtudinal study of the gender gap in mathematics achievement: Evidence from Chile. *Educational Studies in Mathematics*, 107(3), 583–605. 10.1007/s10649-021-10052-1

Merriam, S. B., & Caffarella, R. S. (1999). *Learning in adulthood: A comprehensive guide (2nd.)*. Jossey-Bass Publishers.

Mezirow, J. (1991). *Transformative dimensions of adult learning*. Jossey-Bass.

Miao, F., Holmes, W., Huang, R., & Zhang, H. (2021). *AI and education: A guidance for policymakers*. UNESCO Publishing.

Michael, T., & Jackie, M. (1996). Using Computers in the Nathematics Classroom: The Role of the Teacher. *Mathematics Education Research Journal*, 6(1), 38–57.

Middleton, J. A., & Spanias, P. A. (1999). Motivation for achievement in mathematics: Findings, generalizations, and criticisms of the research. *Journal for Research in Mathematics Education*, 30(1), 65–88. 10.2307/749630

Midgley, C., & Urdan, T. C. (1995). Predictors of middle school students' use of self-handicapping strategies. *The Journal of Early Adolescence*, 15(4), 389–411. 10.1177/0272431695015004001

Mignani, S., Ottaviani, M. G., & Ricci, R. (2014). Teaching statistics at secondary education in Italy: Some issues on large scale standardized test results. *ICOTS9*.

Miheso, M. (2002). Factors Affecting Mathematics Performance Among Secondary Schools Learners, In Nairobi Province, Kenya. [Thesis, Kenyatta University, Nairobi].

Mihindo, W. J., Wachanga, S. W., & Anditi, Z. O. (2017). Effects of Computer-Based Simulations Teaching Approach on Students' Achievement in the Learning of Chemistry among Secondary School Students in Nakuru Sub County, Kenya. *Journal of Education and Practice*, 8(5), 65–75.

Mildenhall, P., & Sherriff, B. (2018). Using multiple metaphors and multimodalities as a semiotic resource when teaching year 2 students computational strategies. *Mathematics Education Research Journal, 30*(4), 383-406. .10.1007/s13394-017-0212-8

Miller, J., Warren, E., & Armour, D. (2018). Mathematical discourse when engaging in pattern generalisation. In R. Hunter, M. Civil, B. Herbel-Eisenmann, N. Planas, & D. Wagner (Eds.), *Mathematical discourse that breaks barriers and creates space for marginalized learners*. Rotterdam: Sense Publishers. https://doi.org/10.1163/9789004378735_011

Miller, T. (2009). *Television studies: The basics*. Routledge. 10.4324/9780203854198

Ministere de l' Education de l' Enseignement Superieure et de la Recherche Scientifiques (MEESRS). (2015). *Curriculum de l'Enseignement fondamental*. MEERS.

Mishra, P., & Koehler, M. (2006). Technological pedagogical content knowledge: A framework for teacher knowledge. *Teachers College Record*, 108(6), 1017–1054. 10.1111/j.1467-9620.2006.00684.x

Mistry, R. S., Benner, A. D., Biesanz, J. C., Clark, S. L., & Howes, C. (2010). Family and social risk, and parental investments during the early childhood years as predictors of low-income children's readiness outcomes. *Early Childhood Research Quarterly*, 25(4), 432–449. 10.1016/j.ecresq.2010.01.002

Mistry, R. S., Biesanz, J. C., Chien, N., Howes, C., & Benner, A. D. (2008). Socioeconomic status, parental investments, and cognitive and behavioral outcomes of low-income children from immigrant and native households. *Early Childhood Research Quarterly*, 23(2), 193–212. 10.1016/j.ecresq.2008.01.002

Mitchell, M., & Gilson, J. (1997). *Interest and Anxiety in Mathematics*. American Educational Research Association.

Mitchell, M. (1993). Situational Interest. *Its Multifaceted Structure in the Secondary School Mathematics Classroom.*, 85(3), 424–436.

Mjenda, M., Mutarutinya, V., & Owiti, D. (2023). Investigating secondary school mathematics teachers' knowledge, preparation and use of assessment rubrics in teaching and learning 3D-geometry in Tanzania. *International Journal of Education and Practice*, 11(3), 557–571. 10.18488/61.v11i3.3434

Mkhize, Z. (2022). 'They are just women, What do they know?'The lived experiences of African women doctoral students in the mathematics discipline in South African Universities. *Transformation in Higher Education*, 7(0), 1–11. 10.4102/the.v7i0.21836124010

Moala, J. G., & Hunter, R. (2019). Developing mathematical resilience among diverse learners: Preliminary progress and problematics. In Hine, G., Blackley, S., & Cooke, A. (Eds.). *Proceedings of the 42nd Annual Conference of the Mathematics Education Research Group of Australasia*. Perth: MERGA.

Monyooe, L. (2017). Reclassifying Township Schools—South Africa's Educational Tinkering Expedition! *Creative Education*, 8(3), 471–485. 10.4236/ce.2017.83036

Moore, D. S. (1997). New pedagogy and new content: The case of statistics. *Internationa statistical review, 65*(2), 123-137.

Moore, P., & Fitz, C. (1993). Gestalt theory and instructional design. *Journal of Technical Writing and Communication*, 23(2), 137–157. 10.2190/G748-BY68-L83T-X02J

Moore, R. L. (2020). Developing lifelong learning with heutagogy: Contexts, critiques, and challenges. *Distance Education*, 41(3), 381–401. 10.1080/01587919.2020.1766949

Morgil, I., Yavuz, S., Oskay, Ö. Ö., & Arda, S. (2005). Traditional and computer-assisted learning in teaching acids and bases. *Chemistry Education Research and Practice*, 6(1), 52–63. 10.1039/B3RP90001K

Morrison, G. R., Ross, S. J., Morrison, J. R., & Kalman, H. K. (2019). *Designing effective instruction*. John Wiley & Sons.

Moshman, D. (1982). Exogenous, endogenous, and dialectical constructivism. *Developmental Review*, 2(4), 371–384. 10.1016/0273-2297(82)90019-3

Mueller, P. A., & Oppenheimer, D. M. (2014). The pen is mightier than the keyboard: Advantages of longhand over laptop note taking. *Psychological Science*, 25(6), 1159–1168. 10.1177/095679761452458124760141

Mugenda, O. M. (1999). *Research methods: Quantitative and qualitative approaches*. African Centre for Technology Studies.

Mugenda, O., & Mugenda, A. (2003). *Research methods: Quantitative and Qualitative methods*. Acts Press.

Muir, T., Beswick, K., Callingham, R., & Jade, K. (2016). Experiencing teaching and learning quantitative reasoning in a project-based context. *Mathematics Education Research Journal*, 28(4), 479–50. 10.1007/s13394-016-0176-0

Mukuka, A., Mutarutinya, V., & Balimuttajjo, S. (2019). Exploring the barriers to effective cooperative learning implementation in school mathematics classrooms. *Problems of Education in the 21st Century, 77*(6), 745–757. 10.33225/pec/19.77.745

Mukuka, A., & Alex, J. K. (2024). Review of research on microteaching in mathematics teacher education: Promises and challenges. *Eurasia Journal of Mathematics, Science and Technology Education*, 20(1), em2381. 10.29333/ejmste/13941

Mukuka, A., Mutarutinya, V., & Balimuttajjo, S. (2020). Data on students' mathematical reasoning test scores: A quasi-experiment. *Data in Brief*, 30, 105546. 10.1016/j.dib.2020.10554632337327

Mukuka, A., Shumba, O., & Mulenga, H. M. (2021). Students' experiences with remote learning during the COVID-19 school closure: Implications for mathematics education. *Heliyon*, 7(7), e07523. 10.1016/j.heliyon.2021.e0752334307946

Mulenga, E. M., & Marbán, J. M. (2020). Is COVID-19 the gateway for digital learning in mathematics education? *Contemporary Educational Technology*, 12(2), ep269. 10.30935/cedtech/7949

Mulenga-Hagane, M., Daka, H., Musango, H., Mwelwa, K., & Kakupa, P. (2019). Formative Assessment as a Means of Improving Learner Achievement: Lessons from Selected Primary Schools of Lusaka, Zambia. *Journal of Lexicography and Terminology, 3*(1), 33-54.

Murchan, D., & Oldham, E. (2017). Exploring the role of computer-based assessment in diagnosing children's mathematical errors in primary education in Ireland. *Irish Educational Studies*, 36(4), 489–510. 10.1080/03323315.2017.1393765

Murphy, K. L., & Cifuentes, L. (2001). Using web tools, collaborating, and learning online. *Distance Education*, 22(2), 285–305. 10.1080/0158791010220207

Mushin, I., Gardner, R., & Munro, J. M. (2013). Language matters in demonstrations of understanding in early years mathematics assessment. *Mathematics Education Research Journal*, 25(3), 415–433. 10.1007/s13394-013-0077-4

Muthivhi, E., & Broom, Y. (2008). Continuities across schooling transition: A case of classroom practices among teachers in Venda, South Africa. *Journal of Educational Studies*, 7(1), 98–121.

Mutodi, P., & Chigonga, B. (2016). Concept map as an assessment tool in secondary school mathematics: An analysis of teachers' perspectives. *Eurasia Journal of Mathematics, Science and Technology Education*, 12(10), 2685–2696. 10.12973/eurasia.2016.2301a

Mwalongo, A. (2011). Teachers' perceptions about ICT for teaching, professional evelopment, administration and personal use. *International Journal of ducation and development using information and communication echnology*, 7(3), 36.

Naimsamphao, P., & Katwibun, D. (2019). Investigation of 10th Grade students'Q agency and authority in a mathematics Problem-based learning classroom. In Hine, G., Blackley, S., & Cooke, A. (Eds.), *Proceedings of the 42nd Annual Conference of the Mathematics Education Research Group of Australasia,* (pp. 524–531). Perth: MERGA.

Nama, S., & Ayalon, M. (2023). Exploring change in secondary mathematics teachers' noticing of argumentation through experiencing peer-assessment strategies. *Journal of Mathematics Teacher Education*, 1–46. 10.1007/s10857-023-09586-3

Namlu, A., & Ceyhan, E. (2002). *Computer anxiety: A study on university students*. Anadolu University Publishing.

Nath, S. R. (2007). Remunerated supplementary tutoring in out-of-school study in rural Bangladesh. BRAC Research Report, Research and Education Division. Dhaka, 1212, Bangladesh.

National Centre for Science and Engineering Statistics. (2019). *Women minorities, and persons with disabilities in science and engneeering (Special Report NSF 19-304)*. National Science Foundation, Directorate for Social, Behavioural and Economic Science.

National Council of Educational Research and Training [NCERT]. (2017). *National Achievement Survey [NAS] 2017: Class III*. Ministry of Human Resource and Development, Government of India.

National Council of Educational Research and Training [NCERT]. (2017). National Achievement Survey [NAS] 2017: Class V. Ministry of Human Resource and Development, Government of India.

National Council of Educational Research and Training [NCERT]. (2017). *National Achievement Survey [NAS] 2017: Class VIII*. Ministry of Human Resource and Development, Government of India.

National Council of Educational Research and Training [NCERT]. (2021). *National Achievement Survey [NAS] National Report 2021 Class III, V, VIII & X. Ministry of Education, Government of India*. NAS. https://nas.gov.in/report-card/2021

National Council of Teachers of Mathematics. (2000). *Principles and standards for school mathematics*. Author.

National Research Council. (2001). *Adding it up: Helping children learn mathematics*. National Academy Press.

NCF. (2005). National Curriculum Framework. National Council of Educational Research and Training, New Delhi, India NEP. (1986). National Policy on Education [NEP]. Ministry of Human Resource and Development, Government of India.

Ndibalema, P. (2021). Online Assessment in the Era of Digital Natives in Higher Education Institutions. *International Journal of Technology in Education, 4*(3).10.1080/0969594X.2015.1108900

Neville, H., Pakulak, E., & Stevens, C. (2015). Family-based training to improve cognitive outcomes for children from lower socioeconomic status backgrounds: Emerging themes and challenges. *Current Opinion in Behavioral Sciences*, 4, 166–170. 10.1016/j.cobeha.2015.05.006

Nicolaidou, M., & Philippou, G. (2003). *Attitudes towards mathematics, self-efficacy and achievement in problem solving. European Research in Mathematics Education III*. University of Pisa.

Niederle, M., & Vesterlund, L. (2010). Explaining the gender gap in math test scores: The role of competition. *The Journal of Economic Perspectives*, 24(2), 129–144. 10.1257/jep.24.2.129

Nimely, D. R., & Nyamu, F. K. (2023). Strategies for Inculcating Gender Equity in Mathematics Classrooms for Enhancement of Learners' Problem-Solving Skills. *International Journal of Research and Innovation in Social Science*, 7(7), 1446–1456. 10.47772/IJRISS.2023.70814

Nishant, R., Kennedy, M., & Corbett, J. (2020). Artificial intelligence for sustainability: Challenges, opportunities, and a research agenda. *International Journal of Information Management*, 53, 102104. 10.1016/j.ijinfomgt.2020.102104

Njoroge, M. C. (2017). The efficacy of using mother tongue language in the teaching of mathematics and science in primary schools: Evidence from grade one classrooms in Kenya. *Journal of Applied Linguistics and Language Research*, 4(8), 133–150.

Nollenberger, N., Rodriguez-Planas, N., & Sevila, A. (2014). The maths gender gap: The role of Culture. *Discussion Paper No. 8379*. Bonn, Germany.

Nova, K. (2023). AI-enabled water management systems: An analysis of system components and interdependencies for water conservation. *Eigenpub Review of Science and Technology*, 7(1), 105–124.

Oakley, A. (1972). *Sex,gender and society*. Temple Smith.

OECD (2012). *Equity and Quality in Education: Supporting Disadvantaged Students and Schools*. OECD Publishing. 10.1787/9789264130852-en

OECD. (2014). *PISA 2012 results: What students know and can do: Students performance in mathematics, reading and science*. OECD.

OECDiLibrary. (2024). *OECDiLibrary.* Indicators B7. https://www.oecd-ilibrary.org/sites/8389c70e-en/index.html?itemId=/content/component/8389c70e-en

Ogembo, J. O. (2017). *Effects of Computer Assisted Learning on Instruction in Organic Chemistry in Public Secondary Schools in Kwale County, Kenya.* [Doctoral dissertation, Kenyatta University, Nairobi].

Oginni, O. I., & Popoola, A. A. (2013). *Effects of Mathematics Innovation and Technology on Students Academic Performance in Open and Distance Learning (ODL).* ODL. oasis.col.org

Ojeda, L., & Flores, L. Y. (2008). The influence of gender, generation level, parents' education level, and perceived barriers on the educational aspirations of Mexican American high school students. *The Career Development Quarterly*, 57(1), 84–95. 10.1002/j.2161-0045.2008.tb00168.x

Ojose, B. (2023). Mathematics literacy: Are we able to put the mathematics we learn into everyday use? *Journal of Mathematics Education*, 4(1).

Oksala, J. (2017). Feminism and Power. In A. Garry, S. Khader, & A. Stone, *The Routledge Companion to feminist Philosophy* (pp. 678-688). Routeledge: New York. 10.4324/9781315758152-55

Okyere, M., & Larbi, E. (2019). *Senior high school mathematics teachers ' perception and use of assessment in the classroom, 15*(2), 43–54.

Olakanmi, E. E., Gambari, I. A., Gbodi, E. B., & Abalaka, E. A. (2016). Promoting intrinsic and extrinsic motivation among chemistry students using computer assisted instruction. *International Journal of Educational Sciences*, 12(2), 155–168. 10.1080/09751122.2016.11890422

Olga, P. (2008). *The effects of computer-assisted instruction on the Achievement, attitudes and retention of fourth grade Mathematics course.* [Thesis, Middle EastTechnical University].

Oliver, R. (2000). *Creating Meaningful Contexts for Learning in Web-based Settings. Proceedings of Open Learning 2000. (Pp.* Learning Network, Queensland.

Olosunde, G. (2019). Influence of globalization on mathematics content, teaching methods, and resource materials in colleges of education in Oyo State. In Tella, A. (Ed.), *Globalized Curriculum Methods for Modern Mathematics Education* (pp. 15–28). IGI Global. 10.4018/978-1-5225-6158-3.ch002

Olsher, S., Yerushalmy, M., & Chazan, D. (2016). How might the use of technology in formative assessment support changes in mathematics teaching? *For the Learning of Mathematics*, 36(3), 11–18. https://www.jstor.org/stable/44382716

Onatsu-Arvillomi, T. P., & Nurmi, J.-E. (2002). The development of achievement strategies and academic skills during the first year of primary school. *Learning and Instruction*, 12(5), 509–527. 10.1016/S0959-4752(01)00029-9

Organisation for Economic Co-operation and Development. (2016). *Excellence and equity in education.* OECD: OECD Publishing.

Organization for Economic Cooperation and Development (OECD). (2014). PISA 2012 results: What students know and can do: Student performance in mathematics,reading and science. *Rev ed Vol 1.*

Ormrod, J. E. (1999). Human learning (3rd ed.). Upper Saddle River, NJ: rentice-Hall

Orora, W., Keraro, F. N., & Wachanga, S. W. (2014). Effects of cooperative e-learning teaching strategy on students' achievement in secondary school Biology in Nakuru County, Kenya. *Sky Journal of Educational Research. 2(1)*, 001-009.

Otach, O. (2008). *Abolishing school fees in Africa: Lessons from Ethiopia. Ghana, Kenya, Malawi and Mozambique.*

Otieno, K. O. (2010). Teaching/learning resources and academic performance in mathematics in secondary schools in Bondo District of Kenya. *Asian Social Science*, 6(12), 126.

Ouyang, F., Xu, W., & Cukurova, M. (2023). An artificial intelligence-driven learning analytics method to examine the collaborative problem-solving process from the complex adaptive systems perspective. *International Journal of Computer-Supported Collaborative Learning*, 18(1), 39–66. 10.1007/s11412-023-09387-z

Ovie-Ekpewu, R. (2011). The sensitivity of science and mathematics teachers to gender issues in the classroom. *Journal of Research in National Development*, 8(2). 10.4314/jorind.v8i2.66848

Owu-Ewie, C. (2006). The language policy of education in Ghana: A critical look at the English-only language policy of education. In *Selected proceedings of the 35th annual conference on African linguistics* (pp. 76-85). Somerville, MA: Cascadilla Proceedings Project.

Ozumba, G. (2005). Gender-sensitivities in Igbo culture: A philosophical reappraisal. *Quidlibet Journal* 7(2).

Page, M. J., McKenzie, J. E., Bossuyt, P. M., Boutron, I., Hoffmann, T. C., Mulrow, C. D., Shamseer, L., Tetzlaff, J. M., Akl, E. A., Brennan, S. E., Chou, R., Glanville, J., Grimshaw, J. M., Hróbjartsson, A., Lalu, M. M., Li, T., Loder, E. W., Mayo-Wilson, E., McDonald, S., & Moher, D. (2021). The PRISMA 2020 statement: An updated guideline for reporting systematic reviews. *BMJ (Clinical Research Ed.)*, (71), n71. Advance online publication. 10.1136/bmj.n7133782057

Paine, L., Aydarova, E., & Syahril, I. (2017). Globalization and teacher education. *The SAGE handbook of research on teacher education, 2*, 1133-1148.

Palloff, R. M., & Pratt, K. (1999). *Building learning communities in cyberspace.* San Francisco, CA.

Papert, S. (1993). *Mindstorms: Children, computers, and powerful ideas* (2nd ed.). Harper Collins Publishers.

Papic, M. (2015). An early mathematical patterning assessment: Identifying young Australian indigenous children's patterning skills. *Mathematics Education Research Journal*, 27(4), 519–534. 10.1007/s13394-015-0149-8

Parish, L. (2016). The power of creativity: A case-study of a mathematically highly capable Grade 5 student. In B. White, M. Chinnappan, & S. Trenholm (Eds.), *Proceedings of the 39th Annual Conference of the Mathematics Education Research Group of Australasia,* (pp. 447–45). Adelaide: MERGA.

Park, H., Byun, S., & Kim, K. (2011). Parental involvement and students' cognitive outcomes in Korea: Focusing on private tutoring. *Sociology of Education*, 84(1), 3–22. 10.1177/0038040710392719

Parmaxi, A., Christou, E., Valdes, J. F., Hevia, D., Perifanou, M., Economides, A., & Manchenko, M. (2023). Gender equality in science, technology, engineering and mathematics: Industrial vis-a-vis academic perspective. *Discover Education*, 3(3), 1–13.

Passey, D., Rogers, C. G., Machell, J., & McHugh, G. (2004). *The Motivational Effect of ICT on pupils*. Department for Education and Skills Research Project.

Patahuddin, S. M., Usman, H. B., & Ramful, A. (2018). Affordances from number lines in fractions instruction: Students' interpretation of teacher's intentions. *International Journal of Science and Mathematics Education*, 16(5), 909–928. 10.1007/s10763-017-9800-z

Patel, D. B. (2013). Construction and effectiveness of computer aided instruction (CAI) programme for the units of science and technology of standard VIII. *International Journal of Research in Education*, 2(1), 112–115.

Paudel, T. (2020). Gender Issue in Teaching and Learning Mathematics. *Proceedings of the Tenth International Mathematics Education and Society*. Research Gate. https://www.researchgate.net/publication/342437701

Pawlak, M. (2022). Research into Individual Differences in SLA and CALL: Looking for Intersections. *Language Teaching Research Quarterly*, 31, 200–233. 10.32038/ltrq.2022.31.14

Pedretti, E., Mayer-Smith, J., & Woodrow, J. (1999). Teaming technology enhanced instruction in the science classroom and teacher professional development. *Journal of Technology and Teacher Education*, 7(2), 131–143.

Pelgrum, W. J. (2001). Obstacles to the integration of ICT in education: Results from a worldwide educational assessment. *Computers & Education*, 37(2), 163–178. 10.1016/S0360-1315(01)00045-8

Pelgrum, W. J. (2005). ICT Integration, Data from International Cooperative Studies. *Computers & Education*, 37.

Pelgrum, W. J., & Plomp, Tj. (1991). *The use of computers m educauon worldwide Results from the lEA "Computers tn Educatton' survey m 19 educattonal systems*. Oxford Pergamon Press.

Pepper, S. C. (1942). *World hypotheses: A study in evidence*. University of California Press.

Pereira-Mendoza, L. (2002). Would you allow your accountant to perform surgery? Implications for the education of primary teachers. In Phillips, B. (Ed.), *Proceedings of the Sixth International Conference on the Teaching of Statistics*. Hawthorn, VIC: International Statistical Institute.

Pfannkuch, M., & Ziedins, I. (2014). A modelling perspective on probability. In Chernoff, E. J., & Sriraman, B. (Eds.), *Probabilistic thinking: resenting plural perspectives (pp. 01–116).* Springer. 10.1007/978-94-007-7155-0_5

Philen, J. (2016). *Implementing Meaningful and Sustainable Project Based Learning Pedagogy.* [Doctoral Dissertation, England].

Piaget, J. (1972). *The principles of genetic epistemology* (Mays, W., Trans.).

Plomp, T.; Pelgrum, W. J. & Law, N. (2007). SITES2006 International comparative survey of pedagogical practices and ICT in education. *Education and Information Technologies, 12*(2), 83- 92.

Plüddemann, P. (2015). Unlocking the grid: Language-in-education policy realisation in post-apartheid South Africa. *Language and Education*, 29(3), 186–199. 10.1080/09500782.2014.994523

Pokrivcakova, S. (2019). Preparing teachers for the application of AI-powered technologies in foreign language education. *Journal of Language and Cultural Education*, 7(3), 135–153. 10.2478/jolace-2019-0025

Poudel, K. (2017). *Women empowerment through English language learning* [Dissertation, Kathmandu University: Kathmadu University and School of Education].

Pratt, D., & Ainley, J. (2014). Chance re-encounters: Computers in probability education revisited. In Wassong, T. (Ed.), *Mit Werkzeugen mathematik and stochastik lernen–using tools for learning mathematics and statistics* (pp. 165–177). Springer Fachmedien. 10.1007/978-3-658-03104-6_13

Premkumar, K. (2016). Use of student response systems for summative assessments. *Creative Education*, 7(13), 1851–1860. 10.4236/ce.2016.713187

Promsawan, S., & Katwibun, D. (2017). 11th grade students' self-regulated learning in a maths problem-based learning (PBL) classroom. In A. Downton, S. Livy, & J. Hall (Eds.), *Proceedings of the 40th Annual Conference of the Mathematics Education Research Group of Australasia,* (pp. 420–427). Melbourne: MERGA.

Quane, K., Chinnappan, M., & Trenholm, S. (2019). The nature of young children's attitudes towards mathematics. In Hine, G., Blackley, S., & Cooke, A. (Eds.), *Proceedings of the 42nd Annual Conference of the Mathematics Education Research Group of Australasia,* (pp. 108–11). Perth: MERGA.

Rahman, M. O., & Islam, M. T. (2009). Universal primary education for all towards millennium development goal 2: Bangladesh perspective. *Journal of Health Management*, 11(2), 405–418. 10.1177/097206340901100209

Rakes, C. R., Stites, M. L., Ronau, R. N., Bush, S. B., Fisher, M. H., Safi, F., Desai, S., Schmidt, A., Andreasen, J. B., Saderholm, J., Amick, L., Mohr-Schroeder, M. J., & Viera, J. (2022). Teaching Mathematics with Technology: TPACK and Effective Teaching Practices. *Education Sciences*, 12(2), 133. 10.3390/educsci12020133

Rakoczy, K., Klieme, E., Leiß, D., & Blum, W. (2017). Formative assessment in mathematics instruction: Theoretical considerations and empirical results of the Co 2 CA project. In Leutner, D., Fleischer, J., Grünkorn, J., & Klieme, E. (Eds.), *Competence assessment in education: Research, models and instruments* (pp. 447–467). Springer. 10.1007/978-3-319-50030-0_26

Ramani, G. B., & Siegler, R. S. (2008). Promoting board and stable improvements in low-income children's numerical knowledge through playing number games. *Child Development*, 79(2), 375–394. 10.1111/j.1467-8624.2007.01131.x18366429

Ramanujam, R., & Subramaniam, K. (2012). *Mathematics Education in India Status and Outlook. Homi Bhabha Centre for Science Education Tata Institute of Fundamental Research.* https://mathedu.hbcse.tifr.res.in/wp-content/uploads/2014/01/INP-Book_Mathematics-Edu-in-India_2012_KS-RR.pdf

Ramanujan, R. (2012). Mathematics Education in India: An Overview. In Ramanujam, R., & Subramaniam, K. (Eds.), *Mathematics Education in India Status and Outlook* (pp. 1–11). Homi Bhabha Centre for Science Education Tata Institute of Fundamental Research. https://mathedu.hbcse.tifr.res.in/wp-content/uploads/2014/01/INP-Book_Mathematics-Edu-in-India_2012_KS-RR.pdf

Rao, K.U, Prema, V and Subramanya, K. N. (2020). Plickers: An ICT Tool for Formative Assessment and Feedback. *Journal of Engineering Education Transformations, 33.*

Rastrollo-Guerrero, J. L., Gómez-Pulido, J. A., & Durán-Domínguez, A. (2020). Analyzing and predicting students' performance by means of machine learning: A review. *Applied Sciences (Basel, Switzerland)*, 10(3), 1042. 10.3390/app10031042

Ravyse, W. S., Seugnet Blignaut, A., Leendertz, V., & Woolner, A. (2017). Success factors for serious games to enhance learning: A systematic review. *Virtual Reality (Waltham Cross)*, 21(1), 31–58. 10.1007/s10055-016-0298-4

Reardon, S. F., & Portilla, X. A. (2016). Recent trends in income, racial, and ethnic school readiness gaps at kindergarten entry. *AERA Open*, 2(3), 1–18. 10.1177/2332858416657343

Republic of South Africa, (1996). Constitution of the Republic of South Africa.*Pretoria: Government Gazette, 378*(17678).

Republic of South Africa. (1996). *South African Schools Act, No. 84*. Government Printers.

Riegle-Crumb, C., Farkas, G., & Muller, C. (2006). The role of gender and friendship in advanced course taking. *Sociology of Education*, 79(3), 206–228. 10.1177/003804070607900302203332 74

Right to Education. (2009). *The Right of Children to Free and Compulsory Education Act, 2009*. Right to Education. https://www.education.gov.in/sites/upload_files/mhrd/files/upload_document/rte.pdf

Ringgold, T. L. (2001). *The evaluation of computer attitudes and the validation of computer self-efficacy scales of minority students at an urban community college in Baltimore City*. Morgan State University.

Rippon, G. (2019). *The gendered Brain: The New Neuroscience that shatters the myth of the female brain.* The Bodley Head.

Ritter, G. X., & Urcid, G. (2021). *Introduction to lattice algebra: With applications in ai, pattern recognition, image analysis, and biomimetic neural networks.* CRC Press. 10.1201/9781003154242

Rittle-Johnson, B., Fyfe, E. R., Hofer, K. G., & Farran, D. C. (2017). Early math trajectories: Low-income children's mathematics knowledge from ages 4 to 11. *Child Development*, 88(5), 1727–1742. 10.1111/cdev.1266227921305

Riyanto, B., Zulkardi, P., & Ratu Ilma Indra, D. (2019). Senior high school mathematics learning through mathematics modeling approach. *Journal on Mathematics Education*, 10(3), 425–444. https://eric.ed.gov/?id=EJ1229897. 10.22342/jme.10.3.8746.425-444

Robinson, J. P., & Lubienski, S. T. (2011). The development of gender achievement gaps in mathematics and reading during elementary and middle school: Examining direct cognitive assessments and teacher ratings. *American Educational Research Journal*, 48(2), 268–302. 10.3102/0002831210372249

Robinson, M., Loch, B., & Croft, T. (2015). Student perceptions of screencast feedback on mathematics assessment. *International Journal of Research in Undergraduate Mathematics Education*, 1(3), 363–385. 10.1007/s40753-015-0018-6

Robitaille, D., & Tavers, K. (1992). International studies of achievement in mathematics. In Grouws, D. (Ed.), *Handbook of research on mathematics education* (pp. 687–709). Macmillan.

Romanelli, F., Bird, E., & Ryan, M. (2009). Learning styles: A review of theory, application, and best practices. *American Journal of Pharmaceutical Education*, 73(1), 9. 10.5688/aj73010919513146

Roos, H. (2023). Students' voices of inclusion in mathematics education. *Educational Studies in Mathematics*, 113(2), 229–249. 10.1007/s10649-023-10213-4

Ross, C. E., & Broh, B. E. (2000). The roles of self-esteem and the sense of personal control in the academic achievement process. *Sociology of Education*, 73(4), 270–284. 10.2307/2673234

Rostvik, C. M., & Fyfe, A. (2018). *Ladies, gentlemen, and scientific publication at the Royal Society, 1945-1990.* Open Library of Humanities. 10.16995/olh.265

Roth, J., & Ames, J. (2014). *Beiträge zum Mathematikunterricht.* WTM-Verlag.

Rothman, D. (2020). *Artificial Intelligence By Example: Acquire advanced AI, machine learning, and deep learning design skills.* Packt Publishing Ltd.

Rozell, E. J., & Gardner, W. L. (1999). Computer-related success and failure: a longitudinal field study of the factors influencing computer- related performance. *Computers in Human Behavior, 15*(1), 1-10.

Rubin, J., & Rajakaruna, M. (2015). Teaching and assessing higher order thinking in the mathematics classroom with clickers. *International Electronic Journal of Mathematics Education*, 10(1), 37–51. 10.29333/iejme/290

Ruhle, R. (2022). Mind the gap-an analysis of gender differences in mathematics and science achievement in South Africa. *Stellenbosch Economic Working Papers: WP04/2022.*

Russell, G., & Bradley, G. (1997). Teachers' computer anxiety: Implications for professional development. *Education and Information Technologies, 2*(1), 17-30.

Russo, J., & Hopkins, S. (2017b). Examining the impact of lesson structure when teaching with cognitively demanding tasks in the early primary years. In A. Downton, S. Livy, & J. Hall (Eds.), *Proceedings of the 40th Annual Conference of the Mathematics Education Research Group of Australasia,* (pp. 450–457). Melbourne: MERGA.

Russo, J. A., & Russo, T. (2019). Teacher interest-led inquiry: Unlocking teacher passion to enhance student learning experiences in primary mathematics. *International Electronic Journal of Mathematics Education*, 14(3), 701–717. 10.29333/iejme/5843

Russo, J., & Hopkins, S. (2017a). Class challenging tasks: Using cognitive load theory to inform the design of challenging mathematical tasks. *Australian Primary Mathematics Classroom*, 22(1), 21–27.

Russo, J., & Hopkins, S. (2017c). How does lesson structure impact teachers' willingness to teach with challenging tasks? *Mathematics Teacher Education and Development*, 19(1), 30–46. 10.1007/s13394-017-0197-3

Russo, J., & Hopkins, S. (2019). Teachers' perceptions of students when observing lessons involving challenging tasks. *International Journal of Science and Mathematics Education*, 17(4), 759–779. 10.1007/s10763-018-9888-9

Salem, S. (2018). Intersectionality and its discontents: Intersectionality as traveling theory. *European Journal of Women's Studies*, 25(4), 403–418. 10.1177/1350506816643999

Sanders, J. (2005). *Gender and technology in education: A research review.* Seattle: Center for Gender Equity.

Sangkaew, S., & Katwibun, D. (2019). 11th grade students' self-beliefs in a mathematics problem-based learning (PBL) classroom. In Hine, G., Blackley, S., & Cooke, A. (Eds.), *Proceedings of the 42nd Annual Conference of the Mathematics Education Research Group of Australasia* (pp. 620–627). Perth: MERGA.

Sanyal, B. C. (2001). New functions of higher education and ICT to achieve education for all. *Paper prepared for the Expert Roundtable on University and Technology-for- Literacy and Education.* Research Gate.

Sarmah, A., & Puri, P. (2014). Attitude towards Mathematics of the students studying in diploma in Engineering Institute (Polytechnic) of Sikkim. *Journal of Research & Methods in Education*, 4(6), 12–18. 10.9790/7388-04630610

Savard, A., & Freiman, V. (2016). Investigating complexity to assess student learning from a robotics-based task. *Digital Experiences in Mathematics Education*, 2(2), 93–114. 10.1007/s40751-016-0016-6

Sawatzki, C. (2016). Insights from a financial literacy task designer: The curious case of problem context. In In B. White, M. Chinnappan, & S. Trenholm (Eds.), *Proceedings of the 39th Annual Conference of the Mathematics Education Research Group of Australasia,* (pp. 576–58). Adelaide: MERGA.

Saxena, S. (2013). *Top 10 Characteristics of a 21st Century Classroom.* EdTech Review. https://www.edtechreview.in/author/saomya-saxena/

Saygin, P. (2020). Gender bias in standard tests: Evidence from a centralised college admission system. *Empirical Economics*, 59(2), 1037–1065. 10.1007/s00181-019-01662-z

Scalise, N. R., Daubert, E. N., & Ramani, G. B. (2020). Benefits of playing numerical card games on head start children's mathematical skills. *Journal of Experimental Education*, 88(2), 200–220. 10.1080/00220973.2019.158172133100376

Scardamalia, M., & Bereiter, C. (1994). Computer support for knowledge-building communities. *Journal of the Learning Sciences*, 3(3), 265–283. 10.1207/s15327809jls0303_3

Schneider, D., Hastings, O. P., & LaBriola, J. (2018). Income inequality and class divides in parental investments. *American Sociological Review*, 83(3), 475–507. 10.1177/0003122418772034

Schoenfeld, A. H. (2015). Summative and formative assessments in mathematics supporting the goals of the common core standards. *Theory into Practice*, 54(3), 183–194. 10.1080/00405841.2015.1044346

Schroeder, L. B., & Dorn, B. (2016). Enabling and integrating online formative assessment in a flipped calculus course. *PRIMUS (Terre Haute, Ind.)*, 26(6), 585–602. 10.1080/10511970.2015.1050619

Selwyn, N. (2012). *Education in a digital world: Global perspectives on technology and education.* Routledge. 10.4324/9780203108178

Serow, P., Callingham, R., & Tout, D. (2016). Assessment of mathematics learning: What are we doing? In Makar, K., Dole, S., Visnovska, J., Goos, M., Bennison, A., & Fry, K. (Eds.), *Research in Mathematics Education in Australasia 2012-2015* (pp. 235–254). Springer. 10.1007/978-981-10-1419-2_12

Sevilla, A., & Cue-Ruiz, P. (2024). How can we reduce gender gaps in mathematics? In *B. Atweh, A. Barton, M. Borba, N. Gough, C. Keitel, V. Vistro-Yu, & R. Vithal, Mathematics Literacy and Globalisation* (pp. 2–17). Springer.

Shakunthala. (2014). Social Inequality and Access to Education A Sociological Study. *The International Journal of Creative Research Thoughts, 2*(1), 732-737.

Shang, B. (2016). *Tackling Gender Inequality: Definitions, Trends, and Policy Designs. WP/22/232.* International Monetary Fund.

Sharma, R. (2003). *Barriers in Using Technology for Education in Developing Countries*. IEEE.

Sharma, R., & Singh, A. (2023). Use of digital technology in improving quality education. In Nadda, V., Tyagi, P. K., Vieira, R. M., & Tyagi, P. (Eds.), *Implementing Sustainable Development Goals in the Service Sector* (pp. 14–26). IGI Global. 10.4018/979-8-3693-2065-5.ch002

Sharma, S., & Sharma, S. (2023). Successful teaching practices for english language learners in multilingual mathematics classrooms: A meta-analysis. *Mathematics Education Research Journal*, 35(4), 821–848. 10.1007/s13394-022-00414-0

Shaughnessy, J. M. (2007). Research on statistics learning and reasoning. *Second handbook of research on mathematics teaching and learning, 2*, 957-1010.

Shaughnessy, J. M. (1992). Research in probability and statistics: Reflections and directions. In Grouws, D. A. (Ed.), *Handbook of Research on Mathematics Teaching and Learning* (pp. 465–494). Macmillan.

Shaughnessy, J. M. (2003). Research on students' understandings of probability. In Kilpatrick, J., Martin, W. G., & Schifter, D. (Eds.), *A Research Companion to Principles and Standards for School Mathematics* (pp. 216–226). National Council of Teachers of Mathematics.

Shaw, B. A. (2004). *The relationship of teacher quality variables to technology integration in K–12 schools*. The University of Texas at El Paso.

Short, D. (2002). *Computer-based mathematics instruction: Automated remedial branching* [Unpublished doctoral dissertation, California State University].

Siddiqui, U., & Khatoon, T. (2013). Teaching physical science: Should we implement teacher-centered CAI or student-centered CAI at secondary school level in India? *European Scientific Journal, ESJ*, 9(10), 136–149.

Siegler, R. S., & Ramani, G. B. (2008). Playing numerical board games promotes low-income children's numerical development. *Developmental Science*, 11(5), 655–661. 10.1111/j.1467-7687.2008.00714.x18801120

Siemens, G. (2004). *A learning theory for the digital age*. eLearning Space. https://www.elearnspace.org/Articles/connectivism.htm

Singh, V. K., Singh, P., Karmakar, M., Leta, J., & Mayr, P. (2021). The journal coverage of Web of Science, Scopus, and Dimensions: A comparative analysis. *Scientometrics*, 126(6), 5113–5142. 10.1007/s11192-021-03948-5

Sinha, S. (2023). *Artificial Intelligence and Mathematics Education: Revolutionizing Learning for the Digital Era*. LinkedIn. https://in.linkedin.com/in/sanjeev-sinha-3474533a?trk=article-ssr-frontend-pulse_publisher-author-card

Sirin, S. R. (2005). Socioeconomic status and academic achievement: A meta-analytic review of research. *Review of Educational Research*, 75(3), 417–453. 10.3102/00346543075003417

Siriwat, R., & Katwibun, D. (2017). Exploring critical thinking in a mathematics problem-based learning classroom. In A. Downton, S. Livy, & J. Hall (Eds.), *Proceedings of the 40th Annual Conference of the Mathematics Education Research Group of Australasia,* (pp. 474–48). Melbourne: MERGA.

Sivin-Kachala, J. (1998). *Report on the effectiveness of technology in schools, 1990-1997.* Software Publishers Association.

Skinner, B. F. (1954). *Teaching Machines and Programmed Learning*. Department of Audiovisual Education, National Education Association.

Skovsmose, O. (2008). Mathematical literacy and globalisation. In *B. Atweh, A. Barton, P. Ernest, N. Gough, C. Keitel, C. Vistro-Yu, & R. Vithal, Internationalisation and Globalisation in Mathematics and Science Education* (pp. 3–18). Springer.

Slack, D., Krishna, S., Lakkaraju, H., & Singh, S. (2023). Explaining machine learning models with interactive natural language conversations using TalkToModel. *Nature Machine Intelligence*, 5(8), 873–883. 10.1038/s42256-023-00692-8

Smaldino, D. E. (2002). The forum report: E-learning adoption rates and barriers. In Rossett, A. (Ed.), *The ASTD e-learning handbook* (pp. 19–23). McGraw-Hill.

Smarkola, C. (2007). Technology acceptance predictors among student teachers and experienced classroom teachers. *Journal of Educational Computing Research*, 37(1), 65–82. 10.2190/J3GM-3RK1-2907-7U03

Smetana, L. K., & Bell, R. L. (2012). Computer simulations to support science instruction and learning: A critical review of the literature. *International Journal of Science Education*, 34(9), 1337–1370. 10.1080/09500693.2011.605182

Snee, R. D. (1993). What's missing in statistical education? *The American Statistician*, 47(2), 149–154. 10.1080/00031305.1993.10475964

Sohn, H., Lee, D., Jang, S., & Kim, T. K. (2010). Longitudinal relationship among private tutoring, student-parent conversation, and student achievement. *KEDI Journal of Educational Policy*, 7(1), 23–41.

Soto, M., & Ambrose, R. (2016). Screencasts: Formative assessment for mathematical thinking. *Technology. Knowledge and Learning*, 21(2), 277–283. 10.1007/s10758-015-9272-6

Sreekanth, Y. (2016). *Students Educational Achievement at School Level: An Analysis across Social Categories.* A paper presented in the International Conference on Social Statistics in India, Patna, Bihar, India.

Srinivasa, K. G., Kurni, M., & Saritha, K. (2022). Harnessing the Power of AI to Education. In *Learning, Teaching, and Assessment Methods for Contemporary Learners: Pedagogy for the Digital Generation* (pp. 311–342). Springer Nature Singapore. 10.1007/978-981-19-6734-4_13

Steele, H. (2006, March). The "gender gap" in computing: Implications for teacher education. In *Society for Information Technology & Teacher Education International Conference* (pp. 811-816). Association for the Advancement of Computing in Education (AACE).

Steinberg, E. R. (1984). *Teaching computers to teach.* Lawrence Erlbaum Associates.

Stephan, M. (2020). Teacher-centered teaching in mathematics education. In Lerman, S. (Ed.), *Encyclopedia of Mathematics Education* (pp. 836–840). Springer International Publishing. 10.1007/978-3-030-15789-0_150

Sternberg, R. J. (2004). *Psychology.* Wadsworth.

Stevenson, D. L., & Baker, D. F. (1992). Shadow education and allocation in formal schooling: Transition to university in Japan. *American Journal of Sociology*, 97(6), 1639–1657. https://www.jstor.org/stable/2781551. 10.1086/229942

Stewart, I. (2017). *The Beauty of Numbers in Nature: Mathematical Patterns and Principles from the Natural World.* MIT Press.

Stoet, G., & Geary, D. C. (2018). The gender-equality paradox in science, technology, engineering, and mathematics education. *Psychological Science*, 29(4), 581–593. 10.1177/095679761774171929442575

Stohl, H. (2005). Probability in teacher education and development. In Jones, G. A. (Ed.), *Exploring probability in school: challenges for teaching and learning* (pp. 345–366). Springer. 10.1007/0-387-24530-8_15

Stoten, D. W. (2022). Navigating heutagogic learning: Mapping the learning journey in management education through the OEPA model. *Journal of Research in Innovative Teaching & Learning*, 15(1), 83–97. 10.1108/JRIT-07-2020-0038

Stratham, D. S., & Torell, C. R. (1996). *Computers in the classroom: The impact of technology on student learning* [Doctoral dissertation, Army Research Institute]. http://www.temple.edu/lss/htmlpublications/spotlights /200/spot206.htm

Sullivan, P., Boreck, C., Walker, N., & Rennie, M. (2016). Exploring a structure for mathematics lessons that initiate learning by activating cognition on challenging tasks. *Journal of Mathematical Behaviour, 41,* 159-170. https://doi.org/.10.1016/j.jmathb.2015.1002

Sun, L., Guo, Z., & Hu, L. (2023). Educational games promote the development of students' computational thinking: A meta-analytic review. *Interactive Learning Environments*, 31(6), 3476–3490. 10.1080/10494820.2021.1931891

Susman, E. B. (1998). Co-operative learning: a review of factors that increase the effectiveness of computer-based instruction. *Journal of Educational Computing Research, 18*(4), 303-322.

Susuwele, B. W. J. (2005). *Classroom Assessment in Malawi : Teachers ' Perceptions and Practices in Mathematics by Classroom Assessment in Malawi : Teachers ' Perceptions and Practices in Mathematics*. Virginia Polytechnic Institute and State University.

Suurtamm, C., Thompson, D. R., Kim, R. Y., Moreno, L. D., Sayac, N., Silver, S. S. E., & Vos, U. S. P. (2016). Assessment in mathematics education. In Kaiser, G. (Ed.), *Assessment in Mathematics Education. ICME-13 Topical Surveys*. Springer. 10.1007/978-3-319-32394-7_1

Syaifuddin, M. (2020). Implementation of authentic assessment on mathematics teaching: Study on junior high school teachers. *European Journal of Educational Research*, 9(4), 1491–1502. 10.12973/eu-jer.9.4.1491

Syyeda, F. (2016). Understanding attitudes towards mathematics (ATM) using modal Model: An exploratory case study with secondary school children in England. *Cambridge Open-Review Educational Research e-Journal 3*, 32-62.

Szeibert, J., Muzsnay, A., Szabó, C., & Bereczky-Zámbó, C. G. (2023). A case study of using test-enhanced learning as a formative assessment in high school mathematics. *International Journal of Science and Mathematics Education*, 21(2), 623–643. 10.1007/s10763-022-10264-8

Tabassum, N., & Nayak, B. S. (2021). Gender Stereotypes and Their Impact on Women's Career Progressions from a Managerial Perspective. *IIM Kozhikode Society & Management Review*, 10(2), 192–208. 10.1177/2277975220975513

Tabrani, Z. A., & Masbur, M. (2016). Islamic Perspectives On The Existence Of Soul And Its Influence In Human Learning (A Philosophical Analysis Of The Classical And Modern Learning Theories). *JURNAL EDUKASI: Jurnal Bimbingan Konseling*, 1(2), 99–112.

Talanquer, V. (2014). Threshold concepts in chemistry: The critical role of implicit schemas. *Journal of Chemical Education*, 92(1), 3–9. 10.1021/ed500679k

Tang, J., Liu, G., & Pan, Q. (2021). A review on representative swarm intelligence algorithms for solving optimization problems: Applications and trends. *IEEE/CAA Journal of Automatica Sinica, 8*(10), 1627-1643.

Tan, S. (2023). Harnessing Artificial Intelligence for innovation in education. In *Learning intelligence: Innovative and digital transformative learning strategies: Cultural and social engineering perspectives* (pp. 335–363). Springer Nature Singapore. 10.1007/978-981-19-9201-8_8

Tanudjaya, C. P., & Doorman, M. (2020). Examining Higher Order Thinking in Indonesian Lower Secondary Mathematics Classrooms. *Journal on Mathematics Education*, 11(2), 277–300. https://eric.ed.gov/?id=EJ1252003. 10.22342/jme.11.2.11000.277-300

Tapalova, O., & Zhiyenbayeva, N. (2022). Artificial Intelligence in Education: AIEd for Personalised Learning Pathways. *Electronic Journal of e-Learning*, 20(5), 639–653. 10.34190/ejel.20.5.2597

Tapia, M., & Marsh, G. (2004). An instrument to measure mathematics attitudes. *Academic Exchange Quarterly*, 8, 1–8.

Tapscott, D. (1998). *Growing up digital: The rise of the Net generation*. McGraw-Hill.

Taub, G. E., Benson, N., & Szente, J. (2014). Improving Mathematics: An Examination of the Effects of Specific Cognitive Abilities on College-age Students' Mathematics Achievement. *International Journal for the Scholarship of Teaching and Learning*, 8(2), 8–15. 10.20429/ijsotl.2014.080208

Tebabal, A., & Kahssay, G. (2011). The effects of student-centered approach in improving students' graphical interpretation skills and conceptual understanding of kinematical motion. *Lat. Am. J. Phy. Edu*, 5(2), 374–381.

Tessa, E. S. C., & Mahzarin, R. B. (2019). Gender in Science, Technology, Engineering, and Mathematics: Issues, Causes, Solutions. *The Journal of Neuroscience : The Official Journal of the Society for Neuroscience*, 39(37), 7228–7243. 10.1523/JNEUROSCI.0475-18.201931371423

Thanheiser, E. (2018). Brief report: The effects of preservice elementary school teachers' accurate self-assessments in the context of whole number. *Journal for Research in Mathematics Education*, 49(1), 39–56. 10.5951/jresematheduc.49.1.0039

Thomas, J. W. (2000). *A review of research on project-based learning*. Bobopearlman. http://www.bobpearlman.org/BestPractices/PBL_Research.pdf

Thomas, M., Clarke, D. M., McDonough, A., & Clarkson, P. C. (2023). Assessing students' understanding of time concepts in Years 3 and 4: Insights from the development and use of a one-to-one task-based interview. *Mathematics Education Research Journal*, 35(S1), 1–22. 10.1007/s13394-023-00451-3

Thuketana, N. S., & Makgabo, M. C. (2022). The use of English to offer learner support and enhance perceptual skills development in South African township schools. *The Journal for Transdisciplinary Research in Southern Africa*, 18(1), 1209. 10.4102/td.v18i1.1209

Touretzky, D., Gardner-McCune, C., & Seehorn, D. (2023). Machine learning and the five big ideas in AI. *International Journal of Artificial Intelligence in Education*, 33(2), 233–266. 10.1007/s40593-022-00314-1

Toyama, K. (2011). *There are no technology shortcuts to good education*. EduTech Debate. http://edutechdebate.org/ict-in-schools/there-are-no-technology-shortcuts-to-good-education/#sthash.nOJ2LCNe.d

Traore, M. (2021). *Guidebook to a successful implementation of digital education*. World Bank. https://docs.edtechhub.org/lib/RTJ2Z8MX

Tudge, J. R. H., & Doucet, F. (2004). Early mathematical experiences: Observing young black and white children's everyday activities. *Early Childhood Research Quarterly*, 91(1), 21–39. 10.1016/j.ecresq.2004.01.007

Turney, S. (2023, November 20). *Systematic review: Definition, example and guide*. Scribbr. https://www.scribbr.com/methodology/systematic-review/

Uddin, M. E. (2015a). Family socio-cultural values affecting early marriage between Muslim and Santal communities in rural Bangladesh. *The International Journal of Sociology and Social Policy*, 35(3/4), 141–164. 10.1108/IJSSP-06-2014-0046

Uddin, M. E. (2015b). Ethnic disparity in family socio-economic status in Bangladesh: Implication for family welfare policy-practice. *Global Social Welfare : Research, Policy & Practice*, 2(2), 29–38. 10.1007/s40609-014-0021-x

Uddin, M. E. (2017). Ecological framework for primary school attainment in the tri-ethnic communities in rural Bangladesh. *Child Indicators Research*, 10(3), 693–713. 10.1007/s12187-016-9401-3

Ukobizaba, F., Nizeyimana, G., & Mukuka, A. (2021). Assessment strategies for enhancing students' mathematical problem-solving skills: A review of literature. *Eurasia Journal of Mathematics, Science and Technology Education*, 17(3), em1945. 10.29333/ejmste/9728

Umar, F. A., Muhoro, S. N., Opondo, F. O., Okumu, J. M., Kasimu, T. M., Wachira, D. W., Odoyo, E. O., Muturi, E. K., Ogari, C. O., Namikoye, E. S., & Wayong'o, S. N. (2018). *Teacher factors influencing the use of mother tongue in teaching of mathematics in lower primary schools in Yola South local government*. [Doctoral dissertation. School of Education. Kenya: Kenyatta University].

Ünal, M. (2017). Preferences of Teaching Methods and Techniques in Mathematics with Reasons. *Universal Journal of Educational Research*, 5(2), 194–202. 10.13189/ujer.2017.050204

UNESCO and Commonwealth of Learning (COL). (2004). *Schoolnet Toolkit.* Bangkok: UNESCO; Vancouver: Commonwealth of Learning.

UNESCO. (2002). *Information and Communication Technology in Education. A Curriculum for Schools and Programme of Teacher Development*. UNESCO.

UNESCO. (2004). *A training manual for sensitising education managers, curriculum and material developers and media professionals to gender concerns*. UNESCO.

UNESCO. (2019). *Open educational resources*. UNESCO. https://www.unesco.org/en/open-educational-resources

UNESCO. (2022). *Challenging gender bias and stereotypes in and through education*. UNESCO. https://unesdoc.unesco.org/ark:/48223/pf0000380827

United Nations Children's Fund. (2006). *Multiple indicator cluster survey manual 2005: Monitoring the situation of children and women*. UNICEF.

United Nations Children's Fund. (2015). *Transforming our world: The 2030 agenda for sustainable development*. UNICEF.

United Nations Scientific and Cultural Organisation (UNESCO). (2016). *Closing the gap in STEM: Drawing more girls and women into Science, Technology, Engineering and Mathematics*. UNESCO.

United Nations. (2006). *United Nations Convention on the Rights of Persons with Disabilities and Optional Protocol*. UN. https://www.un.org/disabilities/documents/convention/convoptprot-e.pdf

Upadyaya, K., & Eccles, J. S. (2014). Gender differences in teachers' perceptions and children's ability self-concepts. InI. Schoon & J. S. Eccles (Eds.), *Gender differences in aspirations and attainment: A life course perspective* (pp. 79–100). Cambridge, UK: Cambridge University Press.

Urrutia, F., & Araya, R. (2022). Do written responses to open-ended questions on fourth-grade online formative assessments in mathematics help predict scores on end-of-year standardized tests? *Journal of Intelligence*, 10(4), 82. 10.3390/jintelligence1004008236278604

Ursini, S., & Sánchez, G. (2008). Gender, technology and attitude towards mathematics: A comparative longitudinal study with Mexican students. *ZDM Mathematics Education*, 40(4), 559–577. 10.1007/s11858-008-0120-1

Valez, J., & Ochoas, S. (2020). Gender sensitivity and Its relation to gender equality. In W. Leal F, A. Azul, L. Brandli, A. Lange S, & T. Wall, *Gender Equality, Encyclopedia of the UN Sustainable Development Goals.* Switzerland: Springer Nature Switzerland. 10.1007/978-3-319-70060-1_46-1

Valle-Cruz, D., Fernandez-Cortez, V., & Gil-Garcia, J. R. (2022). From E-budgeting to smart budgeting: Exploring the potential of artificial intelligence in government decision-making for resource allocation. *Government Information Quarterly*, 39(2), 101644. 10.1016/j.giq.2021.101644

van den Berg, M., Bosker, R. J., & Suhre, C. J. (2018). Testing the effectiveness of classroom formative assessment in Dutch primary mathematics education. *School Effectiveness and School Improvement*, 29(3), 339–361. 10.1080/09243453.2017.1406376

Van Mier, H. I., Schleepen, T. M., & Van den Berg, F. C. (2019). Gender differences regarding the impact of math anxiety on arithmetic performance in second and fourth graders. *Frontiers in Psychology*, 9, 2690. 10.3389/fpsyg.2018.0269030713516

Vanderlinde, R., van Braak, J., & Hermans, R. (2009). Educational technology on a turning point: Curriculum implementation in Flanders and challenges for schools. *Educational Technology Research and Development*, 57(4), 573–584. 10.1007/s11423-009-9126-9

Vandermaas-Peeler, M., Nelson, J., Bumpass, C., & Sassine, B. (2009). Numeracy-related exchanges in joint story-book reading and play. *International Journal of Early Years Education*, 17(1), 67–84. 10.1080/09669760802699910

Vankúš, P. (2021). Influence of game-based learning in mathematics education on students' affective domain: A systematic review. *Mathematics*, 9(9), 986. 10.3390/math9090986

Vasanthan, R., & Ancymol, J. (2021). Literature for gender sensitivity: An Overview. *Journal for Educators. Teachers and Trainers*, 13(1), 61–69.

Vasilyeva, M., Dearing, E., Ivanova, A., Shen, C., & Kardanova, E. (2018). Testing the family investment model in Russia: Estimating indirect effects of SES and parental beliefs on the literacy skills of first graders. *Early Childhood Research Quarterly*, 42, 11–20. 10.1016/j.ecresq.2017.08.003

Veldhuis, M., & van den Heuvel-Panhuizen, M. (2020). Supporting primary school teachers' classroom assessment in mathematics education: Effects on student achievement. *Mathematics Education Research Journal*, 32(3), 449–471. 10.1007/s13394-019-00270-5

Von Glasersfeld, E. (1995). *Radical Constructivism: A Way of Knowing and Learning. Studies in Mathematics Education Series: 6*. Falmer Press, Taylor & Francis Inc.

von Glasersfeld, E. (2014). Piaget and the radical constructivist epistemology. *Costruttivismi, 1*, 94–107. https://www.aippc.it/wp-content/uploads/2019/04/2014.02.094.107.pdf

Votruba-Drzal, E. (2003). Income changes and cognitive stimulation in young children's home learning environments. *Journal of Marriage and Family*, 65(2), 341–355. 10.1111/j.1741-3737.2003.00341.x

Vygotsky, L. (1978). *Mind in Society: The development of higher pyschological processes.* Harvard University Press.

Vygotsky, L. S. (1978). *Mind in society: The development of higher psychological processes.* Harvard university press.

Walick, C. M., & Burns, M. K. (2017). A Proposed Algebra Assessment for Use in a Problem-Analysis Framework. *Assessment for Effective Intervention*, 42(3), 150–159. 10.1177/1534508417690000

Walker, S. P., Wachs, T. D., Gardner, J. M., Wasserman, G. A., Pollitt, E., & Carter, J. A. (2007). Child development in developing countries 2. Child development and risk factors for adverse outcomes in developing countries. *Lancet*, 369(9556), 145–157. 10.1016/S0140-6736(07)60076-217223478

Wangid, M. N., & Widyastuti, P. (2021). Effect of Gender and Department on Motivation, Self-Efficacy, And Perception as Teacher of Pre-Service Teacher Students. *Multicultural Education*, 7(6), 6–8.

Wanjala, M. (2005). *Effectiveness of computer-based instruction versus conventional methods on students' attitude, motivation and achievement in the teaching of mathematics in secondary schools of Bungoma District.* [Unpublished Master's Thesis, Moi University, Eldoret, Kenya].

Wanyonyi, P. W. (2013). *An Assessment of Factors Affecting Students' Performance in Mathematics at KCSE Level in Secondary Schools in Kakamega County, Kenya* [Doctoral dissertation].

Wanyonyi, P. W. (2013). *An Assement of Factors Affecting Students' Perfomonce in Mathematics at K.C.S.E Level in Secondary Schools in Country*. Kenyatta Universty.

Warren, E., & Miller, J. (2016). *Mathematics at the margins*. Springer. 10.1007/978-981-10-0703-3

Wass, R. (2012). *Developing critical thinkers in Higher: A Vygotskian Perspective.* [Thesis, New Zealand: University of Otago].

Watts, T. W., Duncan, G. J., Chen, M., Claessens, A., Davis-Kean, P. E., Duckworth, K., & Susperreguy, M. I. (2015). The role of mediators in the development of longitudinal mathematics achievement associations. *Child Development*, 86(6), 1892–1907. 10.1111/cdev.1241626332124

Watts, T. W., Duncan, G. J., Clements, D. H., & Sarama, J. (2018). What is the long-run impact of learning mathematics during preschool. *Child Development*, 89(2), 539–555. 10.1111/cdev.1271328105650

Watts, T. W., Duncan, G. J., Siegler, R. S., & Davis-Kean, P. E. (2014). What's past is prologue: Relations between early mathematics knowledge and high school achievement. *Educational Researcher*, 43(7), 352–360. 10.3102/0013189X1455366026806961

Way, J., & Thom, J. (2019). Capturing the mathematical drawing process using a digital pen. In Hine, G., Blackley, S., & Cooke, A. (Eds.), *Proceedings of the 42nd Annual Conference of the Mathematics Education Research Group of Australasia,* (pp. 104–107). Perth: MERGA.

Weiland, I. S., Hudson, R. A., & Amador, J. M. (2014). Preservice formative assessment interviews: The development of competent questioning. *International Journal of Science and Mathematics Education*, 12(2), 329–352. 10.1007/s10763-013-9402-3

Wekesa, D. W., Wekesa, E. W., & Amadalo, M. M. (2013). A computer mediated simulation module for teaching cell division in secondary school biology. *International Journal of Educational Research and Development.*, 2(5), 114–130.

Wells, J., Geiger, V., Miller, J., Bruder, R., Roder, U., & Ratnayake, I. (2019). Designing challenging online mathematical tasks for initial teacher education: Motivational considerations. In M. Graven, H. Venkat, A. A. Essien, & P. Vale (Eds.), *Proceedings of the 43rd Conference of the International Group for the Psychology of Mathematics Education.* Pretoria, South Africa: PME.

White, G., Ruther, M., Kahn, J. R., & Dong, D. (2016). Gender inequality amid educational expansion in India: an analysis of gender differences in the attainment of reading and mathematics skills. *Journal of Research in Gender Studies, 6*(2), 153+. https://link.gale.com/apps/doc/A478698526/AONE?u=anon~bec3535f&sid=googleScholar&xid=757f355b10.3389/fpsyg.2021.599199

Wigderson, A. (2019). *Mathematics and computation: A theory revolutionizing technology and science*. Princeton University Press.

Wilkie, K. J., & Ayalon, M. (2019). Learning through critiquing: Investigating students' responses to other graphs of a real-life functional situation. In Hine, G., Blackley, S., & Cooke, A. (Eds.), *Proceedings of the 42nd Annual Conference of the Mathematics Education Research Group of Australasia*. Perth: MERGA.

Wilkie, K. J., & Clarke, D. M. (2016). Developing students' functional thinking in algebra through different visualisations of a growing pattern's structure. *Mathematics Education Research Journal*, 28(2), 223–224. 10.1007/s13394-015-0146-y

Wilkie, K. J., & Sullivan, P. (2018). Exploring intrinsic and extrinsic motivational aspects of middle school students' aspirations for their mathematics learning. *Educational Studies in Mathematics*, 97(3), 235–25. 10.1007/s10649-017-9795-y

Williams, P. (2023). AI, Analytics and a New Assessment Model for Universities. *Education Sciences*, 13(10), 1040. 10.3390/educsci13101040

Windschitl, M. (2002). Framing constructivism in practice as the negotiation of dilemmas: An analysis of the conceptual, pedagogical, cultural, and political challenges facing teachers. *Review of Educational Research, 72*(2), 131–175.

Wolf, S., & McCoy, D. C. (2019). Household socioeconomic status and parental investments: Direct and indirect relations with school readiness in Ghana. *Child Development*, 90(1), 260–278. 10.1111/cdev.1289928708296

Wong, K.-T., Yeop, M. A., & Mazura, M. (2019). Modelling the factor influencing the implementation of mobile-heutagogical practices among teachers: An application of invariance multi-group structural model. *International Journal of Learning. Teaching and Educational Research*, 18(12), 1–16. 10.26803/ijlter.18.12.1

Woods, J., Murzi, H., Greene-Havas, M., Diaz-Strandberg, A., & Vincente, S. (2020). *Role of social interaction in barriers facing the first year -international students in the United States.* American Society for Engineering Education.

World Bank. (2007). Developing Science and Mathematics and ICT education in SSA. World Bank.

World Economic Forum. (2022, June 30). *Online learning: What next for higher education after COVID-19?* WEForum. https://www.weforum.org/agenda/2022/06/online-learning-higher-education-covid-19/

Xie, G., & Liu, X. (2023). Gender in mathematics: How gender role perception influences mathematical capability in junior high school. *The Journal of Chinese Sociology*, 10(10), 10. 10.1186/s40711-023-00188-3

Yaghi, H. M., & Ghaith, G. M. (2002). Correlates of computing confidence among teachers in an international setting. *Computers in the Schools*, 19(1-2), 81–94. 10.1300/J025v19n01_07

Yang, D.-C., & Lin, Y.-C. (2015). Assessing 10-to 11-year-old children's performance and misconceptions in number sense using a four-tier diagnostic test. *Educational Research*, 57(4), 368–388. 10.1080/00131881.2015.1085235

Yang, F. Y., Chang, C. Y., Chien, W. R., Chien, Y. T., & Tseng, Y. H. (2013). Tracking learners' visual attention during a multimedia presentation in a real classroom. *Computers & Education*, 62, 208–220. 10.1016/j.compedu.2012.10.009

Yeager, D. S., & Dweck, C. S. (2012). Mindsets that promote resilience: When students believe that personal characteristics can be developed. *Educational Psychologist*, 47(4), 302–331. 10.1080/00461520.2012.722805

Yeh, C., Cheng, H., Chen, Z., Liao, C., & Chan, T. (2019). Enhancing achievement and interest in mathematics learning through Math-Island. *Research and Practice in Technology Enhanced Learning*, 14(5), 1–19. 10.1186/s41039-019-0100-9

Yeung, W. J., Linver, M. R., & Brooks-Gunn, J. (2002). How money matters for young children's development: Parental investment and family processes. *Child Development*, 73(6), 1861–1879. 10.1111/1467-8624.t01-1-0051112487499

Yildiz, M., & Scharaldi, K. (2015). Cultivating transformative educator model among pre-service teachers while integrating global competencies, critical thinking and 21st century skills: A trans-disciplinary approach. In D. Rutledge & D. Slykhuis (Eds.), *Proceedings of SITE 2015-Society for Information Technology & Teacher Education International Conference* (pp. 1760–1769). Association for the Advancement of Computing in Education (AACE). https://www.learntechlib.org/primary/p/150236/

Yilmaz, N. P. (2011). Evaluation of the technology integration process in the Turkish education system. *Contemporary Educational Technology*, 2(1), 37–54.

Yunus, M. M., & Sukri, S. I. A. (2017). The use of English in teaching mathematics and science: The PPSMI policy vis-à-vis the DLP. *Advances in Language and Literary Studies*, 8(1), 133–142. 10.7575/aiac.alls.v.8n.1p.133

Yushau, B., Mji, A., & Wessels, D. C. J. (2005). The role of technology in fostering creativity in the teaching and learning of mathematics. *Pythagoras*, 62(62), 12–22. 10.4102/pythagoras.v0i62.110

Zadeh, Z. Y., Farnia, F., & Ungerleider, C. (2010). How home enrichment mediates the relationship between maternal education and children's achievement in reading and math. *Early Education and Development*, 21(4), 568–594. 10.1080/10409280903118424

Zawacki-Richter, O., Marín, V. I., Bond, M., & Gouverneur, F. (2019). Systematic review of research on artificial intelligence applications in higher education–where are the educators? *International Journal of Educational Technology in Higher Education*, 16(1), 1–27. 10.1186/s41239-019-0171-0

Zhang, Y., & Xie, Y. (2016). Family background, private tutoring, and children's educational performance in contemporary China. *Chinese Sociological Review*, 48(1), 64–82. 10.1080/21620555.2015.1096193304l6869

Zhou, L., Pan, S., Wang, J., & Vasilakos, A. V. (2017). Machine learning on big data: Opportunities and challenges. *Neurocomputing*, 237, 350–361. 10.1016/j.neucom.2017.01.026

Zobnina, A. (2009). *Glossary of gender-related terms.* Mediterranean Institute of Gender Studies, European Institute for Gender Equality. https://medinstgenderstudies.org/wp-content/uploads/Gender-Glossary-updated_final.pdf

Zreik, M. (2024a). The Transformation of Healthcare Marketing in the New Digital Era: A Comprehensive Analysis with Reference to China. In Djakeli, K. (Ed.), *Modern Healthcare Marketing in the Digital Era* (pp. 42–60). IGI Global. 10.4018/979-8-3693-0679-6.ch003

Zreik, M. (2024b). Cultivating ILCs in China: A Pathway to Culturally Sustaining and Transformative Education. In Adams, S., & Breidenstein, A. (Eds.), *Exploring Meaningful and Sustainable Intentional Learning Communities for P-20 Educators* (pp. 120–140). IGI Global. 10.4018/978-1-6684-7270-5.ch006

Zreik, M. (2024c). Artificial Intelligence and the Future of Chinese Language Pedagogy: An In-Depth Analysis. In Çetin Köroğlu, Z., & Çakır, A. (Eds.), *Fostering Foreign Language Teaching and Learning Environments with Contemporary Technologies* (pp. 1–20). IGI Global. 10.4018/979-8-3693-0353-5.ch001

About the Contributors

Evelyn Njurai is currently a lecturer of Curriculum Studies and Mathematics Education courses in the Department of Curriculum, Instruction, and Media and Coordinator of Teaching Practice in the School of Education at Kisii University, Kenya. She has a wide experience in qualitative research in and out of academia. Her research interests include Mathematics education in the context of language diversity, Curriculum, E-learning, and STEM Education. Evelyn has published internationally and locally in peer-reviewed journals and book chapters and is the author of Dictionary of Mathematics for Secondary Schools, Kenya (2019). She is a co-editor of a handbook for the Competency-Based Curriculum in Kenya and co-author of two book chapters in the book. Evelyn is a member of various professional bodies and is involved in community work through volunteer service in teaching mathematics to marginalized communities in Kenya.

Alphonse Uworwabayeho is a Professor of Mathematics Education at the University of Rwanda-College of Education (UR-CE). He is a member of research committee of African Centre of Excellence for Innovative Teaching and Learning Mathematics and Science (ACEITLMS) based in the UR-CE. He is commissioner on mathematics education for the African Mathematics Union (AMU). Prior joining higher institution of learning, he was a secondary school mathematics teacher. His research interest lies in teacher professional development on enhancing active learning.

***

Jogymol Alex is an Assistant Professor of Mathematics Education at the College of Education in the United Arab Emirates University (UAEU) in UAE. She joined UAEU in January 2024 from Walter Sisulu University, South Africa, where she was a full Professor in Mathematics Education and Head of the Department of Mathematics and Sciences Education. Her research interests concern how mathematics pre-service and in-service teachers and school learners can be empowered through teaching, research, and community projects with local, national, and international partnerships. This has earned her numerous awards in teaching and learning and community engagement.

Sudi Balimuttajjo received his PhD in Mathematics Education from the University of Nevada, Reno in 2010. In the same year, he was appointed and is still serving as Senior Lecturer and head of the Department of Education Foundations and Psychology at Mbarara University of Science and Technology. He has engaged in deep study and published on diverse issues in Mathematics Education.

Chakabwata William holds a PhD in Curriculum and Instruction obtained from University of South Africa on 27 September 2023.He also holds a master degree in Curriculum and arts education which he earned from University of Zimbabwe. He has taught at Midlands State University in Namibia and also at Namibia University of Science and Technology(NUST).

Omosa Elijah Mochama is a lecturer in the Department of Curriculum, Instruction and Educational Media. He holds a PhD in Curriculum, Instruction and Educational Media (Kisii University, 2021), a masters in Physics Education from Moi University (2015) and Bachelor of Education Science (Physics and Mathematics, 1997). He also has a Diploma in Science, Technology, Engineering and Mathematics Education. Omosa has been a District facilitator in SMASSE/SMASE (Physics) project and School Based Teacher Support System (Physics) programme. He wishes every learner creates and maintains interest in STEM education.

Sanjeev Kumar Jha has more than seven years of experience as a school teacher and over four years of research experience at national-level institutions, including the National Council of Educational Research and Training (NCERT), and the National University of Educational Planning and Administration (NUEPA), both in New Delhi, India. He has earned PhD (Education), MSc (Mathematics), MA (Sociology), and MEd. His works include fifteen research papers published in reputable journals, two psychological tests, one edited book, and four chapters included in the edited book. He specialises in higher and school education, psychometrics, educational policy, and mathematics education.

Tebogo Jillian Mampane is a lecturer and researcher at the University of South Africa. Her interests are in Early Childhood Education and Leadership and Management. She has several publications in accredited book chapters. She supervises Master's and PhD students and has trained district officials and curriculum advisors in South Africa.

Marcellina Mjenda is an academician at The University of Dodoma as an assistant lecturer Department of Mathematics and Statistics. She obtained her Bachelor of Science with Education from St John's University of Tanzania and her Master of Science in Mathematics from The University of Dodoma, Tanzania. She is currently pursuing a Ph.D. in Mathematics Education at the University of Rwanda – College of Education-African Centre of Excellence for Innovative Teaching and Learning Mathematics and Sciences (ACEITLMS)

Angel Mukuka is a lecturer of Mathematics Education and Statistics at Mukuba University in Kitwe, Zambia. Dr. Mukuka has a total of 18 years of experience teaching mathematics, both at the secondary and tertiary levels of education. He received a Master of Science in Mathematics Education from Copperbelt University in Zambia and a PhD in Mathematics Education from the African Centre of Excellence for Innovative Teaching and Learning Mathematics and Science, University of Rwanda. His research areas of interest include mathematical reasoning, mathematical problem-solving, technology-based instruction in mathematics classrooms, and mathematics teacher education. Dr. Mukuka is currently undertaking a postdoctoral research fellowship at Walter Sisulu University in South Africa.

Vedaste Mutarutinya is a Professor of Mathematics-Physics and Mathematics Education at the University of Rwanda - College of Education (UR-CE). He pursued his studies in the Faculty of Natural Science of Friendship People's University of Russia (FPUR) in Moscow, where he successively obtained a BSc degree (1994), a Master's degree (1996) in Mathematics and Physics, and a PhD (2000) in Mathematics. His fields of research are Functional Analysis with applications and Mathematics Education. He is the author of several scientific articles in nationally and internationally recognized journals. He is also a co-organizer of different CPD programmes within and outside UR-CE. Prof. Vedaste Mutarutinya is also a member of the Rwanda Mathematics Union and an Associate Member of the African Center of Excellence and Innovation in Teaching/Learning Mathematics and Sciences (ACEITLMS) at the University of Rwanda –College of Education.

Joseph Njiku is an academician in the department of Educational Psychology and Curriculum Studies at the Dar es Salaam University College of Education (DUCE). He graduated his PhD in mathematics education from the University of Rwanda – College of Education. He holds a Master degree in science education and bachelor degree in science majoring in mathematics both from the University of Dar es Salaam. Joseph Njiku believes that changes in education begin with teachers since Human resource is engineered by education and teachers make the heart of it. He is interested in researching how teachers and students navigate and negotiate within the tides of school dynamics. His research and writing are focused on teacher education, school practices, and STEM education.

Benson Njoroge Wanjiru is a teacher trainer and researcher in Mathematics education with over 20 years teaching experience in the University. He is currently serving as a Dean, School of Education at Mount Kenya University. His research interest include but not limited to mathematical language, mathematics resilience, affective issues in Mathematics, pedagogy and ethnomathematics.

B Tatira has been a Senior Lecturer of Mathematics Education in the Faculty of Education at Walter Sisulu University since 2019. He lectures mathematics content courses to undergraduate degree students and also supervises postgraduate students majoring in mathematics education research. His research focus is on mathematics teacher education with a passion in student teachers' conceptualisation of mathematics content knowledge. He has published extensively on teaching and learning undergraduate mathematics and related topics in mathematics education. He is a holder of a PhD degree in mathematics education from the University of KwaZulu-Natal.

Robert Wakhata is a Ph.D. student (in Mathematics Education) at the University of Rwanda, College of Education - African Centre of Excellence for Innovative Teaching and Learning Mathematics and Science, Rwanda. Robert holds a Master of Science degree in Applied Mathematics and a Bachelor's Degree in Education with vast experience in teaching pure mathematics, applied mathematics, and mathematics education.

Mohamad Zreik, a Postdoctoral Fellow at Sun Yat-sen University, is a recognized scholar in International Relations. His recent work in soft power diplomacy compares China's methods in the Middle East and East Asia. His extensive knowledge spans Middle Eastern Studies, China-Arab relations, East Asian and Asian Affairs, Eurasian geopolitics, and Political Economy, providing him a unique viewpoint in his field. Dr. Zreik is a proud recipient of a PhD from Central China Normal University (Wuhan). He's written numerous acclaimed papers, many focusing on China's Belt and Road Initiative and its Arab-region impact. His groundbreaking research has established him as a leading expert in his field. Presently, he furthers his research on China's soft power diplomacy tactics at Sun Yat-sen University. His significant contributions make him a crucial figure in understanding contemporary international relations.

Index

T

V

W

www.ingramcontent.com/pod-product-compliance
Lightning Source LLC
LaVergne TN
LVHW080557300926
849162LV00005B/978

* 9 7 9 8 3 6 9 3 2 8 7 3 6 *